Advanced Multi-Criteria Decision Making for Addressing Complex Sustainability Issues

Prasenjit Chatterjee
MCKV Institute of Engineering, India

Morteza Yazdani
Universidad Loyola Andalucía, Spain

Shankar Chakraborty
Jadavpur University, India

Dilbagh Panchal
Dr. B. R. Ambedkar National Institute of Technology (NIT) Jalandhar, India

Siddhartha Bhattacharyya
RCC Institute of Information Technology Kolkata, India

A volume in the Advances in Environmental Engineering and Green Technologies (AEEGT) Book Series

IGI Global
DISSEMINATOR OF KNOWLEDGE

Published in the United States of America by
IGI Global
Engineering Science Reference (an imprint of IGI Global)
701 E. Chocolate Avenue
Hershey PA, USA 17033
Tel: 717-533-8845
Fax: 717-533-8661
E-mail: cust@igi-global.com
Web site: http://www.igi-global.com

Library of Congress Cataloging-in-Publication Data

Names: Chatterjee, Prasenjit, 1982- editor.
Title: Advanced multi-criteria decision making for addressing complex sustainability issues / Prasenjit Chatterjee, Morteza Yazdani, Shankar Chakraborty, Dilbagh Panchal, and Siddhartha Bhattacharyya, editors.
Description: Hershey, PA : Engineering Science Reference, [2020]
Identifiers: LCCN 2018055860| ISBN 9781522585794 (h/c) | ISBN 9781522585800 (s/c) | ISBN 9781522585817 (eISBN)
Subjects: LCSH: Sustainable engineering. | Green products. | Sustainable development. | Multiple criteria decision making.
Classification: LCC TA170 .A355 2020 | DDC 338.9/27019--dc23 LC record available at https://lccn.loc.gov/2018055860

This book is published in the IGI Global book series Advances in Environmental Engineering and Green Technologies (AEEGT) (ISSN: 2326-9162; eISSN: 2326-9170)

British Cataloguing in Publication Data
A Cataloguing in Publication record for this book is available from the British Library.

All work contributed to this book is new, previously-unpublished material.
The views expressed in this book are those of the authors, but not necessarily of the publisher.

For electronic access to this publication, please contact: eresources@igi-global.com.

Advances in Environmental Engineering and Green Technologies (AEEGT) Book Series

ISSN:2326-9162
EISSN:2326-9170

Editor-in-Chief: Sang-Bing Tsai, University of Electronic Science and Technology of China Zhongshan Institute, China & Ming-Lang Tseng, Lunghwa University of Science and Technology, Taiwan & Yuchi Wang, University of Electronic Science and Technology of China Zhongshan Institute, China

MISSION

Growing awareness and an increased focus on environmental issues such as climate change, energy use, and loss of non-renewable resources have brought about a greater need for research that provides potential solutions to these problems. Research in environmental science and engineering continues to play a vital role in uncovering new opportunities for a "green" future.

The **Advances in Environmental Engineering and Green Technologies (AEEGT)** book series is a mouthpiece for research in all aspects of environmental science, earth science, and green initiatives. This series supports the ongoing research in this field through publishing books that discuss topics within environmental engineering or that deal with the interdisciplinary field of green technologies.

COVERAGE

- Policies Involving Green Technologies and Environmental Engineering
- Green Transportation
- Contaminated Site Remediation
- Pollution Management
- Sustainable Communities
- Waste Management
- Air Quality
- Biofilters and Biofiltration
- Industrial Waste Management and Minimization
- Water Supply and Treatment

IGI Global is currently accepting manuscripts for publication within this series. To submit a proposal for a volume in this series, please contact our Acquisition Editors at Acquisitions@igi-global.com or visit: http://www.igi-global.com/publish/.

The Advances in Environmental Engineering and Green Technologies (AEEGT) Book Series (ISSN 2326-9162) is published by IGI Global, 701 E. Chocolate Avenue, Hershey, PA 17033-1240, USA, www.igi-global.com. This series is composed of titles available for purchase individually; each title is edited to be contextually exclusive from any other title within the series. For pricing and ordering information please visit http://www.igi-global.com/book-series/advances-environmental-engineering-green-technologies/73679. Postmaster: Send all address changes to above address.

Titles in this Series

For a list of additional titles in this series, please visit:
https://www.igi-global.com/book-series/advances-environmental-engineering-green-technologies/73679

Spatial Planning in the Big Data Revolution
Angioletta Voghera (Politecnico di Torino, Italy) and Luigi La Riccia (Politecnico di Torino, Italy)
Engineering Science Reference • ©2019 • 359pp • H/C (ISBN: 9781522579274) • US $195.00

Global Initiatives for Waste Reduction and Cutting Food Loss
Aparna B. Gunjal (Asian Agri Food Consultancy Services Ltd, India) Meghmala S. Waghmode (Annasaheb Magar Mahavidyalaya, India) Neha N. Patil (Annasaheb Magar Mahavidyalaya, India) and Pankaj Bhatt (Dolphin (P.G) College of Biomedical and Natural Sciences Dehradun, India)
Engineering Science Reference • ©2019 • 328pp • H/C (ISBN: 9781522577065) • US $195.00

Green Public Procurement Strategies for Environmental Sustainability
Rajesh Kumar Shakya (The World Bank, USA)
Engineering Science Reference • ©2019 • 228pp • H/C (ISBN: 9781522570837) • US $185.00

Climate Change and Its Impact on Ecosystem Services and Biodiversity in Arid ...
Ahmed Karmaoui (Southern Center for Culture & Sciences (SCCS), Morocco)
Engineering Science Reference • ©2019 • 408pp • H/C (ISBN: 9781522573876) • US $225.00

Bioenergy and the Advanced Application of Bio-Products and Microfluidic Devices
Mohammad Reza Rahimpour (Shiraz University, Iran) Reza Kamali (Shiraz University, Iran) Mohammad Amin Makarem (Shiraz University, Iran) and Mohammad Karim Dehghan Manshadi (Shiraz University, Iran)
Engineering Science Reference • ©2019 • 325pp • H/C (ISBN: 9781522575344) • US $215.00

Handbook of Research on Global Environmental Changes and Human Health
Kholoud Kahime (Cadi Ayyad University, Morocco) Moulay Abdelmonaim El Hidan (Ibn Zohr University, Morocco) Omar El Hiba (Chouaib Doukkali University, Morocco & Cadi Ayyad University, Morocco) Denis Sereno (University of Montpellier, France) and Lahouari Bounoua (NASA, USA)
Engineering Science Reference • ©2019 • 653pp • H/C (ISBN: 9781522577751) • US $265.00

For an entire list of titles in this series, please visit:
https://www.igi-global.com/book-series/advances-environmental-engineering-green-technologies/73679

IGI Global
DISSEMINATOR OF KNOWLEDGE

701 East Chocolate Avenue, Hershey, PA 17033, USA
Tel: 717-533-8845 x100 • Fax: 717-533-8661
E-Mail: cust@igi-global.com • www.igi-global.com

Dr. Prasenjit Chatterjee would like to dedicate this book to his grandparents, father Late Dipak Kumar Chatterjee, his mother Mrs. Kalyani Chatterjee, beloved wife Amrita and his little angel Aheli.

Dr. Morteza Yazdani would like to dedicate the book to his father Mr. Mohammad Yazdani, his mother Mrs. Robabeh Saadati, his grand brother Mani Yazdani, and his nice life partner Dr. Violeta Doval Hernandez.

Dr. Dilbagh Panchal would like to dedicate the book to his father Mr. Karam Singh, his mother Mrs. Sarbati Devi, beloved wife Jyotiand his little angel Evanshi Panchal.

Prof. (Dr.) Shankar Chakraborty would like to dedicate this book to his Late father Phanindra Nath Chakraborty, his Late mother Gouri Chakraborty, his beloved wife Rupa, his son Santonab and daugheter Sirin. A special dedication owes to his Gurudev.

Prof. (Dr.) Siddhartha Bhattacharyya would like to dedicate the book to his Late father Ajit Kumar Bhattacharyya, his Late mother Hashi Bhattacharyya, his beloved wife Rashni, his schoolmates Jayati, Suparna, Joyita, Sukla and Arunima.

Editorial Advisory Board

Table of Contents

Detailed Table of Contents

Dilbagh Panchal, Dr. B. R. Ambedkar National Institute of Technology (NIT) Jalandhar, India
Prasenjit Chatterjee, MCKV Institute of Engineering, India
Morteza Yazdani, Universidad Loyola Andalucía, Spain
Shankar Chakraborty, Jadavpur University, India

The aim of this chapter is to develop a hybrid decision-making framework for studying the risk issues related to failure of an industrial system. On the basis of plant expert's knowledge, failure mode effect analysis (FMEA) sheet has been generated and various failure causes associated with the sub-systems were listed. On the basis of three risk factors, namely probability of occurrence of failure, severity and non-detection, Risk Priority Numbers (RPN) for each failure cause has been tabulated. The demerits of FMEA approach in prioritizing the failure causes has been overcome by implementing fuzzy rule-based tool. The consistency and heftiness of the ranking results have been tested by implementing grey relation analysis (GRA) approach. Comparison of ranking results has been done for effective decision making of ranking results. The accuracy of decision results would be highly useful in developing a planned maintenance policy for the plant. The proposed framework has been tested with its application on a cooling tower system of a thermal power plant located in the northern part of India.

Selection of optimal maintenance strategy for critical systems/machinery is considered as a complex decision-making task that takes into account several available maintenance alternatives that are evaluated in terms of a set of different conflicting qualitative and quantitative factors. In the last few decades, progress has been made in different sustainable-based decision-making problems, where environmental, social, and economic factors played a pivotal role to arrive at the best decision. In this chapter, a hybrid artificial intelligence (AI)-based conceptual decision-making model is described by taking advantages of both expert system and case-based reasoning methodology to solve sustainable maintenance strategy selection problems. Adding to this, a flowchart of the model is suitably described by hypothetical examples of a sustainable maintenance strategy selection program.

This chapter aims to select the best biodiesel for a diesel power generator considering sustainable criteria. The study proposes the application of an almost unexplored prospect theory-based multi-criteria decision-making (MCDM) method, popularly known as Tomada de Decisao Interativa Multicriterio (TODIM). Several conflicting criteria including calorific value, cetane number, density, viscosity, flash point, and pour point are considered as the most predominant criteria, while pongamia oil, jatropha oil, cotton seed oil, linseed oil, madhuca indica oil, olive oil, and sunflower oil are among the considered biodiesel alternatives. Results show that madhuca indica oil scored the highest value followed by others. The study is also complemented by an analysis of the sensitivity of the numerical results obtained to show the robustness of the proposed method.

Preface

Sustainability is a broad discipline, which insights into the global aspects of business, technology, environment and the social sciences. In last few decades, researches on modeling sustainability issues for different engineering applications have witnessed a huge evolution. Sustainability is an epoch-making enabler to remain competitive in the global business scenario. Sustainability concept mainly focuses on three basic attributes like environmental, economic and social performance. It aims to preserve energy and natural resources to ensure negligible effect on the environment and surroundings. Sustainability mainly focuses on accomplishing the requirements of the present without restraining future generations to meet their needs. Since a long time, environmental impact had been the focus for many organizations craving a sustainability philosophy. The topic of sustainability constantly appears in Media, News, Articles, and daily communication as in publics. We have a world with limited resources that are not recoverable sometimes. Therefore, it is essential for project planners and developers to think, evaluate and select such technologies that will reduce the technological and human impact of the project on the environment, organize and balance energy demand, and focus on supplying renewable energy sources. Hence, the practice of green activities has become mandatory to balance these conflicts; even manufacturing processes cannot make an exception. Among the public consciousness, environmental concerns have attracted more attention in contemporary globalization. Especially, in engineering and business areas, it is an advantage to use methodologies that help and support decision-making, reducing the complexity of the problems and the risks inherent in this process. Many firms and organizations may face the problem of evaluating and choosing between several options. Each kind of decision-making problem is designed in the context of different variables, goals and constraints. In a strategic decision environment, policy makers to finding the optimal solution have to rely on the optimality and effectiveness of the outcomes. The question is the way of utilizing these tools together and participating in an effective decision procedure. In real problems often it is required the consideration of a large number of conflicting criteria that affect the final consequences. Therefore, decision makers (DMs) need systematic mathematical approaches to conduct those analyses. The growing complexity in modern social-economics or engineering environments or systems has forced the researchers to solve a complicated problem by using multi-criteria decision-making (MCDM) approaches. Therefore, the use of MCDM met5hods has gained considerable attention in both academia and practice to solving such real time industrial problems. Over time, intelligent decision-making approaches were grown and developed in depth. This is the subject that the book dedicates widely.

Among operation research (OR) sub disciplines, decision analyses have gained crucial importance in economics and are widely recognized as a sound prescriptive theory. Precisely, MCDM is most directly characterized by a set of multiple criteria methods and is a famous sub-discipline of the operation research. It analyzes and selects alternatives under several criteria in real world applications. It is a systematic quantitative approach, which aims to support DMs in order to make rational and efficient decisions by considering important objectives and criteria. The concern of the experts and researchers is that each method might not be convenient to every problem, and so how to handle such condition and based on what regulations. MCDM can be broadly divided into two classes, i.e. multi-attribute decision-making (MADM) and multi-objective decision-making (MODM). The fundamental difference between MODM and MADM problems is that the former concentrates on a continuous decision space subject to problem specific constraints, while the later aids to evaluate alternatives in a discrete decision space consisting of predetermined set of alternatives. MADM usually consists of identification of the alternatives, specification of each alternative with respect to each evaluating criterion, indicating the relative preference of each alternative over another and finally, scoring the alternatives according to the decision criteria. While, MADM methods select the best alternative among a finite number of choices, in MODM the best alternative is designed with multiple objectives based on continuous decision variables subject to various constraints.The background of this book is based on the assumption that the success of decision-making problems relies on the assessment and integration of different design and selection objectives. The varying design issues and available alternatives call for different areas of expertise to be involved in the decision-making process. This makes it difficult to evaluate the overall goodness of a proposed alternative. MADM is basically an approach to rank the feasible alternatives with respect to different attributes. This is achieved on the basis of the impact of the alternatives on certain criteria, may be conflicting in nature, in view of the objective of optimization. In all MADM methods, taking decision regarding the best alternative amongst all the available choiceis resolved using mathematical treatments to form a decision matrix. MADM also largely depends to a high degree on subjective preferences stated by the decision makers as the methodology deals with criteria that are difficult to quantify or assign with a numerical value.Many decision-making approaches have been developed and designed by the researchers for measuring and analysis of multi-dimensional sustainability goals and the complexity of socio-economic and biophysical systems. Therefore, based on the above foundations, we invited authors to submit innovative research articles to explore new proposals related to MCDM methods to rationalize the process of optimal decision-making to address sustainability modelling. Thus, this book intended to present some comprehensive chapters, employing different MCDM methods to address the burning themes of sustainability in the presence of mutually conflicting attributes, objectives, alternatives and involvement of cognitive state of uncertainties and fuzziness indecision-making situations to explore the opportunities and challenges in modelling sustainability aspects of real time applications. This book makes an effort to discuss and address the challenges in conceptualization and implementation of decision-making techniques in the context of green, lean and sustainable engineering, factor identification, quantification, comparison, selection, simulation modeling and analysis, system approach to manufacturing, supply chain, transportation, operations, maintenance strategy, biodiesel, agriculture, laboratory footprints, e-commerce and financial sectors to name a few.

ORGANIZATION OF THE BOOK

The book is organized into 13 chapters. A brief description of each of the chapters is presented below:

Chapter 1 develops a hybrid decision-making framework for studying the risk issues related to failure of an industrial system. Failure Mode Effect Analysis (FMEA) sheet has been generated and various failure causes associated with the sub-systems are listed. Limitations of the traditional FMEA approach in prioritizing the failure causes has been overcome by implementing fuzzy rule base tool. Grey Relation Analysis (GRA) is also implemented for checking the consistency and heftiness of the ranking results.

Chapter 2 proposes a combined application of preference ranking organization method for enrichment of evaluations (PROMETHEE) and geometrical analysis for interactive aid (GAIA) approaches to evaluate the performances of smart cities in Indian context based on several critical criteria. A GIS (geographic information system) method and an HSV (hue-saturation-value) color coding scheme-based on cartographic principles are also employed to identify the influence of individual criterion on the overall rank of the smart cities.

Chapter 3 presents a holistic model for the selection of battery-operated electric vehicles (BEVs) based on a newly developed combined compromise solution

(CoCoSo) and criteria importance through inter criteria correlation (CRITIC) methods. Performance of the proposed model has also been compared with some of the popular MCDM methods.

In Chapter 4, a hybrid artificial intelligence (AI) based conceptual decision-making model is described by taking advantages of both Expert System and Case-Based Reasoning methodology to solve sustainable maintenance strategy selection problems. Adding to this, flowchart of the model is suitably described by hypothetical examples of sustainable maintenance strategy selection program.

Chapter 5 proposes the application of an almost unexplored prospect theory-based TOmada de Decisao Interativa Multicriterio (TODIM) method for selecting select the best biodiesel for a diesel power generator considering sustainable criteria. Several conflicting criteria including calorific value, cetane number, density, viscosity, flash point and pour point are considered, while pongamia oil, jatropha oil, cotton seed oil, liseed oil, madhucaindica oil, olive oil and sunflower oil are among the considered biodiesel alternatives. Madhucaindica oil emerges out as the most appropriate biodiesel alternative for the considered decision-making problem.

Chapter 6 emphasizes on modeling and optimization of performance and emissions parameters of a biodiesel run diesel engine using an integrated MCDM approach using entropy with multi criteria ranking analysis (MCRA) methods.

Chapter 7 examines the applicability and feasibility of a modified similarity-based approach for selection of cotton fiber which in turn assists in achieving sustainable design characteristics. Two illustrative examples from past researches and one existing cotton fiber selection problem of a small textile organization are solved applying the adopted method to validate its veracity and robustness.

Chapter 8 focuses on the application of a preference function based outranking method in sustainable agriculture for prioritization of different farming issues including policies, equipment and process.

Chapter 9 explores the agricultural hazards factors in the farming sectors by the use of literature and standard nordic questionnaires. Subsequently, Best Worst Method (BWM) is used to rank the different hazards based on the risk factors involved. Finally, an ergonomic evaluation is made by using both Rapid Entire Body Assessment (REBA) and Ovako Working posture Assessment. System (OWAS) ergonomic tools, considering different postures taken by farmers in selected farming activities, and accordingly the corrective measures are recommended.

Chapter 10 makes an attempt to help the physicians in order to reduce the number of tests required for diagnosis. The analogy between concentration of different contaminants in drinking water and risk analysis based on IFSs have been investigated. The findings will not only help in diagnosis but also offer cost-cutting and eco friendly strategy by avoiding certain required laboratory examinations.

Chapter 11 put forwards a hybrid MCDM model consisting of entropy and Vlse Kriterijuska Optimizacija I Komoromisno Resenje (VIKOR) methods for performance parameter optimization of Vertical axis wind turbine (VAWT). A real life case on optimization of performance parameters of Savonius type VAWT (SVAWT) is considered.

Chapter 12 focuses on the product prediction and recommendation in sustainable e-commerce using association rule mining and k-means clustering techniques. It addresses the performance issues by scaling up the neighborhood formation process through the use of clustering techniques for business sustainability.

In Chapter 13, a novel method based on the technique for order preference by similarity to ideal solution (TOPSIS) and quality function deployment (QFD) is proposed for the first time to evaluate the sustainable efficiency of banking operations. QFD is employed to provide importance to the customers' needs with respect to banks' sustainability, and subsequently the priority weights of the considered criteria of sustainability principles are estimated. TOPSIS method is applied to rank the alternatives considered.

Prasenjit Chatterjee
MCKV Institute of Engineering, India

Morteza Yazdani
Universidad Loyola Andalucía, Spain

Shankar Chakraborty
Jadavpur University, India

Dilbagh Panchal
Dr. B. R. Ambedkar National Institute of Technology (NIT) Jalandhar, India

Siddhartha Bhattacharyya
RCC Institute of Information Technology Kolkata, India

Acknowledgment

The editors would like to express their gratitude to the many people who cooperated through this book; to all those who provided support, talked things over, read, wrote, offered comments, and assisted in the editing, proofreading and design to make this book possible. No undertaking like an edited volume uniting a number of scholarly persons from around the world can be done by one person. This is certainly no exception.

First, the editors owe a lot to the Almighty and their family members for their endless support, motivation, guidance and love all through their lives.

Second, the editors wish to acknowledge the valuable contributions of the authors regarding the improvement of quality, coherence, and content presentation of the chapters. Many of the authors also served as referees; the editors highly appreciate their double task.

Third, words are not adequate to express gratitude, appreciation and much more, to all the individuals who dedicated their considerable time and expertise to the book by serving as editorial board members and reviewers. Their generous contribution is deeply admired.

The editors are very much thankful to the entire IGI Global team for keeping faith on them and showing the right path to create this book. Their guidance, motivation, positive responses and resources ultimately led the foundation of this book.

Last but definitely not the least, the editors take this opportunity to thank all the readers for their trust and it is expected that this book will continue to inspire and guide them for their future endeavour.

Prasenjit Chatterjee
MCKV Institute of Engineering, India

Morteza Yazdani
Universidad Loyola Andalucía, Spain

Shankar Chakraborty
Jadavpur University, India

Dilbagh Panchal
Dr. B. R. Ambedkar National Institute of Technology (NIT) Jalandhar, India

Siddhartha Bhattacharyya
RCC Institute of Information Technology Kolkata, India

Chapter 1
A Hybrid MCDM Approach-Based Framework for Operational Sustainability of Process Industry

Dilbagh Panchal
Dr. B. R. Ambedkar National Institute of Technology (NIT) Jalandhar, India

Prasenjit Chatterjee
https://orcid.org/0000-0002-7994-4252
MCKV Institute of Engineering, India

Morteza Yazdani
https://orcid.org/0000-0001-5526-8950
Universidad Loyola Andalucía, Spain

Shankar Chakraborty
Jadavpur University, India

ABSTRACT

The aim of this chapter is to develop a hybrid decision-making framework for studying the risk issues related to failure of an industrial system. On the basis of plant expert's knowledge, failure mode effect analysis (FMEA) sheet has been generated and various failure causes associated with the sub-systems were listed. On the basis of three risk factors, namely probability of occurrence of failure, severity and non-detection (O_f, S , and O_d), Risk Priority Numbers (RPN) for each failure cause has been tabulated. The demerits of FMEA approach in prioritizing the failure causes has been overcome by implementing fuzzy rule-based tool. The consistency and heftiness of the ranking results have been tested by implementing grey relation analysis (GRA) approach. Comparison of ranking results has been done for effective decision making of ranking results. The accuracy of decision results would be highly useful in developing a planned maintenance policy for the plant. The proposed framework has been tested with its application on a cooling tower system of a thermal power plant located in the northern part of India.

DOI: 10.4018/978-1-5225-8579-4.ch001

INTRODUCTION

Repairable systems due to their complexity in operation are difficult to keep in upstate condition. Unavailability of a plant system is a big threat for the operational sustainability of the particular industry. Even a minor failure in the system's operation may result in plant breakdown due to which there is increase in the production loss (Sharma and Sharma, 2012; Panchal et al., 2018). Production loss for an industry directly affects its sustainability and it become difficult to sustain itself in the market. According to a survey in Europe, for a group of heavy process industries the maintenance cost varies from 15-40% of the total production cost which is a great challenge for their sustainability (Wang et al., 2007; Panchal & Kumar 2016; Panchal et al., 2018; Panchal et al., 2017). Thus to lower down the production cost it is essential to deal with the risk associated with operational behavior of various component of the system. Further, minimization of failure risk by minimizing or eliminating the various failure causes associated with various equipment/components of the system is of supreme importance for improving the operational sustainability of the plant. Hence, for improving the availability by minimizing the risk of sudden failure of the system the study of risk issues is presented in the present work.

LITERATURE BACKGROUND

In the past various studies related to risk of sudden failure has been performed by various authors in different field. Different mathematical models based on qualitative approaches were developed by these researches and are implemented on different systems of different process industries. Sharma et al., (2005) developed FMEA approach based framework for studying the risk issues of a system in a paper plant located in northern part of India. Tay and Lim (2006) developed a new FMEA approach for prioritizing the failure of an industrial system. Application of FMEA approach was presented with guided rule reduction method. Sharma et al., (2007) again developed a fuzzy decision support system for analyzing the risk issues of a system of paper industry. Hekmatpanah and Ravichandran (2011) presented the application of FMEA approach in oil industry for studying the risk issues associated with its operation. Kumru and Kumru (2013) presented the application of fuzzy FMEA approach in a hospital for improving its purchase process. Panchal and Kumar (2015) presented the application of FMEA approach based framework for performing the risk analysis of Power Generating System of thermal power plant located in Northern part of India. Panchal and Kumar (2016) developed a fuzzy methodology based framework for studying the failure risk of water treatment plant in thermal power Industry. Adar et al., (2017) expounded the application of fuzzy FMEA approach based framework for identifying the critical components of water gasification system in a sewage treatment plant. Similarly, applying the same framework, risk analysis of compressor house unit in a thermal power industry has been done and risky component of the considered system has been identified (Panchal & Kumar, 2017). Panchal and Srivastva (2018) expounded the application FMEA approach based framework for studying the risk issues of CNG dispensing system. Panchal et al., (2019) developed a new FMEA approach based framework for analyzing the risk aspects of chlorine gas plant of a chemical industry. In the above reported studies fuzzy methodology has been incorporated in order to consider the uncertainty involved in the expert's judgment. Consideration of expert's judgment uncertainty using fuzzy methodology results in making the above mentioned framework more sound which further result in quality/accurate decision making. The framework is also useful in overcoming the limitations of the traditional FMEA approach like similar ranking to different failure causes and different ranking for the causes which has been defined by same linguistic terms etc. This ability of framework to overcome the drawbacks of traditional FMEA approach would further strength its application to study the risk issues of complex industrial system. Also, from reviewed literature it has been noted that the application of proposed framework has not yet been found for studying the risk issues of cooling tower of the thermal power plant. Thus, the application of fuzzy methodology based framework is presented in the next section of this manuscript.

PROPOSED DECISION-MAKING FRAMEWORK

The framework implemented for studying the risk issues of the considered system is shown in figure 1.

With the proposed framework in the first stage qualitative information has been collected and on the basis of experts knowledge a FMEA sheet was organized. On the basis of experts feedback a scale has been generated and the scores related to three risk factors, namely O_f, S and O_d, has been assigned. RPN score $\left(O_f \times S \times O_d\right)$ has been tabulated and the failure causes were prioritized. Further, to overcome the demerits of FMEA approach fuzzy methodology based framework has been applied within FMEA approach and fuzzy RPN score were tabulated. In the second stage, for the stability of ranking results GRA approach was applied and the ranking results are compared for effective decision making of the failure causes.

Figure 1.

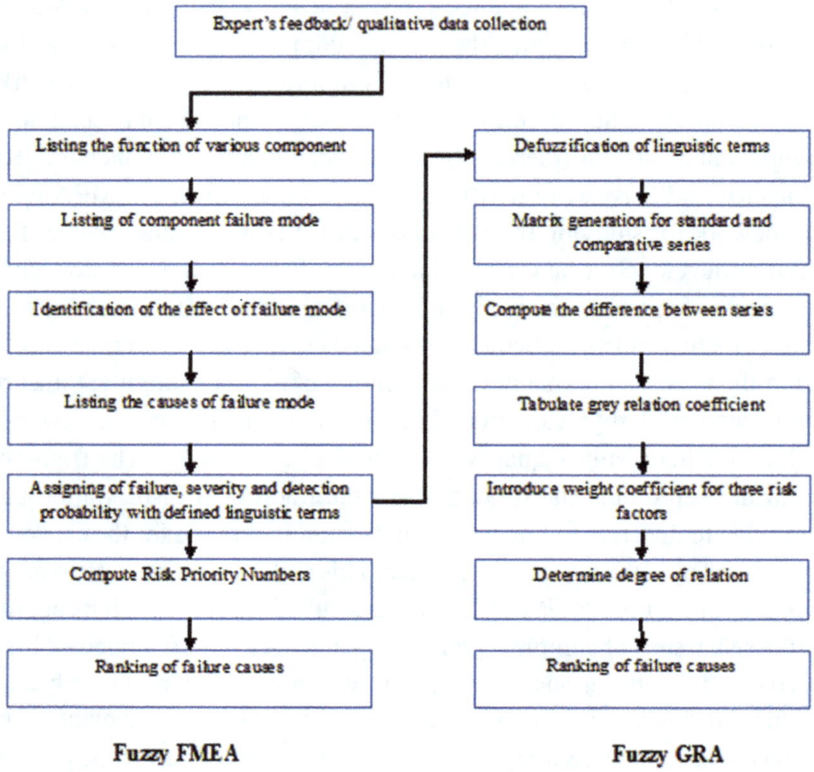

DECISION-MAKING APPROACHES

FMEA and Fuzzy Rule Base

FMEA is one of the systematic and efficient approach for analyzing the risk issues for the various systems of different process industries. Under FMEA approach all the components/subsystems were listed with complete information and the three risk factors were given a rating on the basis of a scale and RPN values were tabulated. The importance of risk factor is that they contribute in judging the risk priority on the basis of RPN values. As this tool allows the experts to list the complete information related to system operation with an ease therefore in the current work FMEA has been implanted in the current work. However, there are some demerits which are associated with FMEA approach such as i) similar RPN values for more than one failure cause although their defined linguistic terms are different ii) defined linguistic terms are same for more than one failure cause but different RPN values for the failure causes.

To overcome these drawbacks fuzzy methodology based FMEA approach has been applied by various researches in different filed (Sharma & Sharma, 2012; Panchal & Kumar, 2017; Ebeling, 2001; Tay & Lim, 2006; Panchal et al., 2018; Panchal & Srivastva, 2018). In the present work fuzzy rule base MAT-LAB toolbox has been applied to overcome the drawbacks and its equation for generating the IF-THEN rules is represented by equation 1 as (Sharma & Sharma, 2012):

$$R_1 : if\, x\, is\, S_i\, then\, y\, is\, T_i \quad where\, i = 1, 2, 3 \ldots\ldots\ldots.n. \tag{1}$$

Grey Relation Approach

Grey relation approach proposed by Deng in 1986 is another powerful decision making tool. The ability of considering the multiple inputs makes this tool more useful. Also, the consideration of three risk factors weights make this approach more sound for quality decision making of ranking results. In the past it has been implemented by various researchers in different process industries (Sharma & Sharma, 2012; Panchal & Kumar, 2016; Panchal & Kumar, 2017). The various steps of GRA approach are discussed as follows:

Step 1: The fuzzified linguistic terms are converted into defuzzified values by using the relation given by Chen and Klien's method (Chen & Klien, 1997)

Step 2: The comparative series representing the crisp set of values of various linguistic terms for the set of failure causes under a system/component is represented as:

$$p = \begin{bmatrix} p_1 \\ p_2 \\ \vdots \\ p_n \end{bmatrix} = \begin{bmatrix} p_1^1 & p_1^2 & \cdots & p_1^1 \\ p_2^1 & p_2^1 & \cdots & p_2^1 \\ \vdots & \vdots & \vdots & \vdots \\ p_n^1 & p_n^1 & \cdots & p_n^1 \end{bmatrix}. \tag{2}$$

where,

n = Number of failure modes; p_i^1, p_i^2, p_i^3 = Crisp numbers for the risk factors

Step 3: The standard series used to reflect the various linguistic terms is represented by using equation 3.

$$q = \begin{bmatrix} q_1^1 & q_1^2 & q_1^3 \\ q_2^1 & q_2^2 & q_2^3 \\ \cdots & \cdots & \cdots \\ q_n^1 & q_n^1 & q_n^1 \end{bmatrix}. \tag{3}$$

where; $q_i^1, q_i^2 \, and \, q_i^3$ = crisp number of lowest level of risk factor

Step 4: The difference between the two series is computed by using equation 4 as:

$$R = \begin{bmatrix} r_i^j & r_i^j & r_i^j \\ r_i^j & r_i^j & r_i^j \\ \cdots & \cdots & \cdots \\ r_i^j & r_i^j & r_i^j \end{bmatrix}. \tag{4}$$

where, $r_i^j = \left| p_i^j - q_i^j \right|$, for any $i \in \left\{1, 2, \ldots\ldots, n\right\} and \, j \in \left\{1, 2, 3\right\}$.

Step 5: Grey relation coefficient for comparing the decision factor with standard series is tabulated by using relation as:

$$\gamma_I^J = \frac{R_{min} + ¶R_{max}}{r_i^j + ¶R_{max}}. \tag{5}$$

where, $R_{min} = \begin{matrix} min \; min \\ i \quad j \end{matrix} \, r_i^j$ and $R_{max} = \begin{matrix} max \; max \\ i \quad j \end{matrix} \, r_i^j$ = minimum and maximum value

$\zeta \in \left[0, 1\right]$ = an identifier which affects the relative value of risk and in the present study on expert's feedback assumed to be 0.5.

Step 6: The equation for tabulation of degree of relation is represented as:

$$\Gamma = \beta_1\gamma_i^1 + \beta_2\gamma_i^2 + \beta_3\gamma_i^3, \text{ With } \sum_{k=1}^{n}{}^{2}{}_{k} = 1. \tag{6}$$

where,

(β_k) = weighting coefficient for tree risk factors

Step 7: On the basis of grey output ranking of failure causes is done ascending order.

APPLICATION OF PROPOSED FRAMEWORK

Case Study

For the application of fuzzy methodology, cooling tower system of thermal power plant has been considered. Cooling tower consists of various sub-systems namely: Cooling water pump (CP_1), cooling water intake tunnel (CW_2), Condenser tube box (CB_4) and cooling water discharge tunnel (CD_4). Due to flow of water in this system there is a big problem in maintaining the sustainable operation of the plant. Hence, for sustainable plant operation risk analysis has been performed with the application of proposed framework. The various steps involved in the proposed framework are presented in the discussion below:

FMEA and Rule Base FMEA Application

A detailed FMEA sheet (Table 1) has been generated on the basis of plant's maintenance expert's knowledge. On the basis of a designed scale (Table 2) scores were allocated by the experts to three risk factors (O_f, S, O_d) and Risk Priority Number (RPNs) were calculated for each risk factors. Priorities were allocated to the failure causes (Table 1) and it has been noticed that failure causes CP_1 and CP_3 of Cooling water pump system have different RPN (140 and 112) with same set of linguistic terms (*Moderate, High, Moderate*) and ranked as 3rd and 5th respectively in Table 4. Similar observation has been made for causes CW_1 and CW_3 of the Cooling water intake tunnel of the Cooling tower system. (ii) Also the cause CT_1 and CT_3 of the Cooling tower Bottom system are represented by different set of linguistic terms (*High, High, Moderate and Moderate, High, High*) but have same RPN values (245) and same rank is given to these causes which could be misleading as far as the system analyst is concerned.

Table 1. FMEA sheet

Components	Function	Potential failure mode	Potential effect of failure	Potential cause of failure	O_f	S	O_d	RPN
Cooling tower	To cool the water	Non uniform water pattern	Decrease in cooling efficiency	Spray nozzle blockage **[CT_1]**	**7**	**7**	**5**	**245**
				Fill pack blockage **[CT_2]**	**6**	**5**	**5**	**150**
				Improper water flow rate **[CT_3]**	**5**	**7**	**7**	**245**
				Irregular air flow **[CT_4]**	**5**	**8**	**5**	**200**
Cooling water pump	To carry cool water to cooling intake tunnel	Bearing seizure	Operation loss	Foreign particle in water **[CP_1]**	**4**	**7**	**5**	**140**
				Scanty lubrication **[CP_2]**	**5**	**5**	**5**	**125**
				Scale formation **[CP_3]**	**4**	**7**	**4**	**112**
		Vibration	Operational efficiency loss	Improper alignment **[CP_4]**	**7**	**8**	**7**	**392**
				Scrap on the shaft **[CP_5]**	**9**	**7**	**6**	**378**
Cooling water intake tunnel	To deliver cool water to condenser	Uneven air flow	Operational efficiency loss	Inexact pressure range **[CW_1]**	**5**	**8**	**6**	**240**
				Lack of air flow**[CW_2]**	**4**	**4**	**4**	**64**
				Inexact temperature range **[CW_3]**	**4**	**7**	**4**	**112**
Condensers tube box	To increase water temperature	leakage	Operational efficiency decrease	Scaling of joint **[CB_1]**	**8**	**7**	**6**	**336**
				Improper cleaning of tube**[CB_2]**	**6**	**7**	**7**	**294**
		Improper vacuum pressure	Operational efficiency decrease	Sump over flow **[CB_3]**	**6**	**6**	**6**	**216**
Cooling water discharge tunnel	To deliver water to the cooling tower	Valve failure	Operational efficiency decrease	Improper water flow rate**[CB_4]**	**4**	**7**	**5**	**140**

Table 2. FMEA input scale (Panchal and Kumar, 2016; Panchal and Kumar, 2017)

Linguistic terms	Score/rank No.	MTBF	Occurrence rate (%)	Severity effect	Likelihood of non-detection (%)
Very low	1	> 5 years	< 0.01	Not noticed	0-5
Low	2/3	2-5 years	0.01-0.1	minor infuriation to operator	6-15/16-25
Moderate	4/5/6	1-2 years	0.1-0.5	minor fall in system performance	26-35/36-45/46-55
High	7/8	0.5-1year	0.5-1	Considerable deterioration in system performance	56-65/66-75
Very high	9/10	< 6 months	> 1	Power generation loss	76-85/86-100

Further, to overcome the limitations of traditional FMEA approach fuzzy rule base FMEA and GRA approaches have been implemented within FMEA approach. The scores for three risk factors were used in the fuzzy inference system with Trapezoidal fuzzy numbers as input for defined linguistic terms where IF-THEN rules were applied. Trapezoidal membership function is chosen because of its ability to consider the wide range of uncertainty in the raw data. Further, with the set of three five linguistic terms total 125 rules were generated. Now by eliminating the common rules we left with twenty five rules. With these rules and FMEA risk factor values, fuzzy outputs in the form Triangular fuzzy numbers are obtained and Fuzzy Risk Priority Number (FRPN) values (crisp value) as output were noted (Table 4).

GRA Application

Further, fuzzified values of defined linguistic terms were defuzzified by using the relation as per cited article (Chen and Klien, 1997). Using equation 2-3 comparison and standard series were generated for each set of failure causes. Differences between the two equations for the set of failure causes were tabulated using equation 4. Using equation 5-6 grey relation coefficient and grey output values were calculated for each failure cause shown in Table 3.

Here for computing grey output weight values $\left(\beta_k\right)$ for three risk factors were considered on the basis of expert's feedback and the values are 0.20, 0.56 and 0.24. Table 4 shows the comparison of the ranking results obtained with the application of three approaches used in the proposed framework.

Table 3. Degree of coefficient and Grey output values

Failure cause	γ_f	γ_s	γ_d	Grey output
CT_1	0.4647	0.4647	0.5680	0.4894
CT_2	0.5680	0.5680	0.5680	0.5680
CT_3	0.5680	0.4647	0.4647	0.4854
CT_4	0.5680	0.4647	0.5680	0.5102
CP_1	0.5680	0.4647	0.5680	0.5102
CP_2	0.5680	0.5680	0.5680	0.5680
CP_3	0.5680	0.4647	0.5680	0.5102
CP_4	0.4647	0.4647	0.4647	0.4647
CP_5	0.4017	0.4647	0.5680	0.4769
CW_1	0.5680	0.4647	0.5680	0.5101
CW_2	0.5680	0.5680	0.5680	0.5680
CW_3	0.5680	0.4647	0.5680	0.5101
CB_1	0.4647	0.4647	0.5680	0.4894
CB_2	0.4647	0.5680	0.5680	0.5473
CB_3	0.5680	0.5680	0.5680	0.5680
CB_4	0.5680	0.4647	0.5680	0.5102

RESULT DISCUSSION

Table 4 clearly indicated that with the implementation of FMEA approach causes CP_1 and CP_3 with RPN score 140 and 112 are prioritized as 3rd and 5th respectively but as per fuzzy RPN and GRA approach outputs these causes are as ranked 3rd which means same priority for both these causes. In other case causes CT_1 and CT_3 were obtained same RPN score under FMEA approach but they are represented by different set of linguistic terms as per Table 3. Since both these causes are represented by same RPN value so both should be ranked 1st, but with fuzzy RPN and grey output the values obtained are different so these causes are ranked accordingly. In this case due to the weighting coefficient and a high probability of failure occurrence, CT_1 is given higher priority Table 4 also indicate that for most of the failure causes same ranking is obtained under the application of three decision making approach which proves the robustness and stability of the decision results.

Table 4. Comparison of traditional FMEA, Fuzzy FMEA and Grey ranking

Failure codes	RPN output	RPN ranking	FRPN output	Fuzzy ranking	Gray output	Gray ranking
CT_1	245	1	7.33	1	0.4894	2
CT_2	150	3	4.75	3	0.5680	4
CT_3	245	1	6.00	2	0.4854	1
CT_4	200	2	4.50	4	0.5102	3
CP_1	140	3	4.50	3	0.5102	3
CP_2	125	4	4.50	3	0.5680	4
CP_3	112	5	4.50	3	0.5102	3
CP_4	392	1	5.64	1	0.4647	1
CP_5	378	2	5.00	2	0.4769	2
CW_1	240	1	4.75	1	0.5101	1
CW_2	64	3	4.50	2	0.5680	2
CW_3	112	2	4.50	2	0.5101	1
CB_1	336	1	6.28	1	0.4894	1
CB_2	294	2	5.83	2	0.5473	2
CB_3	216	3	5.21	3	0.5680	3
CB_4	140	4	4.5	4	0.5102	4

CONCLUSION

Risk analysis for the operational sustainability of the cooling tower of thermal power industry has been done. Critical failure causes were identified using hybrid decision based framework. Fuzzy set theory based concepts were incorporated for considering the vagueness involved in the information collected from experts. The comparison of ranking results indicates towards the stability of the decision results which are useful for the maintenance personal in fixing the optimal maintenance interval for the considered plant. Operational sustainability of the considered system could be maintained with the implementation of the analysis results. The overall cost of plant operation may also be reduced. Further, the correctness of analysis results are totally based on the available information obtained from the plant sources. However, plant manager has shown his interest with the analysis result but the suitability of the results would be tested once the management of the plant took decision about its implementation.

REFERENCES

Adar, E., Ince, M., Karatop, B., & Bilgili, M. S. (2017). The risk analysis by failure mode and effect analysis (FMEA) and Fuzzy FMEA of supercritical water gasification system used in the sewage sludge treatment. *Journal of Environmental Chemical Engineering*, *5*(1), 1261–1268. doi:10.1016/j.jece.2017.02.006

Ebeling, C. (2000). An Introduction to Reliability and Maintainability Engineering. Tata McGraw-Hill Company Ltd.

Hekmatpanah, M., Shahin, A., & Ravichandran, N. (2011). The application of FMEA in the oil industry in Iran: The case of four litre oil canning process of Sepahan Oil Company. *African Journal of Business Management*, *5*(7), 3019.

Kumru, M., & Kumru, P. Y. (2013). Fuzzy FMEA application to improve purchase process in public hospital. *Applied Soft Computing*, *13*(1), 721–733. doi:10.1016/j.asoc.2012.08.007

Meng Tay, K., & Peng Lim, C. (2006). Fuzzy FMEA with a guided rules reduction system for prioritization of failures. *International Journal of Quality & Reliability Management*, *23*(8), 1047–1066. doi:10.1108/02656710610688202

Panchal, Chattergi, Shukla, Choudhury, & Temostallin. (2017). Integrated fuzzy AHP-CODAS framework for maintenance decision in urea fertilizer industry. *Econ Comput Econ Cyb.*, *51*(3), 179-196.

Panchal, D., & Kumar, D. (2016). Stochastic behaviour analysis of power generating unit in thermal power plant using Fuzzy methodology. *Opsearch.*, *53*(1), 16–40. doi:10.100712597-015-0219-4

Panchal, D., & Kumar, D. (2016). Integrated framework for behavior analysis in a process plant. *Journal of Loss Prevention in the Process Industries*, *40*, 147–161. doi:10.1016/j.jlp.2015.12.021

Panchal, D., & Kumar, D. (2017). Risk analysis of compressor house unit in a thermal power plant using integrated fuzzy FMEA and GRA approach. *Int. J. Industrial and Systems Engineering.*, *25*(2), 228–250. doi:10.1504/IJISE.2017.081519

Panchal, D., & Kumar, D. (2017). *Stochastic behaviour analysis of real industrial system. Int J Syst Assur Eng Manag*. doi:10.100713198-017-0579-7

Panchal, D., Mangla, S. K., Tyagi, M., & Ram, M. (2018). Risk analysis for clean and sustainable production in a urea fertilizer industry. *International Journal of Quality & Reliability Management*, *35*(7), 1459–1476. doi:10.1108/IJQRM-03-2017-0038

Panchal, D., Singh, A. K., Chatterjee, P., Zavadkas, E. K., & Ghorabaee, M. K. (2018). A new fuzzy methodology- based structured framework for RAM and risk analysis. *Applied Soft Computing*, *74*, 242–254. doi:10.1016/j.asoc.2018.10.033

Panchal, D., & Srivastava, P. (2018). *Qualitative analysis of CNG dispensing system using fuzzy FMEA–GRA integrated approach. Int J Syst Assur Eng Manag*. doi:10.100713198-018-0750-9

Sharma, R., Kumar, D., & Kumar, P. (2005). Systematic failure mode and effect analysis using fuzzy linguistic modeling. *International Journal of Quality & Reliability Management*, *22*(9), 986–1004. doi:10.1108/02656710510625248

Sharma, R. K., Kumar, D., & Kumar, P. (2007). Modeling system behaviour for risk and reliability analysis using KBARAM. *Quality and Reliability Engineering International*, *23*(8), 973–998. doi:10.1002/qre.849

Sharma, R. K., & Sharma, P. (2012). Integrated framework to optimize RAM and cost decision in process plant. *Journal of Loss Prevention in the Process Industries*, *25*(6), 883–904. doi:10.1016/j.jlp.2012.04.013

Tay, K. M., & Lim, C. P. (2006). Fuzzy FMEA with a guided rules reduction system for prioritization of failures. *International Journal of Quality & Reliability Management*, *23*(8), 1047–1066. doi:10.1108/02656710610688202

Wang, L., Chu, J., & Wu, J. (2007). Selection of optimum maintenance strategies based on a fuzzy analytic hierarchy process. *International Journal of Production Research*, *107*(1), 151–163. doi:10.1016/j.ijpe.2006.08.005

Chapter 2
Performance Evaluation of Sustainable Smart Cities in India:
An Adaptation of Cartography in PROMETHEE-GIS Approach

Rajeev Ranjan
Doon Business School, India

Prasenjit Chatterjee
https://orcid.org/0000-0002-7994-4252
MCKV Institute of Engineering, India

Dilbagh Panchal
Dr. B. R. Ambedkar National Institute of Technology (NIT) Jalandhar, India

Dragan Pamucar
https://orcid.org/0000-0001-8522-1942
University of Defence in Belgrade, Serbia

ABSTRACT

Indian cities have seen accelerated economic and social growth, attracting more and more people from all parts of the country. Growth achieved by cities is linked to their ability to address issues related to urbanization and associated social, environmental, and economic issues in a holistic manner, while making the most of future opportunities. In this chapter, using PROMETHEE and GAIA (geometrical analysis for interactive aid) approaches, an attempt is made to evaluate the performances of 20 smart cities in Indian context based on 10 critically important criteria. A GIS (geographic information system) method and an HSV (hue-saturation-value) color coding scheme-based on cartographic principles are also employed to identify the influence of individual criterion on the overall rank of the smart cities. This analysis would help the decision makers to identify the strengths and deficiencies of Indian smart cities with respect to considered criteria conditions so that proper promotional and growth actions can be implemented.

DOI: 10.4018/978-1-5225-8579-4.ch002

INTRODUCTION

As India's population continues to raise, more citizens are moving to cities. Professionals expect that about 25-30 people are migrating every minute to major Indian cities from rural areas in search of better living and better lifestyles. It is estimated that by the end of the year 2050, the number of people living in Indian cities will touch 843 million. To accommodate this massive urbanization, India must try to find smarter ways to manage problems, reduce expenses, increase efficiency and improve the quality of life. A smart city is a town that practices diverse electronic data collection sensors to supply information which is use to manage assets and resources efficiently. This includes data collected from citizens, devices, and assets that is processed and analyzed to monitor and manage traffic and transportation systems, power plants, water supply networks, waste management, law enforcement, information systems, schools, libraries, hospitals, and other community services. The smart city concept integrates information and communication technology (ICT) (Yigitcanlar and Baum 2008; Caragliu et al. 2011), and various physical devices connected to the network (the internet of things or IOT) to enhance the efficiency of city operations and services and connect to citizens. Smart city technology allows city administrators to interact directly with both community and city infrastructure and to monitor what is happening in the city and how the city is sprouting. ICT is used to improve quality and performance of urban services, to reduce costs and resource consumption and to improve the contact between citizens and government. Smart city applications are established to manage urban streams and allow for real-time responses. A smart city may therefore be more equipped to respond to challenges than one with a simple transactional relationship with its citizens. Yet, the term itself remains unclear to its essentials and therefore, open to many interpretations. Other terms that have been used for similar concepts include cyber villages, digital city, electronic communities, flex city, information city, intelligent city, knowledge-based city, mesh city, Telecity, Teletopia, ubiquitous city, wired city. The notion of smart city is relatively new and can be seen as a successor of information city, digital city and sustainable city (Yigitcanlar 2006).

Major technological, economic and environmental changes have encouraged interest in smart cities, including climate change, economic reformation, the move to online retail and entertainment, mature populations, urban population growth and pressures on public finances. European union (EU) has devoted constant efforts to developing a strategy for achieving 'smart' urban growth for its metropolitan city-regions. In 2010, it emphasized its attention on solidification of origination and investment in ICT services for the purpose of refining public services and quality of life. Examples of smart city technologies and programs have been implemented in Singapore, Dubai, Milton Keynes, Southampton, Amsterdam, Barcelona, Madrid,

Stockholm, china and New York. Smart city has evolved since its initial arrival in 1996 to the developing market and to a multi-discipline scientific domain. Recently, most standardization bodies around the globe have introduced several competitive standards in their attempt to clarify the smart city and corresponding industrial products. Though, standardization has left out so far smart services modelling as well as corresponding policy making. Such policy making is mainly supported by decision making tools like multi-criteria decision-making (MCDM) but it has not been modeled yet. Sustainability and sustainable urban development concepts generate awareness of the production and use of resources required for residential, industrial, transportation, commercial or recreational processes (Yigitcanlar et al. 2007; Pietrosemoli and Monroy 2013; Goonetilleke et al. 2014; Yigitcanlar and Kamruzzaman 2014, 2015)

In order to achieve a correct description and explanation of the concept of smart city, it is required to first analyze the topic through a specific framework. The framework is divided into five main dimensions: technology framework, human framework, institutional framework, energy framework and data management framework. A technological smart city is not just one concept but there are different amalgamations of technological infrastructure that shape a concept of smart city like digital city, virtual city, information city, intelligent city and cognitive smart city. Human framework is crucial alliance for city development. It includes creative city, learning city, human city and knowledge city. Institutional framework, since 1990s the smart community's movement took shape as a strategy to extend the base of users involved in information technology (IT). Members of these communities are people that share their interest and work in a partnership with government and other institutional organizations to push the use of it to improve the quality of daily life as a consequence of different worsening in daily actions. Smart cities use data and technology to improve sustainability, create economic development, and enhance quality of life factors for people living and working in the city. It also means that the city has a smarter energy infrastructure. Smart city services a combination of data collection, processing, and spreading technologies in combination with networking and computing technologies and data security and privacy measures encouraging application innovation to boost the overall quality of life for its citizens and covering dimensions that include: utilities, health, transportation, entertainment and government services. Especially during the last two decades, metropolitan areas around the world have been engaged in initiatives to improve urban infrastructure and services, aiming at a better environment, social and economic conditions, improving the attractiveness and competitiveness of cities (Lee et al. 2008). Additionally, the concept of smart city goes beyond the definitions of information cities, digital cities, and intelligent cities, because it contextualizes technology to be used in favor of systems and services for people (Jong et al. 2015).

LITERATURE REVIEW

Anand et al. (2017) has developed a model related to the importance of various criteria for sustainability in a smart city, which has determined by using fuzzy and fuzzy-analytic hierarchy process (FAHP) method. The sustainability indicators for designing a smart city in a developing country has been identified. Data envelopment analysis (DEA) model was further adopted to determine the relative efficiency of each of the sustainability indicators for a smart city in the context of input and output criteria. Tiwari and Mishra (2017) has proposed an effective solution-based technique for order preference by similarity to an ideal solution (TOPSIS) approach to help planner, developers, and designers, working for smart city project having a need to evaluate their strategies. It has also been suggested to integrate solid waste generation, population growth, economic and social factor assessment procedures to support decision-making in the context of smart city. Anthopoulos and Giannakidis (2017) had proposed a model by task-based modelling method (TBM) and focuses on policy-making process standardization for smart cities. Rondini et al. (2016) conceptualized smart cities as an integration of human and social capital, linked with traditional and modern communication, infrastructure, economic growth, quality life and proper management of natural resources. In order to facilitate the ranking and the prioritization of the ideas, two-step assessment method based on importance performance analysis (IPA) was developed in the area of product-service system. From the review of the existing literature, it is quite evident that till date, very few researches have yet been conducted in order to evaluate the performance of smart cities, and their conditions and characteristic features in the Indian context. In this chapter, performance of Indian smart cities is evaluated based on several identified parameters using a combined preference ranking organization method for enrichment of evaluations (PROMETHEE), geometrical analysis for interactive aid (GAIA) and geographic information system (GIS) methods.

PROMETHEE-GAIA METHOD

PROMETHEE is an outranking MCDM approach developed by Bans and Vincke (1985). It can be applied to various decisions-making circumstances where the best alternative needs to be selected from a finite set of feasible alternatives based on several conflicting criteria. The basic principle of outranking approach is that if one alternative performs better than another in most of the criteria and it does not present worse performance in others, then it is the most preferred choice. The following decision matrix is the initial step of PROMETHEE method:

$$\begin{bmatrix} g_1(a_1) & g_2(a_1) & \cdots & g_j(a_1) & \cdots & g_n(a_1) \\ g_1(a_2) & g_2(a_2) & \cdots & g_j(a_2) & \cdots & g_n(a_2) \\ \cdots & \cdots & \cdots & \cdots & \cdots & \cdots \\ g_1(a_i) & g_2(a_i) & \cdots & g_j(a_i) & \cdots & g_n(a_i) \\ \cdots & \cdots & \cdots & \cdots & \cdots & \cdots \\ g_1(a_m) & g_1(a_m) & \cdots & g_j(a_m) & \cdots & g_n(a_m) \end{bmatrix} \quad (1)$$

where $g_j(a_i)$ shows the performance of ith alternative on j^{th} criterion, m is the number of alternatives and n is the number of criteria.

The PROMETHEE method application requires some additional information, e.g. the relative importance or weight of the considered criteria and the decision makers' (DMs') preference function for comparing the contribution of the alternatives with respect to each criterion (Brans et al. 1986). Among several methods for criteria weight estimation, Shannon's entropy method has been considered here due to its simplicity and objective nature (Ranjan et al. 2015). The preference structure of PROMETHEE method is based on pair-wise comparison and these preferences are real numbers varying between 0 and 1. It means that for each criterion, the DM considers the following preference function:

$$P_j(a,b) = F_j[d_j(a,b)] \qquad \forall a,b \in A \quad (2)$$

where $d_j(a,b) = [g_j(a) - g_j(b)]$, $\quad 0 \le p_j(a,b) \le 1$ and A is a set containing the feasible alternatives.

For beneficial criteria where, higher the better criteria are always required, this function gives the preference of 'a' over 'b' for the observed deviations between their evaluations on criterion $g_j(.)$. The preference equals to 0 when the deviations are negative. For non-beneficial criteria where lower the better values are always preferable, the preference function can be overturned as follows:

$$P_j(a,b) = F_j[-d_j(a,b)] \quad (3)$$

A generalized criterion $\{g_j(.), P_j(a,b)\}$ is associated to criterion $g_j(.)$ and it is well-defined for each criterion. For the enable selection of a specific preference function, six basic types of preference functions, i.e. level criterion, usual criterion, V-shaped criterion, U-shaped criterion, V-shape with indifference criterion and Gaussian criterion are anticipated to the DM. In some of these preference functions, different threshold parameters (p, q or s) need to be specified by the DM. As the evaluation matrix, $g_j(.)$ is recognized, and the relative importance (w_j) and generalized criterion, $\{g_j(.), P_j(a,b)\}$ are defined, PROMETHEE method prepared for implementation. PROMETHEE method is basically based on pair-wise comparisons among different alternatives where the aggregated preference indices are defined as follows (Brans and Mareschal, 1994):

$$\begin{cases} \pi(a,b) = \sum_{j=1}^{n} p_j(a,b)w_j \\ \pi(b,a) = \sum_{j=1}^{n} p_j(b,a)w_j \end{cases} \quad (4)$$

where $\pi(a, b)$ expresses the degree with which 'a' is preferred to 'b' over all the criteria and $\pi(b, a)$ represents how 'b' is preferred to 'a'. In most of the cases, there are criteria for which 'a' is better than 'b', and criteria for which 'b' is better than 'a', and consequently, $\pi(a, b)$ and $\pi(b, a)$ values are usually positive.

When the values of $\pi(a, b)$ and $\pi(b, a)$ are computed for each pair of alternatives in the evaluation matrix, a complete outranking graph, including two arcs between each pair of nodes (alternatives) is developed. In this method, each alternative 'a' faces (m - 1) number of other alternatives in the evaluation matrix. Now, the following two outranking flows can be defined as:

$$\text{Positive outranking flow, } \varphi^{+}(a) = \frac{1}{m-1}\sum_{x \in A} \pi(a,x) \quad (5)$$

$$\text{Negative outranking flow, } \varphi^{-}(a) = \frac{1}{m-1}\sum_{x \in A} \pi(x,a) \quad (6)$$

The positive outranking flow expresses how an alternative 'a' outranks all other alternatives. The higher the value of $\varphi^{+}(a)$, the better is the alternative. The negative outranking flow expresses how an alternative 'a' is being outranked by the others. Lower value of $\varphi^{-}(a)$ signifies better alternatives. In PROMETHEE I method, a partial ranking (P^I,I^I,R^I) is obtained from the positive and negative outranking flow values where P^I,I^I and R^I respectively stands for preference, indifference and incomparability relations. Both the flows do not usually induce the same rankings. PROMETHEE II method can provide a complete rank preorder of the alternatives by using a net flow, though it loses much information of preference relations. In this method, there is a balance between the positive and negative outranking flows. The net outranking flow for each alternative can be obtained using the following equation:

$$\varphi(a) = \varphi^{+}(a) - \varphi^{-}(a) \tag{7}$$

The higher the value of $\varphi(a)$, the better is the alternative. Thus, the best alternative is the one having the highest $\varphi(a)$ value.

The involvement of various criteria is shown in The GAIA plane considered with alternatives. The GAIA plan of the alternatives and criteria, which clearly shows the best alternative and the alternative for which a criterion is the best, distinguishes it from the other MCDM approaches. From the positive and negative outranking flow values, as expressed by Eqns. (5) and (6) respectively, the net outranking flow value can be estimated, as given below:

$$\varphi(a) = \varphi^{+}(a) - \varphi^{-}(a) = \frac{1}{m-1} \sum_{j=1}^{n} \sum_{x \in A} [P_j(a,x) - P_j(x,a)] w_j \tag{8}$$

Consequently,

$$\varphi(a) = \sum_{j=1}^{n} \varphi_j(a) w_j \tag{9}$$

$$\varphi_j(a) = \frac{1}{m-1} \sum_{x \in A} [P_j(a,x) - P_j(x,a)] \tag{10}$$

where $\varphi_j(a)$ is the single criterion net flow obtained when only criterion $g_j(.)$ is considered. It expresses how an alternative 'a' is outranking ($\varphi_j(a) > 0$) or outranked ($\varphi_j(a) < 0$) by all other alternatives on criterion $g_j(.)$. From Eqn. (9), it is observed that the global net flow of an alternative is the product between the vector of the criteria weights and the profile vector of that alternative. This property is primarily used for developing the GAIA plane. This can be defined based on the single criterion net flows of all the alternatives, as shown by the matrix M.

$$M = \begin{bmatrix} \phi_1(a_1) & \phi_2(a_1) & \ldots & \phi_j(a_1) & \ldots & \phi_n(a_1) \\ \phi_1(a_2) & \phi_2(a_2) & \ldots & \phi_j(a_2) & \ldots & \phi_n(a_2) \\ \ldots & \ldots & \ldots & \ldots & \ldots & \ldots \\ \phi_1(a_i) & \phi_2(a_i) & \ldots & \phi_j(a_i) & \ldots & \phi_n(a_i) \\ \ldots & \ldots & \ldots & \ldots & \ldots & \ldots \\ \phi_1(a_m) & \phi_1(a_m) & \ldots & \phi_j(a_m) & \ldots & \phi_n(a_m) \end{bmatrix} \tag{11}$$

The information included in matrix M is more exhaustive than that of the decision matrix in Eqn. (1), because the degree of reference given by the generalized criteria are taken into consideration in M, moreover, the $g_j(a_i)$ values are expressed on their own scale, while the $\varphi_j(a_i)$ values are dimensionless. In addition, the matrix M is not dependent on the weights of the considered criteria. Therefore, the set of m alternatives can be represented as a cloud of m points in an n-dimensional space. As the number of criteria is usually larger than two, it is not possible to obtain a clear view of the relative positions of the points with respect to the criteria. The GAIA plane is found by projection of this on a plane such that as few information may get vanished. In this plane, alternatives $(a_1,a_2,\ldots,a_m)$ are represented by points and the criteria $(c_1,c_2,\ldots,c_n)$ by axes. From Eqn. (9), it can be said that the PROMETHEE net flow of a_i is the projection of vector of its single criterion net flows on w. Therefore, the relative positions of the projections of all the alternatives on w provide the PROMETHEE ranking of the alternatives. Clearly, the vector w plays a pivotal role. It is represented in GAIA plane by the projection of the unit vector of the weights. This projection is referred to as the PROMETHEE decision axis (π). This axis shows the direction of the compromise solution resulting from the weights allocated to the criteria. It is clear that the π-axis will coincide with the axis of that criterion in GAIA plane, when the weights are concentrated on one criterion. the π-axis appears as a weighted resultant of all the criteria axes, when the weights are distributed over all the criteria. If π is long, the PROMETHEE decision axis has a strong decision power and the DM is requested to select alternatives as far as possible on its direction. On the other hand, when π is short, it has no strong decision power, which means that according to the weights, the criteria are contradictory and the selection of a good compromise solution is a tough problem (Ranjan & Chakraboty, 2015). In this plane, criteria expressing alike preferences on the evaluation data are oriented in the same direction, and the contradictory criteria are pointing in the opposite directions. When the weights are modified, the positions of the alternatives and criteria remain unchanged in GAIA plane. The weight vector appears as a decision stick that the DM can change according to the preference in favor of a particular criterion. When a sensitive analysis is performed by changing the criteria weights, the PROMETHEE decision stick (w) and the π-axis move in such a way that the consequences for decision-making can be easily detected in GAIA plane.

CARTOGRAPHY AND HSV COLOR CODING SCHEME

In order to create a better visual analysis, a GIS-based approach is also applied in this paper. The GIS can be considered as a system that permits the user to store, manage, analyze and present spatially referenced data (Chakraborty et al. 2014). It delivers various means to represent data in the form of maps, tables and diagrams. The geographical view obtained from GIS application is useful for finding the flow and position (both absolute and relative) of the alternatives. The HSV color coding scheme, based on cartography and geographical representations, is one such GIS technique which contains two main features, i.e. color and size, for differentiating between various alternatives (Lidouh et al., 2011). Color can again be disintegrated into three components, e.g. color hue, color saturation (relative darkness or lightness of color relative to a grayscale) and color brightness (value to refer to the perceived intensity of reflecting objects). By applying these concepts, the HSV color coding scheme helps to find different profiles of every alternative. The three components of color, i.e. hue, saturation and value are represented by a cylinder. By superimposing the GAIA plane on this color chart, the color connected with each alternative can be easily identified. Thus, alternatives lying close to one another on GAIA plane and thereby, having a similar profile, fall in similar color regions. Once the color connected with each alternative is properly identified, they need to be represented on the geographical map. For this purpose, instead of coloring the entire region, the colors can be represented within circles of variable diameters. The size of the circle, used for representing an alternative on the geographical map serves as an indicator of the alternative's overall performance. A bigger diameter corresponds to a higher net flow value and a smaller one to a lower value. Different colors are assigned to the alternatives to identify their individual weak and strong points (Lidouh et al., 2011).

PROMETHEE-GAIA-BASED PERFORMANCE ASSESSMENT OF INDIAN SMART CITIES

The current scenario requires smart cities to find ways to manage new challenges. Cities worldwide have started to look for solutions which enable transportation linkages, mixed land uses, and high-quality urban services with long-term positive effects on the economy. For instance, high-quality and more efficient public transport that responds to economic needs and connects labor with employment is considered a key element for smart city growth. Many of the new approaches related to urban services have been based on harnessing technologies, including ICT, helping to create smart cities. The concept of the smart city is far from being limited to the application of technologies to cities. In fact, the use of the term is booming in many sectors with no agreed upon definitions. This has led to confusion among urban policy makers, hoping to institute policies that will make their cities smart. This chapter seeks to advance state-of-the-art knowledge on what a smart city is, what its key dimensions are, and how its performance can be evaluated. It is based on a review of the literature, including peer reviewed papers published. In particular, it is structured as follows. First, the main definitions of smart city are reviewed, highlighting the different meanings given to this concept and the several perspectives through which it has been studied; next, it analyzes the key dimensions of a best smart city; then it focuses on the measures of performance of a smart city, reports on the experiences of so called, smart cities; finally closing with a discussion of the main findings of the study. Proper understanding of the nature, significance and importance of resources are thus essential for the DMs at all levels to make appropriate plans for improving the quality of life in all the Indian smart cities. Although Indian cities has developed in education, health, transportation, tourism, jobs, population, pollution control etc but still required a more development in these areas. India is bounded by the Indian Ocean on the south, the Arabian Sea on the south-west and the Bay of Bengal on the south-east. Geographically, India consists of 30 states and 6 union territories with 98 smart cities but only 20 smart cities are considered in this paper for their performance analysis. In order to achieve the objectives of this paper, at first, several evaluation criteria are shortlisted based on important field required to make a city perfect. A city relatively large dense permanent settlement of heterogeneous individuals and groups of individuals organized to perform, or to facilitate the performance of locality relevant functions in an integrated manner and to insure in integration with system of which the city is the part. A review of the published literature reveals that there have been a few studies to measure the performance competitiveness among the Indian smart cities. The major attributes which contribute to the performance evaluation of best smart cities are health, education, population, area, population density, male female ratio of the city, pollution, tourist visited to particular city, crime, job opportunity in the city, roads condition, airport facility and railway station connectivity, situational conditions, management of the destination, attractions and natural resources, climate, culture, architectural heritage, museums, festivals, hotels, transport and entertainment (Gooroochurn and Sugiyarto, 2004; Crouch and Ritchie, 1999; Kozak and Remmington, 1999; Melián-Gonzáles and García-Falcón, 2003). Based on the available information in various

Indian smart cities related websites, the evaluation of 20 smart Indian cities is performed with respect to 10 criteria, as enlisted in Table 1. The total Indian population is found as 1.22 billion. In this case, Ahmedabad city is the highest population (55,77,940) and Kakinada (312538) is lowest one. In the case of total area, the total area of India as per 2011census considered 29 states and 6 territories is found 32,87,240 sq. km. Vishakhapatnam tops the list with maximum area of 513.61 sq. km, while new Delhi is at the bottom of the list (42.74 sq. km). The next criterion is Slum population, the current slum population of India is 65 million as per the 2011 census. This criterion is taken as non-beneficial criteria, so the lower value is the better one. Among all the alternatives kochi is the lowest one and Jabalpur is the highest one. The next criterion is the population density, which is correlated to population and can be one of the important resources for the smart city. City population and especially area are, however, heavily dependent on the definition of "urban area" used: densities are almost invariably higher for the central city area than when suburban settlements and the intervening rural areas are included, as in the areas of agglomeration or metropolitan area, the latter sometimes including neighboring cities. Chennai is the one of the highest populated density cities per unit area, while Vishakhapatnam and Guwahati are the lowest one. A total of 10 different criteria, as shown in Table 1, are considered for this evaluation purpose since all these 10 criteria are predominantly important for the current appraisal system and are entirely uncorrelated. Literacy rate is an extremely important parameter which can be used as an indicator of the status of education in individual city. It is defined as the total percentage of population of an area at a particular time aged seven years or above who can read and write with understanding. According to that report, Kochi is the most literate smart city where the literacy rate is 97.36%. The second most literate smart city in the country is Guwahati with a literacy rate of 91.47%. On the other hand, Kakinada is the most illiterate in considered smart city in India. However, literacy rate alone cannot serve as a guideline towards determining the educational performance of individual smart city, since a host of other factors need to be taken into consideration. The total number of schools, colleges and other educational institutions present in a smart city act as a direct indicator of the scope of education present in that city. A smart city having lower number of schools and colleges inherently has a major disadvantage. However, simply counting the number of educational institutions would not serve the basic purpose because the overall area of every smart city is different from each other. For this purpose, the numbers of schools and colleges present in a smart city are divided by its geographical area to obtain two parameters, i.e. number of schools and number of colleges, which can be used as two important criteria. But in this both criteria have merged and the criterion become Total number of school and colleges. In this criterion Delhi is the top and Solapur is the lowest one. The next very important criterion for the measure of smart city is poverty ratio. The head count ratio (HCR) is the proportion of a population that exists, or lives, below the poverty line. The Poverty headcount ratio at national poverty line (percentage of population) in India was last reported at 21.9% in 2011-12. In this criterion, Delhi (0.87) has the lowest poverty ratio smart city and Belagavi (32.33) is the highest one. Income per capita is a measure of the amount of money earned per person in a certain area. It can apply to the average per-person income for a city, region or country, and is used as a means of evaluating the living conditions and quality of life in different areas. It can be calculated for a country by dividing the country's national income by its population. In the case of per capita income among the smart cities, Delhi is the highest per capita income almost Rs 112510 and Udaipur is the lowest with Rs. 24,135. One of the important criteria for any smart city is the unemployment rate, which directly get the benefit of the young generation and GDP of any country are correlated with it. This criterion is non-beneficial type, where lowest value would be the better and highest value would be the worst. Surat is the lowest (0.29%) and Jabalpur is the highest (10.40%). To become smart city, the health of the citizen is very important. For the healthy life, availability of hospitals in the city is very important. The next criterion is the number of hospitals available in the city. In this case, in Delhi, there is highest number of hospitals available while in Kakinada, a smaller number of hospitals is available.

Table 1. Criteria for performance evaluation of smart cities

Criteria	Symbol	Criteria Type	Weight
Total Population	C_1	Non-Beneficial	0.0797
Slum Population (in% age)	C_2	Non-Beneficial	0.1028
Area of City (sq. km)	C_3	Beneficial	0.1041
Population Density (persons/sq. km)	C_4	Non-Beneficial	0.1212
Literacy Rate (%age)	C_5	Beneficial	0.1589
Poverty Ratio(%age)	C_6	Non-Beneficial	0.0835
Per Capita Income (Rs)	C_7	Beneficial	0.1375
Unemployment Rate (%age)	C_8	Non-Beneficial	0.0756
Number of Hospitals in Particular city	C_9	Beneficial	0.0933
Number of Schools and Colleges in the city	C_{10}	Beneficial	0.0429

In order to evaluate the relative performance of the considered 20 Indian smart cities, Visual PROMETHEE 1.4 software is used here. The initial input window of Visual PROMETHEE software is shown in Figure 1. where the user or the DM can easily select the number of alternatives, number of criteria, number of scenarios for group decision-making problems, type of each criterion (beneficial or non-beneficial), priority weights for the criteria, unit for each criterion and type of the preference function (usual or U-shape or V-shape or level or linear or Gaussian). The weights for various criteria are calculated using Shannon's entropy method as $C_1 = 0.0797$, $C_2 = 0.1028$, $C_3 = 0.1041$, $C_4 = 0.1212$, $C_5 = 0.1589$, $C_6 = 0.0835$, $C_7 = 0.1375$, $C_8 = 0.0756$, $C_9 = 0.0933$ and $C_{10} = 0.04293$, which are then used for the PROMETHEE-GAIA method-based analyses. Even though there are mainly six types of preference functions in PROMETHEE method-based analysis, most of them require definition of some preference parameters, like preference and indifference thresholds. But, in many real time circumstances, it may be difficult for the DM to exactly specify which specific form of preference function is fit for each criterion and

Table 2. Decision matrix for performance evaluation of Indian smart cities

Sl. No.	Smart cities	C_1	C_2	C_3	C_4	C_5	C_6	C_7	C_8	C_9	C_{10}
A_1	Bhubaneswar	840834	19.5	135	6228	91.89	4.7	33312	4.27	46	240
A_2	Pune	3124458	22.1	276.4	11304	89.56	2.73	88341	3.29	108	1396
A3	Jaipur	3046163	10.62	484.64	6285	83.33	5.92	30461	4.09	44	684
A_4	Surat	4467797	10.46	335.82	13304	87.89	5.61	38301	0.29	50	250
A_5	Kochi	602046	0.86	107.13	5620	97.36	4.05	63599	4.45	18	148
A_6	Ahmedabad	5577940	4.49	468.92	11894	88.29	6.64	45773	0.86	64	668
A_7	Jabalpur	1055525	45.82	152.53	6920	87.39	16.93	38968	10.4	34	169
A_8	Vishakhapatnam	1728128	44.61	513.61	3365	81.79	6.35	50580	3.95	29	223
A_9	Solapur	951558	18.43	178.57	5329	82.8	19.29	45859	1.65	28	81
A_{10}	Davanagere	434971	13.79	77.12	5640	84.9	23.76	30219	3.64	27	113
A_{11}	Indore	1964086	30.05	172.39	11393	85.87	2.8	52501	2.07	40	290
A_{12}	Delhi	2578030	7.76	42.74	6032	89.83	0.87	112510	2.03	172	2740
A_{13}	Kakinada	312538	35.85	57.36	5449	80.62	4.64	37712	4.94	12	91
A_{14}	Belagavi	488157	10.76	99.61	4900	89.82	32.33	28856	6.5	27	87
A_{15}	Bhopal	1798218	26.68	285.88	6290	83.47	9.72	47214	1.53	49	387
A_{16}	Udaipur	451100	14.36	56.92	7925	89.66	15.97	24135	0.3	17	160
A_{17}	Guwahati	957352	2.69	219.06	4370	91.47	9.09	43278	4.32	15	246
A_{18}	Chennai	4646732	28.89	175	26553	90.18	2.34	57706	2.32	34	631
A_{19}	Ludhiana	1618879	15.08	159.37	10158	85.77	9.51	51633	0.8	38	272
A_{20}	Coimbatore	1050721	12.29	105.6	9950	91.3	3.66	65781	3.21	126	473

also to determine the parameter values. To overcome these difficulties and make the related computation easier, the simplest form of the preference function, i.e. usual criterion is adopted here. Now, the alternative Indian smart cities are compared in pairs for each criterion based on the given preference function. Based on the weighted sum of single criterion preferences, the positive and negative outranking flows are calculated as a measure of dominance of the alternatives.

Based on the positive and negative outranking flows, a partial preorder of the alternatives is defined according to PROMETHEE I method. The net outranking flow is also calculated to avoid incomparability's and define a complete preorder of

Figure 1. Input window of Visual PROMETHEE for performance evaluation of 20 Indian smart cities

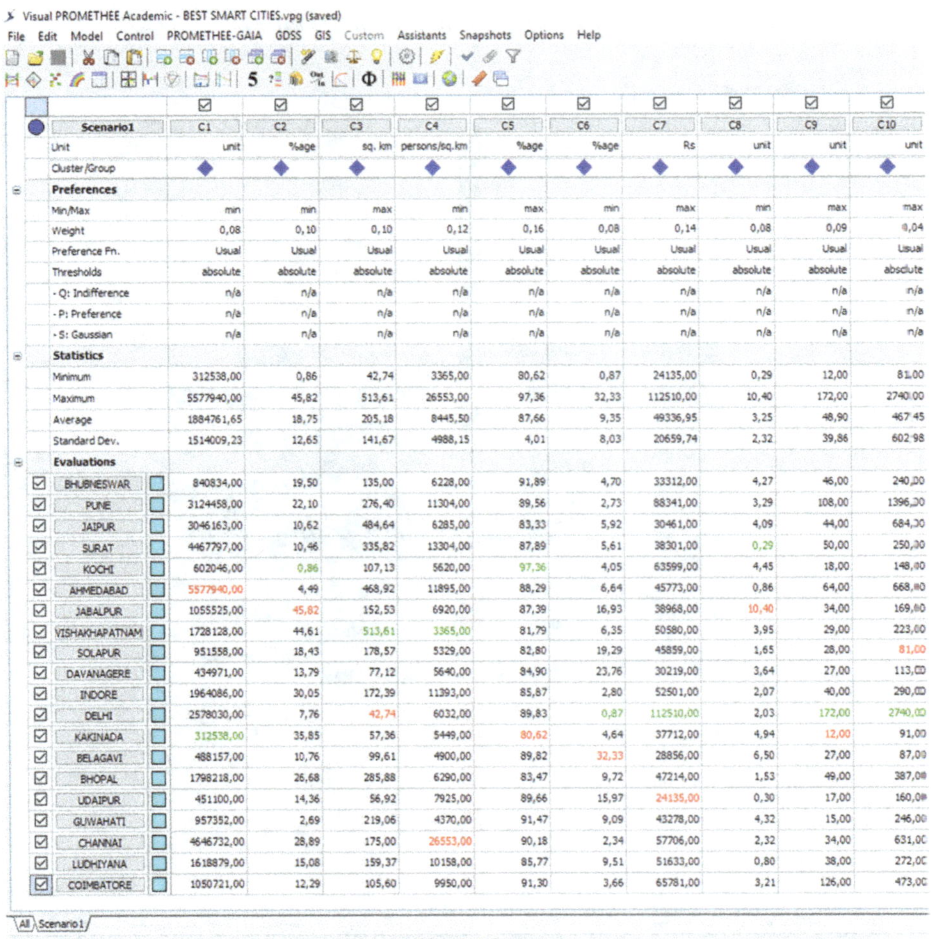

Scenario1	C1	C2	C3	C4	C5	C6	C7	C8	C9	C10
Unit	unit	%age	sq. km	persons/sq.km	%age	%age	Rs	unit	unit	unit
Cluster/Group										
Preferences										
Min/Max	min	min	max	min	max	min	max	min	max	max
Weight	0,08	0,10	0,10	0,12	0,16	0,08	0,14	0,08	0,09	0,04
Preference Fn.	Usual	Usual	Usual	Usual	Usual	Usual	Usual	Usual	Usual	Usual
Thresholds	absolute	absolute	absolute	absolute	absolute	absolute	absolute	absolute	absolute	absolute
- Q: Indifference	n/a	n/a	n/a	n/a	n/a	n/a	n/a	n/a	n/a	n/a
- P: Preference	n/a	n/a	n/a	n/a	n/a	n/a	n/a	n/a	n/a	n/a
- S: Gaussian	n/a	n/a	n/a	n/a	n/a	n/a	n/a	n/a	n/a	n/a
Statistics										
Minimum	312538,00	0,86	42,74	3365,00	80,62	0,87	24135,00	0,29	12,00	81,00
Maximum	5577940,00	45,82	513,61	26553,00	97,36	32,33	112510,00	10,40	172,00	2740,00
Average	1884761,65	18,75	205,18	8445,50	87,66	9,35	49336,95	3,25	48,90	467,45
Standard Dev.	1514009,23	12,65	141,67	4988,15	4,01	8,03	20659,74	2,32	39,86	602,98
Evaluations										
BHUBNESWAR	840834,00	19,50	135,00	6228,00	91,89	4,70	33312,00	4,27	46,00	240,00
PUNE	3124458,00	22,10	276,40	11304,00	89,56	2,73	88341,00	3,29	108,00	1396,00
JAIPUR	3046163,00	10,62	484,64	6285,00	83,33	5,92	30461,00	4,09	44,00	684,00
SURAT	4467797,00	10,46	335,82	13304,00	87,89	5,61	38301,00	0,29	50,00	250,00
KOCHI	602046,00	0,86	107,13	5620,00	97,36	4,05	63599,00	4,45	18,00	148,00
AHMEDABAD	5577940,00	4,49	468,92	11895,00	88,29	6,64	45773,00	0,86	64,00	668,00
JABALPUR	1055525,00	45,82	152,53	6920,00	87,39	16,93	38968,00	10,40	34,00	169,00
VISHAKHAPATNAM	1728128,00	44,61	513,61	3365,00	81,79	6,35	50580,00	3,95	29,00	223,00
SOLAPUR	951558,00	18,43	178,57	5329,00	82,80	19,29	45859,00	1,65	28,00	81,00
DAVANAGERE	434971,00	13,79	77,12	5640,00	84,90	23,76	30219,00	3,64	27,00	113,00
INDORE	1964086,00	30,05	172,39	11393,00	85,87	2,80	52501,00	2,07	40,00	290,00
DELHI	2578030,00	7,76	42,74	6032,00	89,83	0,87	112510,00	2,03	172,00	2740,00
KAKINADA	312538,00	35,85	57,36	5449,00	80,62	4,64	37712,00	4,94	12,00	91,00
BELAGAVI	488157,00	10,76	99,61	4900,00	89,82	32,33	28856,00	6,50	27,00	87,00
BHOPAL	1798218,00	26,68	285,88	6290,00	83,47	9,72	47214,00	1,53	49,00	387,00
UDAIPUR	451100,00	14,36	56,92	7925,00	89,66	15,97	24135,00	0,30	17,00	160,00
GUWAHATI	957352,00	2,69	219,06	4370,00	91,47	9,09	43278,00	4,32	15,00	246,00
CHANNAI	4646732,00	28,89	175,00	26553,00	90,18	2,34	57706,00	2,32	34,00	631,00
LUDHIYANA	1618879,00	15,08	159,37	10158,00	85,77	9,51	51633,00	0,80	38,00	272,00
COIMBATORE	1050721,00	12,29	105,60	9950,00	91,30	3,66	65781,00	3,21	126,00	473,00

candidate Indian smart cities according to PROMETHEE II method. The performance of 20 Indian smart cities with respect to their potential is displayed in a comparative ranking diagram, as shown in Figure 2.

In the central of Figure 2, there is a scale having a range of values from -1 to 1. The upper half of this scale corresponds to positive net outranking flow values, whereas, the lower half represents negative net outranking flow values. As mentioned earlier, PROMETHEE II method ranks the alternatives by calculating net outranking flow values. Therefore, the alternative with the highest net outranking flow value is the best compromised solution. Figure 2 also shows that Delhi is the best performing Indian smart city, while, Kakinada lies at the lowest of the ranking list of Indian smart cities. This ranking signifies that Delhi has excellent performance with respect to all the considered criteria, whereas, Kakinada is the worst performing smart

Figure 2. PROMETHEE II ranking of 20 Indian smart cities

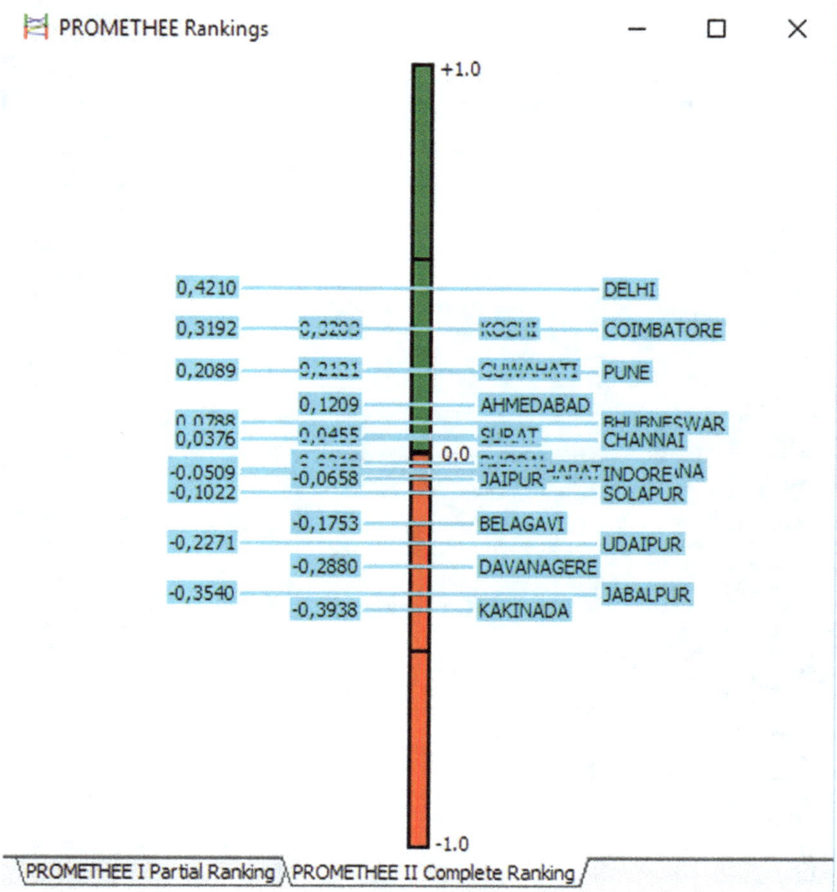

city with respect to the designated parameters for performance evaluation. Besides Delhi, the performances of Kochi, Coimbatore and Pune are also satisfactory. The performances of Kakinada and Jabalpur are relatively poor in the race of smart city. From Figure 2, it is further observed that 9 Indian smart cities have positive net outranking flows, while the remaining 11cities have negative net outranking flows, and it can be concluded that the State as well as Central Governments should emphasize on developing these smart cities accompanied by a detailed study of their characteristics in all criteria.

GAIA plane is a useful tool to assess a given decision-making problem. In order to attain a better understanding of the considered smart cities evaluation problem, the developed GAIA plane is exhibited in Figure 3. The positions of the alternative smart cities are shown in light green while the criteria are shown in blue. Figure 3 indicates that Delhi, Kochi, Coimbatore and Pune have the farthest distances from the origin in the direction of π axis indicating the dominance of these smart cities over the others. Moreover, Delhi is the farthest from the origin in the direction of π axis, which makes it the best performing smart city. According to this plane, criteria

Figure 3. GAIA plane for performance evaluation of 20 Indian smart cities

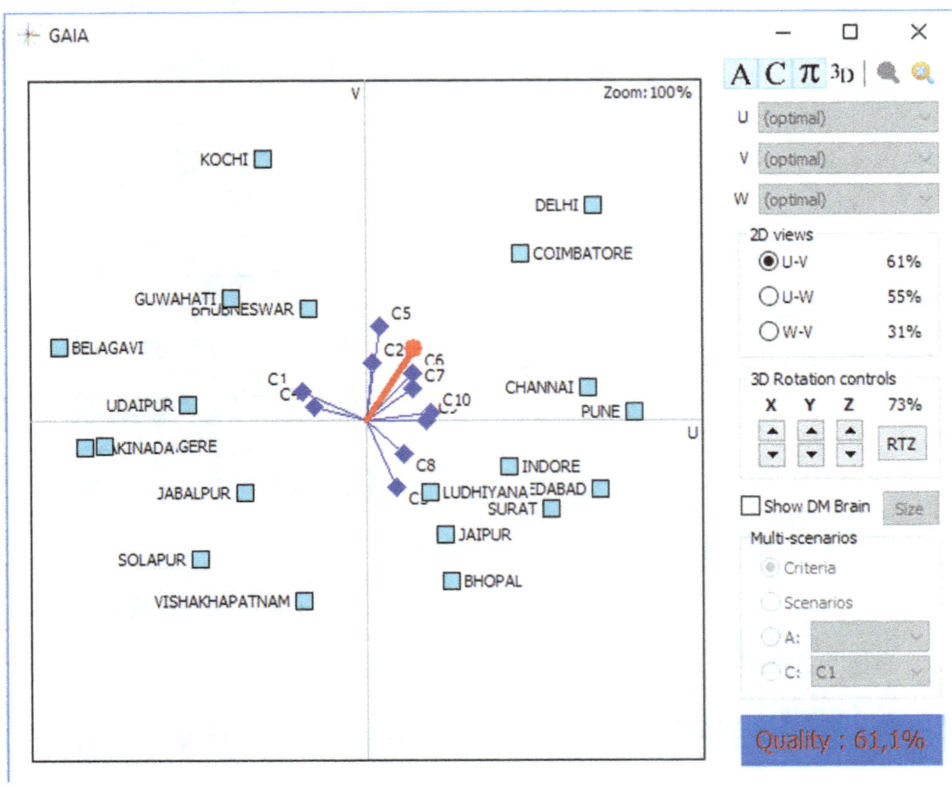

of slum population (C_2), literacy rate (C_5), poverty ratio (C_6), per capita income (C_7), number of hospitals (C_9) and number of schools and colleges (C_{10}), form one cluster, while area of city (C_3) and unemployment rate (C_8) are in the second cluster. The criteria in both the clusters have the same effects in this evaluation process and show similar preferences. In this figure, it is also observed that the positions of all the non-beneficial criteria are symmetrically opposite in the GAIA plane. The criterion total population (C1) and population density (C4) are non-beneficial criteria. Their position is totally opposite in the GAIA plane. The quality of solution for this performance evaluation problem of 20 Indian smart cities is 61.1%, which suggests that PROMETHEE-GAIA method can solve it proficiently.

The PROMETHEE network is another presentation of PROMETHEE II ranking. The ranking of 20 Indian smart cities from the best to the worst is shown with the help of nodes and arrows in PROMETHEE network in Figure 4. The nodes are

Figure 4. PROMETHEE network of 20 Indian smart cities

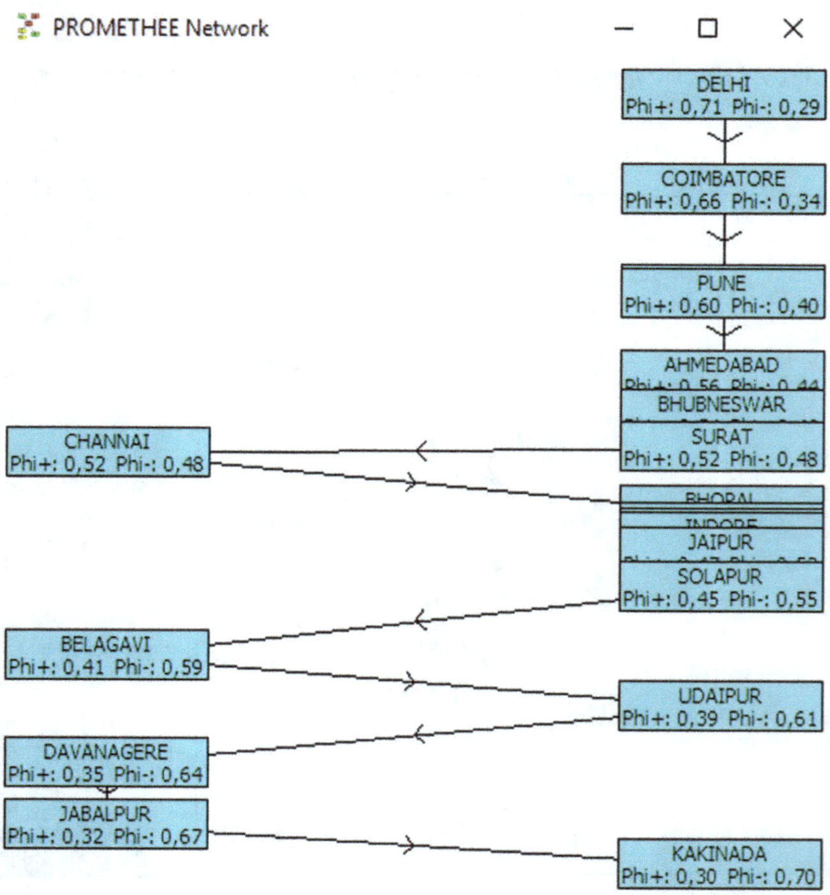

positioned at relative positions corresponding to PROMETHEE diamond so that the contiguities between the flow values appear clearly. It is observed from this figure that Delhi is the best performer among all the smart cities, while Kakinada and Jabalpur stay at the bottom of the ranking list. The PROMETHEE rainbow diagram for performance evaluation of 20 Indian smart cities is shown in Figure 5. The PROMETHEE rainbow diagram is a disaggregated view of PROMETHEE II net flow values. The alternatives are displayed from left to right according to their PROMETHEE II rankings. For each alternative, a colored vertical bar is drawn which is composed of different slices. Each slice within the bar corresponds to the contribution of one criterion in the computation of net flow values. Its height is equal to the uni-criterion net flow value multiplied by the corresponding criterion weight. Thus, the multi-criteria net flow value is the sum of all slices (positive ones minus negative ones). Larger positive slices (most important good features of the alternatives) are on top and larger negative slices (most important weaknesses) are at the bottom of the slice. The PROMETHEE rainbow makes it possible to visualize the characteristic profiles of the alternatives, taking into account the criteria weights. It is observed that the alternative at the leftmost position in Figure 5 got the highest rank, whereas, the rightmost alternative comes last in the ranking. So, Delhi is the top performer among all the smart cities, whereas, Kakinada and Jabalpur are poor performers. Now, when the top two ranked smart cities, i.e. Kochi and Coimbatore are compared between themselves, it is observed from Figure 6 that Kochi outperforms Coimbatore with respect to criteria total population (C_1), slum population (C_2), area

Figure 5. PROMETHEE rainbow for performance evaluation for 20 Indian smart cities

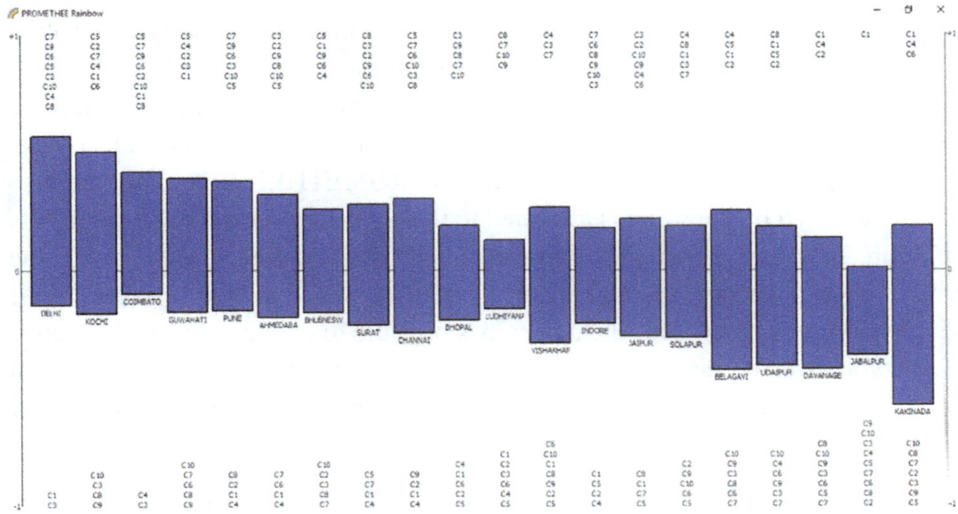

Figure 6. PROMETHEE rainbow between Kochi and Coimbatore

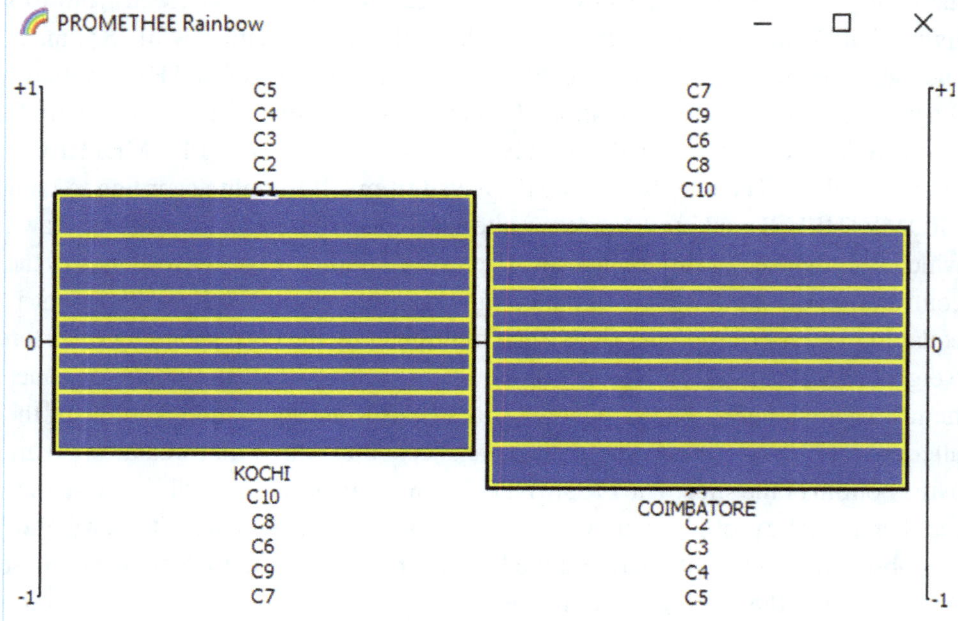

of city (C_3), population density (C_4) and literacy rate (C_5) whereas, Coimbatore performs better than Kochi with respect to the remaining criteria. For example, the population density which is a non-beneficial criterion, is only 5620 in Kochi, whereas, it is 9950 in Coimbatore. While, literacy rate (C_5), a beneficial criterion, in Kochi is 97.36%, the same for Coimbatore is 91.30%. These two criteria values signify that Kochi excels over Coimbatore. As the sum of weights of criteria total population, slum population, area of city population density and literacy rate is dominating over that for other criteria, so Kochi comes out as the best performer as compared to Coimbatore. According to PROMETHEE rainbow diagram, Kakinada has the weakest performance amongst the 20 Indian smart cities with respect to all criteria.

As compared to any MCDM method, the PROMETHEE II rankings are also directly affected by the weights allocated to the considered criteria and the results of Figure 7 are based on a predetermined set of criteria weights. Therefore, if the criteria weights change, the ranking of alternatives would likely to change. Thus, with a good sensitivity analysis, it becomes possible to obtain more interpretative results which would enhance the DM's understanding of the decision-making problem and assess whether the derived solutions are sensitive to parametric changes. A special feature of Visual PROMETHEE software, called 'walking weights' allows the DM to amend the values of criteria weights and visualize the resulting changes in PROMETHEE II ranking. It is basically an interactive weight sensitivity analysis tool. Changing the weights would only change the position of the π-axis, while the positions of the alternatives and criteria would remain unchanged. Figure 7 shows the best to worst ranking orders of the Indian smart cities when a different combination of weights is allocated to the evaluation criteria. It is observed that at the new criteria weight combination, two different clusters are formed in Figure 7. Delhi, Kochi, Coimbatore, Pune, Guwahati, Ahmedabad, Bhubaneswar, Surat and Chennai form the first cluster. This cluster of smart cities have positive flow values which signify that these cities are good performers among all smart cities. While in the second cluster, Bhopal, Ludhiana, Vishakhapatnam, Indore, Jaipur, Solapur, Belagavi, Udaipur, Davanagre, Jabalpur and Kakinada have negative flow values, which signify their poor performance with respect to all the considered criteria in Smart Cities.

Figure 7. Walking weights for 20 Indian Smart cities

The Visual PROMETHEE software can also provide analysis on the stability of criteria and sub-criteria weights, the results of this weight stability analysis are given in Figure 8. It can be used for further analysis with respect to modification of the criteria weight coefficients. However, it should be noted that the considered criteria and their weights have a decisive influence on determining the ranking of the smart cities. Figure 8 represents the visual stability interval with respect to C_1 (total population) criterion weight which exhibits how the net flow values change as a function of C_1 criterion weight. In this figure, the horizontal axis represents the weight of criterion C_1 and the vertical axis measures the net flow values. For each Smart City, a line is drawn showing the net flow value as a function of C_1 criterion weight. At the right edge of this figure, the weight of C_1 criterion is 100%, and the Indian smart cities are ranked here only with respect to C_1 criterion. The position

Figure 8. Visual stability interval for criterion "total population" (C_1)

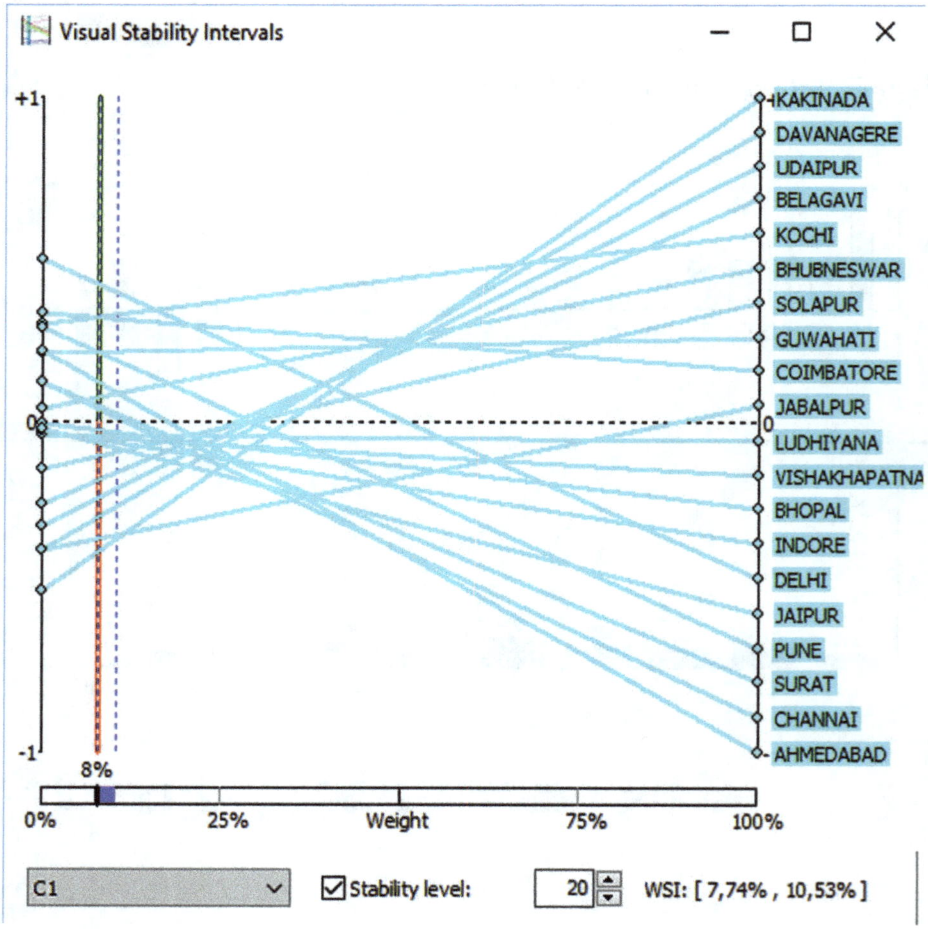

of the vertical red bar corresponds to the current weight of criterion C_1 at 8%. The intersections of different lines with this vertical bar provide the present complete ranking of the Indian smart cities. It is observed that the weight stability interval for criterion C_1 is 7.74% to 10.53%, which indicates that within this range, there would be no change in the ranking of the top ranked smart cities. In case of Kakinada, it remains as an underperformer in the lower range of C_1 criterion weight, but in the higher weight for C_1 criterion, Jabalpur supersedes Kakinada to be the worst performer in the Indian smart cities. Similarly, the weight stability intervals can also be determined for the remaining evaluation criteria.

The GAIA web as developed for this smart cities' performance evaluation problem with respect to the best recital smart city to Delhi is portrayed in Figure 9 where instead of displaying various criteria at arbitrary angles, the position of criteria axes in the GAIA plane are uses as a reference. Those criteria which are

Figure 9. Developed GAIA plane for Smart City Delhi

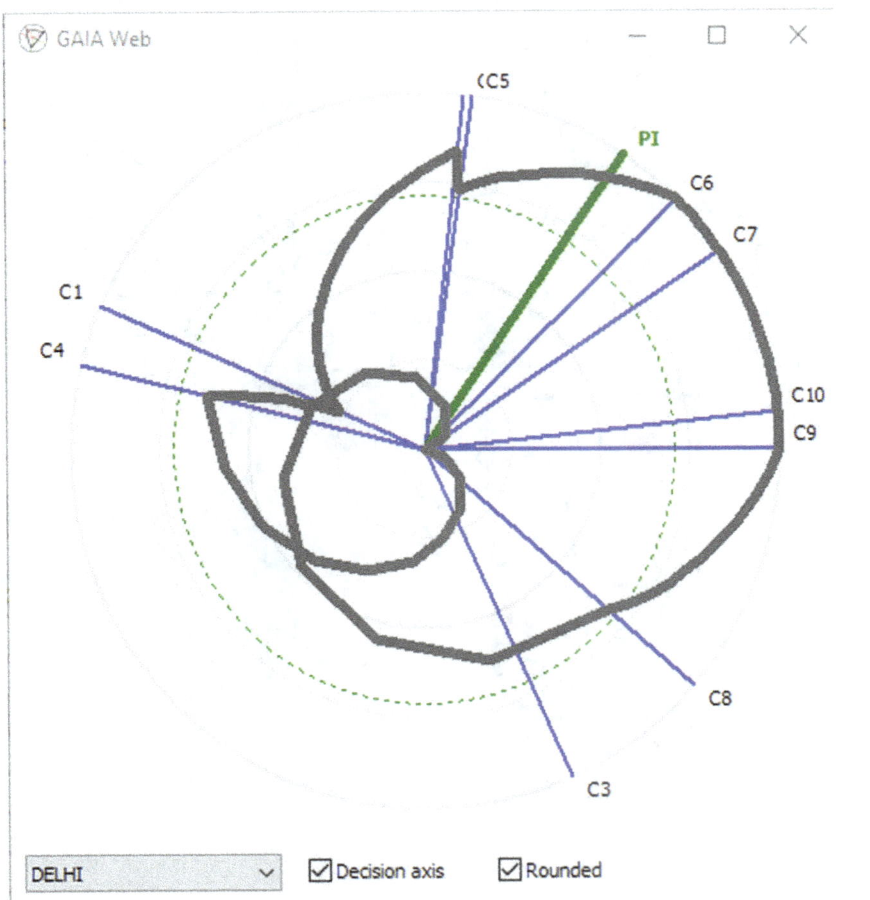

strongly correlated are close to each other in this web. Uni-criterion net flows are represented on the web, -1 are drawn at the center of the web while +1 values are the positioned on the outer circle. A polygon is finally drawn connecting different criteria. From this figure, it is very clear that Delhi, criteria C_6, C_7, C_2, C_5, C_9 and C_{10} are highly correlated. These observations are quite predictive in nature and show the right direction toward which Delhi is moving to one of the good examples of smart city. The PI axis is also oriented toward criteria C_6 and C_7, as already observe in established GAIA plane.

In Figure 10, the geographical map of India is provided on which the profiles of different smart cities with their performance are portrayed based on GIS and HSV color coding scheme. It would be very easy for DMs to compare those smart cities

Figure 10. Profiles of 20 Indian smart cities using GIS and HSV color coding scheme

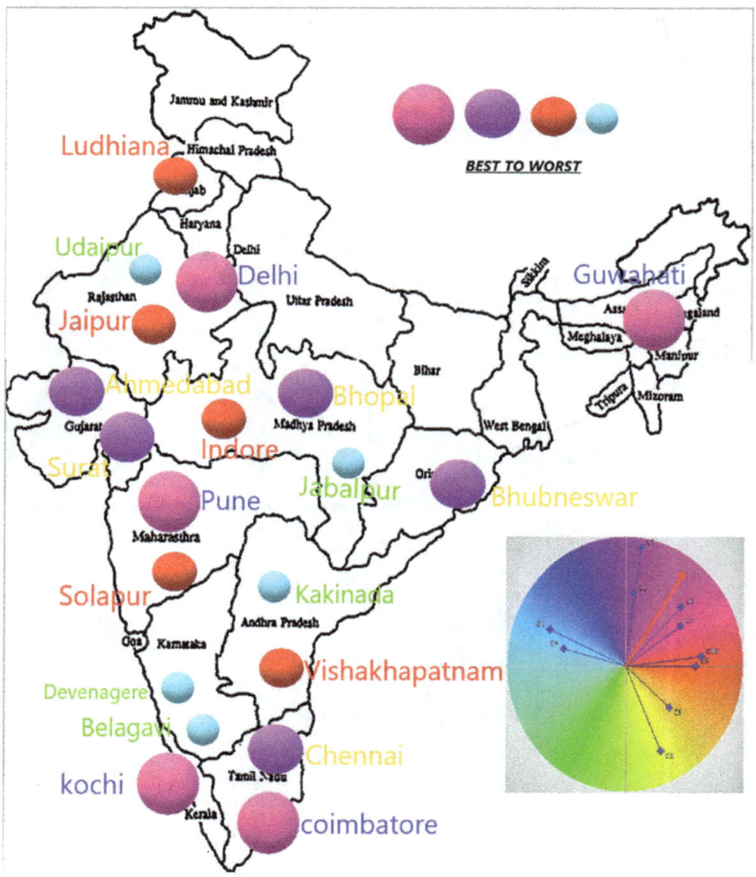

profiles, and identifies their strength and weaknesses in the GAIA referential. The best performer smart cities like Delhi, Kochi, Coimbatore, Pune and Guwahati have large circle in the map. The circle for Delhi, Kochi and Coimbatore is drawn in cyan, which shows that they are the strong performer in the criterion C_6 (poverty ratio) and C_7 (per capita income). On the other hand, light blue color circle for Bhopal, Surat, Bhubaneswar, Ahmedabad and Chennai proves its superiority in the criterion of C_1 (total population) and C_4 (population density). The smaller circles are the worst performer smart cities like Kakinada, Jabalpur, Belagavi, Devenagare and Udaipur in different color also display their potentialities in respective areas. The performance of Kakinada and Jabalpur is very strong in the criterion of C_1. Similarly, the red color circle for Jaipur, Ludhiana, Indore, Vishakhapatnam and Solapur proves its strength in the criterion C_9 (number of hospitals) and C_{10} (number of school and colleges). The orientation of PI axis towards the cluster of criteria C_6 and C_7 in the GAIA plane is the reason behind the supremacy of Delhi over the other Smart Cities. The medium size circles drawn in different color shows the intermediate ranking of the remaining smart cities. To have better understanding on the performance of 20 smart cities, when a detailed zonal analysis is carried out, it discovered that all the smart cities in the southern zone of India i.e. Devenagere, Belagavi, Kakinada, and Solapur are the underperformers. The central and state government should have to take appropriate measure step to develop these lagging Cities.

CONCLUSION

In this chapter, the current performance of 20 Indian smart cities conditions has been evaluated with respect to 10 different criteria using an integrated PROMETHEE-GIS methodology. It is observed that Delhi and Kochi are the top two Performing smart cities, where as Kakinada and Jabalpur are the worst performer among all 20 smart cities. It can be recommended that the total financial outlay for the development of some of the smart cities like Kakinada, Jabalpur, Belagavi, Devenagere and Udaipur needs to be increased. This analysis would help the DM to identify the strength and deficiencies of each Indian smart cities with respect to their criteria so that proper promotional and growth actions can be implemented. The key issues identified are of immense help to the policy makers for having detailed insight for further improving the deficiencies of smart cities.

REFERENCES

Ali, N. H., Sabri, I. A. A., Noor, N. M. M., & Ismail, F. (2012). Rating and ranking criteria for selected Islands using fuzzy analytic hierarchy process (FAHP). *International Journal of Applied Mathematics and Informatics*, *6*(1), 57–65.

Anand, A., Rufuss, D. D. W., Rajkumar, V., & Suganthi, L. (2017). Evaluation of sustainability Indicators in smart cities for India using MCDM approach. *Energy Procedia*, *141*, 211–215. doi:10.1016/j.egypro.2017.11.094

Anthopolous, L., & Giannakidis, G. (2017). Policy Making in Smart City: Standardizing City's Energy Efficiency with Task-Based Modelling. *Journal of ICT*, *4*(2), 111–146.

Brans, J. P., & Mareschal, B. (1994). The PROMCALC and GAIA decision-support system for multi criteria decision aid. *Decision Support Systems*, *12*(4-5), 297–310. doi:10.1016/0167-9236(94)90048-5

Brans, J. P., & Vincke, P. (1985). Preference ranking organization method: The PROMETHEE method for MCDM. *Management Science*, *31*(6), 647–656. doi:10.1287/mnsc.31.6.647

Brans, J. P., Vincke, P., & Mareschal, B. (1986). How to select and how to rank project: The PROMETHEE method. *European Journal of Operational Research*, *24*(2), 228–238. doi:10.1016/0377-2217(86)90044-5

Caragliu, A., Del Bo, C., & Nijkamp, P. (2011). Smart cities in Europe. *Journal of Urban Technology*, *18*(2), 65–82. doi:10.1080/10630732.2011.601117

Chakraborty, S., Ranjan, R., & Mondal, P. (2014). A state-wise performance appraisal of the Indian roads using PROMETHEE-GIS approach. *Benchmarking: An International Journal*, *9*(25), 1–19.

Crouch, G. I., & Brent Ritchie, J. R. (1999). Tourism, Competitiveness, and Societal Prosperity. *Journal of Business Research*, *44*(3), 137–152. doi:10.1016/S0148-2963(97)00196-3

Formica, S. (2000). *Dissertation Thesis, Destination Attractiveness as a Function of Supply and Demand Interaction,* Virginia Tech.

Goonetilleke, A., Yigitcanlar, T., Ayoko, G., & Egodawatta, P. (2014). *Sustainable urban water environment: Climate, pollution and adaptation*. Cheltenham, UK: Edward Elgar. doi:10.4337/9781781004647

Gooroochurn, N., & Sugiyarto, G. (2004). *Discussion Paper, Measuring Competitiveness, TTRI*. Nottingham University Business School, University of Nottingham.

Jong, M., Joss, S., Schraven, D., Zhan, C., & Weijnen, M. (2015). Sustainable–smart–resilient–low carbon–eco–knowledge cities; making sense of a multitude of concepts promoting sustainable urbanization. *Journal of Cleaner Production*, *109*, 25–38. doi:10.1016/j.jclepro.2015.02.004

Kozak, M., & Rimmington, M. (1999). Measuring tourist destination competitiveness: Conceptual considerations and empirical findings. *International Journal of Hospitality Management*, *18*(3), 273–283. doi:10.1016/S0278-4319(99)00034-1

Lee, J. H., Hancock, M. G., & Hu, M. C. (2014). Towards an effective framework for building smart cities: Lessons from Seoul and San Francisco. *Technological Forecasting and Social Change*, *89*, 80–99. doi:10.1016/j.techfore.2013.08.033

Lee, S. H., Han, J. H., Leem, Y. T., & Yigitcanlar, T. (2008). Towards Ubiquitous City: Concept, Planning, and Experiences. *Igi Global*, *2*, 148–169.

Lidouh, K., Smet De, Y., & Zimányi, E. (2011). An adaptation of the ´ GAIA visualization method for cartography. *Procs. of the IEEE Symposium on Computational Intelligence in Multicriteria Decision-Making*, 29-35.

Melián-González, A., & García-Falcón, J. M. (2003). Competitive potential of tourism in destinations. *Annals of Tourism Research*, *30*(3), 720–740. doi:10.1016/S0160-7383(03)00047-1

Pietrosemoli, L., & Monroy, C. R. (2013). The impact of sustainable construction and knowledge management on sustainability goals. A review of the Venezuelan renewable energy sector. *Renewable & Sustainable Energy Reviews*, *27*, 683–691. doi:10.1016/j.rser.2013.07.056

Ranjan, R., & Chakraboty, S. (2015). Performance Evaluation of Indian Technical Institutions Using PROMETHEE-GAIA Approach. *Informatics in Education*, *1*(14), 103–125. doi:10.15388/infedu.2015.07

Ranjan, R., Chatterjee, P., & Chakraborty, S. (2014). Evaluating performance of engineering departments in an Indian University using DEMATEL and compromise ranking methods. *Opsearch*, *2*(52), 307–328.

Ranjan, R., Chatterjee, P., & Chakraborty, S. (2015). Performance evaluation of Indian states in tourism using an integrated PROMETHEE-GAIA approach. *Opsearch*, *1*(53), 63–84.

Rondini, A., Lagorio, A., Pezzotta, G., & Pinto, R. (2016). A multi-criteria decision-making approach for prioritising product-service systems implementation in smart cities. *International Journal of Management and Decision Making*, *17*(4), 415–446. doi:10.1504/IJMDM.2018.095729

Tiwari, A., & Mishra, N. (2017). Sustainable and smart city project: An overview of the application of multi criteria decision making techniques and approaches for Indian context. *International Journal of Management and Applied Science*, *3*(7), 1–6.

Yigitcanlar, T. (2006). Australian local governments' practice and prospects with online planning. *URISA Journal*, *18*(2), 7–17.

Yigitcanlar, T., & Baum, S. (2008). Benchmarking local e-government. In Electronic Government: Concepts, Methodologies, Tools, and Applications. IGI Global. doi:10.4018/978-1-59904-947-2.ch033

Yigitcanlar, T., Dodson, J., Gleeson, B., & Sipe, N. (2007). Travel self-containment in master planned estates: Analysis of recent Australian trends. *Urban Policy and Research*, *25*(1), 129–149. doi:10.1080/08111140701255823

Yigitcanlar, T., & Kamruzzaman, M. (2014). Investigating the interplay between transport, land use and the environment: A review of the literature. *International Journal of Environmental Science and Technology*, *11*(8), 2121–2132. doi:10.100713762-014-0691-z

Yigitcanlar, T., & Kamruzzaman, M. (2015). Planning, development and management of sustainable cities: A commentary from the guest editors. *Sustainability*, *7*(11), 14677–14688. doi:10.3390u71114677

Chapter 3
An Integrated Methodology for Evaluation of Electric Vehicles Under Sustainable Automotive Environment

Tapas Kumar Biswas
MCKV Institute of Engineering, India

Željko Stević
https://orcid.org/0000-0003-4452-5768
University of East Sarajevo, Bosnia and Herzegovina

Prasenjit Chatterjee
https://orcid.org/0000-0002-7994-4252
MCKV Institute of Engineering, India

Morteza Yazdani
https://orcid.org/0000-0001-5526-8950
Universidad Loyola Andalucía, Spain

ABSTRACT

In this chapter, a holistic model based on a newly developed combined compromise solution (CoCoSo) and criteria importance through intercriteria correlation (CRITIC) method for selection of battery-operated electric vehicles (BEVs) has been propounded. A sensitivity analysis has been performed to verify the robustness of the proposed model. Performance of the proposed model has also been compared with some of the popular MCDM methods. It is observed that the model has the competency of precisely ranking the BEV alternatives for the considered case study and can be applied to other sustainability assessment problems.

DOI: 10.4018/978-1-5225-8579-4.ch003

INTRODUCTION

Transport sectors are the major consumer of world oil output, as well as a leading source of greenhouse gas (GHG) emissions Worldwide, mainly the carbon dioxide (CO_2). The transportation sector generates 28% of total GHG in the United States (US) (USEPA, 2014; Knittel, 2012). Objective of the Paris Agreement set of limiting the increase in the global average temperature to well below 2°C above preindustrial levels and pursuing efforts to limit the temperature increase to 1.5°C above pre-industrial levels (UNFCCC, 2015b). Reducing use of fossil fuel energy and GHG emissions from transport sectors could consequently play an important role in solving both the global fossil fuel reduction and climate change challenges which the whole world faces. Battery electric vehicles (BEVs) are continually regarded as an important means of solving both problems. BEVs have the potential to make an important contribution to reduce greenhouse gases because of their potential to run with zero emissions when operating with electricity from renewable sources (United Nations, 2014; United Nations,1992, April 30-May 9). Global climate change has strengthened the need for sustainable development over the last few decades. Battery Electric vehicles will reduce the CO_2 footprint, pollution level and help challenge anthropogenic climate change (Dijk et al., 2013). Therefore, human factors research has focused more intensively on the interaction between humans and BEVs to better understand relevant factors i.e., range related user experience (Franke et al., 2015; Rauh et al., 2015), acceptance of BEVs (Bühler et al., 2014) and ways to increase and optimize the usage of BEVs. A sustainable balance is thus struck between limited resources and socio-environmental demands. With zero tailpipe emissions in the case of battery electric vehicles, EVs also offer a clean alternative to vehicles with internal combustion engines by helping to reduce exposure to air pollution resulting from fuel combustion and limiting noise. This is especially relevant in urban areas and along major transportation access. The relevance of EVs for the reduction of air pollution and noise is well demonstrated by the leading role that cities assume in promoting EV deployment, in 2015, nearly a third of global electric car sales took place in just 14 cities (Hall et al., 2017).

Alternate fuel vehicles are classified as EV, Solar powered vehicle, Hydrogen fuel vehicle, Ethanol and biodiesel fuel vehicle, liquefied petroleum gas (LPG) and compressed natural gas (CNG) vehicles. EVs are further classified as Battery EV (BEV), Hybrid EV (HEV), Plug-in HEV (PHEV), Range Extender EV (REXEV) to name a few (Ajanovic & Haas, 2016). Vehicles run with hydrogen fuel by modified internal combustion (IC) engines. Bio-diesel is another source of sustainable fuel for IC engines and diesel engines run by biodiesel fuel. But in all the cases the combustion process would still cause some pollutants. However, BEVs are the zero emission vehicles at the point of use considering the dependence of fossil fuels in the other forms of EVs. The low carbon impact of EVs will be possible if electricity is generated from sustainable sources. As the fuel cells and solar vehicle also produce zero emission than that of IC engines it is likely that fuel cells and solar energy will eventually become the chosen option.

BACKGROUND

An EV is a vehicle propelled by an electric motor-generator set, instead of an IC engine and the batteries delivered the power by which the motor run. The batteries have to be charged frequently by plugging into any main (120 V or 240 V) supply. EVs are known as zero emissions vehicles and are much environment friendly than gasoline, diesel, CNG or LPG-powered vehicles. The benefits of the electric vehicles are, it has fewer moving parts, maintenance is also minimal like no oil changes due to absence of the engine, no tune-ups or timing and there is no exhaust. EVs are more energy efficient than IC engines and they are very silence in operation. Advancement of battery technology, system integration, aerodynamics, new light materials and research and development by major vehicle manufacturers has led to the producing of electric vehicles that will play a practical role on present and future automobile. In EVs, electrical energy is stored in a storage battery and this electrical energy is used to power an electric motor, which then turns the wheels and provides propulsion. However, EVs have a serious disadvantage in terms of limited range and acceleration with lower speed. Due to the limited range, EVs can only travel in the order of more or less 80 km or so on a single charge, although modification in battery technology can enhance the vehicle range. There are some new battery technologies that offer good promise in terms of range, but none has the desired combination of fast charging/discharging (high power density), large storage capacity (high energy density), and low cost. Traditional lead-acid batteries have been improved some over the years, but their power and energy densities remain insufficiently low. Presently maximum EVs use Lithium ion battery instead of lead acid batteries. Main components of an EV are motor, controller, converter, battery. For accurate identification, a motor should be identified by name and their specification. Following are the commonly used motors in EVs: Series Wound Brushed DC Motors, BLDC motor, Induction motor etc. Batteries typically account for one third or more of vehicle weight and one fourth or more of the life-cycle cost of an electric vehicle. Major improvements in batteries are expected because, presently effort has been put into designing and building batteries of the size needed for vehicles. The list of electric vehicle battery such as Lead acid, Nickel Cadmium/Nickel Iron (Ni-Cd)/(Ni-Fe), NiMH, NaS, Lithium ion, Lithium Cd, Zinc air, Aluminium air etc. (Schalkwijk and Scrosati, 2007) All the batteries use solid, liquid or gaseous electrolytes; replaceable metals, and replaceable liquids. Lithium ion batteries or lead acid batteries have been recognized as a clean alternative to conventional vehicles for reducing the GHG emissions from transportation sector. But the advantages of Lithium ion batteries are capable of fast charging (up to 80% charge in 30 minutes) compared with lead-acid batteries (7-8 hours for a full charge). Nowadays, Lithium batteries are extensively used due to the low electro negativity and low equivalent weight of lithium. The largest part of rechargeable lithium batteries available today are lithium ion batteries. The anode consists of lithiated graphite (LiC6). The most common cathode materials are lithium metal oxides. Vehicle range, and vehicle speed depends upon the quality of the battery. Main auxiliaries in an EV Chargers a good charger is crucial to EV performance. Presently all the EVs charging station takes more or less same time for charging. Modern chargers can sense the level of charge in the battery pack, there are differences between 220 V and 110 V input chargers. A 220 V will charge the pack faster, but it is bulkier and heavier, not really suitable for on-board mounting. A 110 V charger will charge more slowly, but it is small and light enough to be mounted onboard so the driver can take advantage of opportunity charging anywhere there is 110 V power. However, new strategies also become important like the efficient usage of regenerative braking (saving energy during deceleration and restore into battery) (Cocron et al., 2013). The successful usage of eco-driving strategies has the potential to reduce energy consumption of a BEV by more than 25% (Helmbrecht et al., 2013). Among EU countries, the Netherlands is one of the leaders in terms of sales of EVs in 2014 (Lutsey, 2015). It is expected that by 2020,200,000 EVs will add to its roads in Netherland (Mouli et al., 2016). In Norway fleet of vehicles contained 146006 battery electric vehicles at the end of 2017 (Lambert, 2017), out of these, 139474 where passenger vehicles (BEVs), which is 5.1% of the total passenger vehicle fleet. In India, exhaust from different transport is currently one of the biggest sources of pollution in cities. In 2013, Ministry of Heavy Industries (MoHI), Government of India launched National Electric Mobility Mission Plan (NEMMP) with a target of 6 to 7 million EVs on Indian roads by 2022. European Union aimed at a replacement of 10% of the conventional fuels in the road transport sector before the year 2020 to decrease GHG emissions and improve the security of energy supply (VROM, 2007).20% reduction of greenhouse gasses in 2020 compared to the levels of 1990 was proposed by the European Union in 2007. The transport sector accounts for about 31% of European energy use (EEA, 2008) and 25% of the European CO_2 emissions. A big part of the CO_2 emission reduction can be achieved by introducing alternative fuels and drive trains. International Energy Agency (IEA 2017) has forecast that GHG emissions from transportation will by 120% from 2000 to 2050 as a result demand of electric vehicle increase worldwide. Some industry and advocacy groups have set global deployment target of 100 million electric cars and 400 million electric motorcycles and scooters by 2030.

An original equipment manufacturer (OEM) is a company that produces parts and equipment that may be marketed by another manufacturer. For example, BMW, Chevrolet (General Motors), Ford, Volvo, Tesla etc. manufacturing company are the largest OEM Company in the world by both scale and revenue. Table 1 indicates the various announcements made by these organizations in terms of sales targets of EVs.

Even if some new Battery electric car and electric two wheelers are rising for EVs market, they are focused on the economic side with integrating social and environmental dimensions into the core of the business. Sustainability is still seen as an enhancement outside the scope of business models, which leads to various forms of missing value opportunities. BEVs seem to be a relatively clean and efficient way of using energy in comparison with other fuels (Mierlo et al., 2006). Therefore, there is a great potential of saving energy and reduce emissions when the BEV is being introduced on large scale. The efficiency as well as the emissions of the BEV already have been subject of research and numerous can be found in literature. The development of battery technology and the future costs of advanced batteries are uncertain but are utmost important for the success of the BEV. The battery technology is considered to be the most critical factor in the commercialization of

Table 1. Announcements of some EV manufacturing companies with respect to sales targets

Sl. No.	OEM	Announcement
1	BMW	0.1 million electric car sales in 2017 and 15-25% of the BMW group's sales by 2025. (Lambert, 2017).
2	Chevrolet (GM)	30 thousand annual electric car sales by 2017. (Loveday,2017)
3	Chinese OEMs	4.52 million annual electric car sales by 2020 (CNEV, 2017)
4	Daimler	0.1 million annual electric car sales by 2020 (Daimler, 2016)
5	Ford	13 new EV models by 2020 (Ford, 2017)
6	Honda	Two-thirds of the 2030 sales to be electrified vehicles (including hybrids, PHEVs, BEVs and FCEVs) (Honda, 2016)
7	Renault-Nissan	1.5 million cumulative sales of electric cars by 2020. (Cobb, 2015)
8	Tesla	0.5 million annual electric car sales by 2018 1 million annual electric car sales by 2020 Goliya and Sage (2016). (Tesla, 2017)
9	Volkswagen	2-3 million annual electric car sales by 2025 (Volkswagen, 2016)
10	Volvo	1 million cumulative electric car sales by 2025*.

(**Sources:* www.media.volvocars.com/global/en-gb/media/pressreleases/).

the BEV. Delucchi et al. (1989) already researched the lifecycle costs, performance and battery technology of EVs in 1989. The research predicted a large technology improvement of the battery and a commercial breakthrough of the EV at the turn of the century. Therefore, prediction the price and performance of the battery are important for the commercialization of the BEV. Over the last few years different car manufacturing company produced BEVs, because BEVs constitutes a sustainable innovation that is more energy efficient than conventional gasoline vehicles and emits no local pollutants. Current government policies and subsidies in many countries are seeking to stimulate the new beginning of EVs adoption and help it gain market traction. EV technology, which is claimed to be sustainable, holds the promise to electrify vehicles and ultimately, achieve a zero-emission transport system. However, several challenges including the high costs compared with conventional cars, the range anxiety and long charging time still impede the market penetration of EVs (Xuewu et al., 2012). In present era, there are many fastest growing BEV industries in this world. Today's BEV industry plays an important role in consumer life. Consumers search for the best BEV with the most beneficial features at a reasonable cost. Every year, various models of BEVs with latest technology and excellent characteristics are introduced in the market. Due to this, consumers face difficulties in selecting the best car among the available alternatives according to their needs. The involvement of a number of mutually conflicting criteria in the BEV selection process makes it multi-criteria decision-making (MCDM) problem. Recently many studies have focused on market analysis of BEVs and usage advantages. Researchers have already used various methods for the decision-making process when selecting the most suitable potential vehicles. For example Biswas and Das (2018a) presented a decision support system for hybrid vehicle selection using an entropy- multi-attributive border approximation area comparison (MABAC) method. Biswas and das (2018b) proposed a fuzzy analytic hierarchy process (AHP)-based MABAC approach for electric vehicle selection problems. Behnam et al. (Behnam et al., 2011) considered the problem of alternative fuel buses selection using two novel fuzzy MCDM methods. Ma et al. (Ma et al., 2019) combined AHP and logit regression model for market forecasting of new energy vehicles.

From the literature survey as presented above, it has been observed that padt researchers have not presented any methodology for evaluation of BEVs. Thus, the present chapter takes this opportunity to propose an integrated systematic and logical model combining the newly developed Combined Compromise Solution (CoCoSo) method with Criteria Importance Through Inter criteria Correlation (CRITIC) under sustainable automotive environment in order to bridge this gap and to help the decision makers (DMs) in taking appropriate decision for evaluation and selection of a group of battery-operated EVs.

PROPOSED INTEGRATED DECISION-MAKING APPROACH

This section presents the mathematical formulations of the considered CRITIC and COCOSO methods which are subsequently applied for the evaluation of EVs.

CRITIC Method

CRITIC (CRiteria Importance Through Intercriteria Correlation) method is one of the weighting methods which determine objective weights of criteria. CRITIC method was proposed by Diakoulaki et al. in 1995. This method includes the intensity of the contrast and the conflict in the structure of the decision making problem (Diakoulaki et al., 1995). It uses correlation analysis to find out the contrasts between criteria (Yılmaz and Harmancıoglu, 2010). In this method the decision matrix is evaluated and the standard deviation of normalized criterion values by columns and the correlation coefficients of all pairs of columns are used to determine the criteria contrast (Madic and Radovanovic, 2015). In the literature there are many applications of CRITIC method. Various researcher applied this method such as pharmaceutical industries (Diakoulaki et al. 1995), water resource management model (Yılmaz and Harmancıoglu, 2010), logistic firms (Cakır and Percin, 2013), an evaluation index system of city's soft power (Guo et al., 2013), financial statement of stock exchange (Kazan and Ozdemir,2014), non-traditional machining process (Madic and Radovanovic, 2015).

CRITIC method is presented in the following steps (Madic and Radovanovic, 2015). It is assumed that there is a set of *m* feasible alternatives A_i (i = 1,2,…,m) and *n* evaluation criteria C_j (j = 1,2,…,n).

Step 1: Development of the decision matrix (*X*), expressed as follows.

$$x_{ij} = \begin{bmatrix} x_{11} & x_{12} & \cdots & x_{1n} \\ x_{21} & x_{22} & \cdots & x_{2n} \\ \cdots & \cdots & \cdots & \cdots \\ x_{m1} & x_{m2} & \cdots & x_{mn} \end{bmatrix}; \ i = 1,2,...,m; \ j = 1,2,...,n. \tag{1}$$

where (i = 1,2…,m and j = 1,2,…,n)

The elements (x_{ij}) of the decision matrix (*X*) represent the performance value of ith alternative on j^{th} criterion.

Step 2: Normalization of original decision matrix using the following equations:

$$r_{ij} = \frac{x_{ij} - \min_i x_{ij}}{\max_i x_{ij} - \min_i x_{ij}}; \text{ for benefit criterion;} \quad (2)$$

$$r_{ij} = \frac{\max_i x_{ij} - x_{ij}}{\max_i x_{ij} - \min_i x_{ij}}; \text{ for cost criterion} \quad (3)$$

Step 3: Calculation of symmetric linear correlation matrix (m_{ij}):

A linear correlation coefficient between the each pair of criteria is estimated using the following equation to quantify the conflict resulted among different criteria. It can be seen that the more discordant the scores of the alternatives in two criteria i and j, the lower the value m_{ij}.

$$m_{ij} = \frac{\sum_{i=1}^{m} (r_{ij} - \overline{r}_j)(r_{ik} - \overline{r}_k)}{\sqrt{\sum_{i=1}^{m} (r_{ij} - \overline{r}_j)^2 \sum_{i=1}^{m} (r_{ik} - \overline{r}_k)^2}} \quad (4)$$

Step 4: Determination of the objective weight of a criterion using CRITIC method also requires the estimation of both standard deviation of the criterion and its correlation with other criteria. In this regard, the weight of the jth criterion (w_j) is obtained using Equation (5).

$$W_j = \frac{C_j}{\sum_{j=1}^{n} C_j} \quad (5)$$

where, C_j is the amount of information contained in the criterion j and is determined as follows:

$$C_j = \sigma \sum_{j'=1}^{n} 1 - m_{ij} \tag{6}$$

where σ is the standard deviation of jth criterion and is the correlation coefficient between the two criteria. CRITIC method provides higher weight to the criterion with higher value of σ and low correlation with the other criteria. A higher value of C_j signifies greater amount of information contained in a particular criterion, hence it is provided with higher weight value.

Combined Compromise Solution (CoCoSo) Method

Combined Compromise Solution (*CoCoSo*) method (Yazdani et al. 2018) is based on the integration of two most popular MCDM methods namely Simple Additive Weighting (SAW) and Exponentially Weighted Product (MEP). The *CoCoSo* method consists of the following three easy steps and it is implemented after obtaining the criteria weights through the application of CRITIC method.

Step 1: Estimation of sum of weighted comparability (S_i) sequence and power weighted comparability sequences (P_i) for each alternative respectively:

$$S_i = \sum_{j=1}^{n} (w_j r_{ij}). \tag{7}$$

$$P_i = \sum_{j=1}^{n} (r_{ij})^{w_j} \tag{8}$$

Step 2: Computation of relative weights of the alternatives:

In this step, three aggregated appraisal scores are used to generate relative performance scores of the alternatives, using the following equations:

$$\text{a) } k_{ia} = \frac{P_i + S_i}{\sum_{i=1}^{m} (P_i + S_i)} \tag{9}$$

b) $k_{ib} = \frac{S_i}{\min_i S_i} + \frac{P_i}{\min_i P_i}$ (10)

c) $k_{ic} = \frac{\lambda(S_i) + (1-\lambda)(P_i)}{(\lambda \max_i S_i + (1-\lambda)\max_i P_i)}; \ 0 \le \lambda \le 1.$ (11)

Equation (9) basically expresses the arithmetic mean of sums of WSM and WPM scores, while Equation (10) signifies the sum of relative scores of WSM and WPM compared to the best alternative. Finally, Equation (11) computes a balanced compromise score of WSM and WPM models. In Equation (11), the value of λ (usually the threshold $\lambda = 0.5$) ranges from 0 to 1 and is chosen by the decision-maker.

Step 3: The final ranking of the alternatives is determined based on k_i, values:

Higher k_i values indicate better position of the alternatives in the ranking pre-order.

$$k_i = \left(k_{ia} k_{ib} k_{ic}\right)^{\frac{1}{3}} + \tfrac{1}{3}\left(k_{ia} + k_{ib} + k_{ic}\right) \tag{12}$$

CASE STUDY

Now, it is high time to introduce the integrated decision-making model for the performance evaluation and selection of EVs while considering specific criteria which may be conflicting in nature. Here 12 US BEV models including BMW i3, Chevy Bolt, Chevy Spark, Fiat 500e, Ford Focus Electric, Mitsubishi i-MiEV, Nissan leaf, Fisker Emotion, Tesla model S, VW e Golf, Tesla Original Roadster and Tesla Model X have been considered as alternative EVs for selection and ranking under passenger car category. Some EV selection attributes such as EPA rated combined fuel economy (C_1), battery range (C_2), top speed on flat road (C_3), accelerating time (C_4) and vehicle price (C_5) have been considered. Brief description of the criteria is given in the following Table 2. Out of five criteria, first three are beneficial type, i.e., higher the better type, while the remaining two are non-beneficial criteria. A summary of the available specifications of the considered EV models as available in the manufacturers' websites is presented in Table 2.

Table 2. List of criteria with description

Sl. No.	Criteria	Description
1	Combined fuel economy (C_1)	This criterion is expressed as Mile per gallon gasoline equivalent (MPG_e). Combined fuel economy is a weighted average of highway and city values. It indicates that how much mile the vehicle can go by using a quantity of fuel with the same energy content as a litre of gasoline.
2	Battery range (C_2)	Battery range means that the maximum distance the car can travel between two subsequent charging and is expressed in Mile.
3	Top speed (C_3)	It indicates the maximum speed at which the vehicle can runs on good road. It is expressed in the Mile per hour.
4	Accelerating time (C_4)	It represents that how much time is required to accelerate the car from 0 to 60 Mile per hour. It is expressed in the time(in seconds)
5	Vehicle price (C_5)	It is the selling price of vehicle in US. It is important criteria from customers' perspective.

At first, the criteria weights for the BEV selection case study are estimated using CRITIC method. As the initial step, the decision matrix of Table 3 is normalized by using Eqs. (2) and (3), respectively for beneficial and cost criteria. The normalized decision matrix is shown in Table 4. The last row of Table 4 shows the values of standard deviations (σ) for the considered criteria. The values of correlation coefficient are then calculated using Equation (4) and shown in Table 5. Finally the

Table 3. Decision matrix for selection of BEVs

Sl. No.	BEV	C_1	C_2	C_3	C_4	C_5
1	BMW i3 (EV_1)	137	81	93	7.2	42400
2	Chevy Bolt (EV_2)	119	238	91	7	37495
3	Chevy Spark (EV_3)	119	82	90	8	21375
4	Fiat 500e (EV_4)	122	87	88	9	31800
5	Ford Focus Electric (EV_5)	110	100	84	10	36630
6	Mitsubishi i-MiEV (EV_6)	112	126	81	15	22995
7	Nissan leaf (EV_7)	129	151	90	6.3	30680
8	Fisker Emotion (EV_8)	100	400	125	6.3	129000
9	Tesla model S (EV_9)	107	315	120	2.3	68000
10	VW e Golf (EV_{10})	126	125	102	9.2	28995
11	Tesla Original Roadster (EV_{11})	99	245	130	3.7	109000
12	Tesla Model X (EV_{12})	93	256	120	3.8	88000

(Source: Manufacturer website and http://www.comparehybridcars.net (Dated 13/8/2018)

Table 4. Normalized decision matrix

BEV	C_1	C_2	C_3	C_4	C_5
EV_1	1.0000	0	0.2449	0.6142	0.8046
EV_2	0.5909	0.4922	0.2041	0.6299	0.8502
EV_3	0.5909	0.0031	0.1837	0.5512	1.0000
EV_4	0.6591	0.0188	0.1429	0.4724	0.9031
EV_5	0.3864	0.0596	0.0612	0.3937	0.8583
EV_6	0.4318	0.1411	0	0	0.9849
EV_7	0.8182	0.2194	0.1837	0.6850	0.9135
EV_8	0.1591	1.0000	0.8980	0.6850	0
EV_9	0.3182	0.7335	0.7959	1.0000	0.5668
EV_{10}	0.7500	0.1379	0.4286	0.4567	0.9292
EV_{11}	0.1364	0.5141	1.0000	0.8898	0.1858
EV_{12}	0	0.5486	0.7959	0.8819	0.3810
Standard deviation (σ)	0.3019	0.3288	0.359	0.2664	0.3361

Table 5. Correlation coefficient values of paired criteria

Criteria	C_1	C_2	C_3	C_4	C_5
C1	1	-0.6853.	-0.6777.	-0.3466	0.7367.
C2	-0.6853.	1	0.8064.	0.5987	-0.8413
C3	-0.6777.	0.8064.	1	0.7684	-0.9084
C4	-0.3466	0.5987	0.7684	1	-0.6137
C5	0.7367.	-0.8413	-0.9084	-0.6137	1

criteria weights of Table 6 are determined using Eq. (5). According to the values of corrected calculation of Table 6, C_5 is the most important while C_4 is least important.

Next step to solve this BEV selection problem is to follow the mathematical steps of CoCoSo method. At first, the sum of weighted comparability sequence (S_i) and

Table 6. Weights of the BEV selection criteria

Criteria	C_1	C_2	C_3	C_4	C_5
C_j	1.501	1.355	1.440	0.957	1.891
w_j	0.210	0.190	0.202	0.134	0.265

power weight of comparability sequences (P_i) are computed using Equations. (7) and (8) respectively, as exhibited in Table 7. As stated earlier, for CoCoSo method, different ranking scores are computed and ultimately an accumulated index produces ranking of the alternative BEVs. Those formulas are introduced through Eqs. (9), (10) and (11) respectively and the results are also shown in Table 7.

From the ranking preorder, as obtained according to the descending order of the k values (Table 7), it is observed that EV_9 (Tesla model S) is the most favorite candidate while EV_6 (Mitsubishi i-MiEV) is the worst one among others. Figure 1 shows sensitivity analysis based on the varying λ values in a range of 0 to 1. From this figure, it is clearly seen that for the different λ values, there is no change in the position of the best ranked alternative and EV_9 (Tesla model S) remains the best one throughout the analysis, thus establishing its superior performance and acceptance over other alternatives. Some minor changes in the ranking orders of some intermediate alternatives have been observed in an acceptable range.

Figure 2 shows the second part of sensitivity analysis which represents comparing with three other MCDM methods: SAW (MacCrimmon, 1968), WASPAS (Zavadskas et al. 2012) and MABAC method (Pamučar & Ćirović 2015; Chatterjee et al. 2017).

As can be seen from Figure 2 that alternative EV_9 (Tesla model S) is the best as given by all the considered methods, while other alternatives have placed in different

Table 7. Calculated score values in CoCoSo method

BEV	S_i	P_i	k_{ia}	k_{ib}	k_{ic}	k	Rank
EV_1	0.5547	3.6340	0.0791	2.9065	0.8063	1.8342	**7**
EV_2	0.5681	4.3934	0.0937	3.2427	0.9551	2.0926	**3**
EV_3	0.5003	3.8644	0.0824	2.8539	0.8402	1.8414	**6**
EV_4	0.4732	3.9402	0.0834	2.8122	0.8496	1.8324	**8**
EV_5	0.3848	3.8170	0.0794	2.5296	0.8088	1.6848	**11**
EV_6	0.3782	2.5239	0.0548	2.0000	0.5587	1.2653	**12**
EV_7	0.5841	4.3463	0.0931	3.2665	0.9491	2.0972	**2**
EV_8	0.4958	3.6087	0.0775	2.7408	0.7901	1.7545	**9**
EV_9	0.6504	4.5446	0.0981	3.5202	1.0000	2.2411	**1**
EV_{10}	0.5773	4.3522	0.0931	3.2506	0.9489	2.0907	**4**
EV_{11}	0.4961	4.1644	0.0880	2.9616	0.8971	1.9317	**5**
EV_{12}	0.4834	3.6053	0.0772	2.7066	0.7871	1.7382	**10**

Figure 1. Sensitivity analysis for BEV selection case study

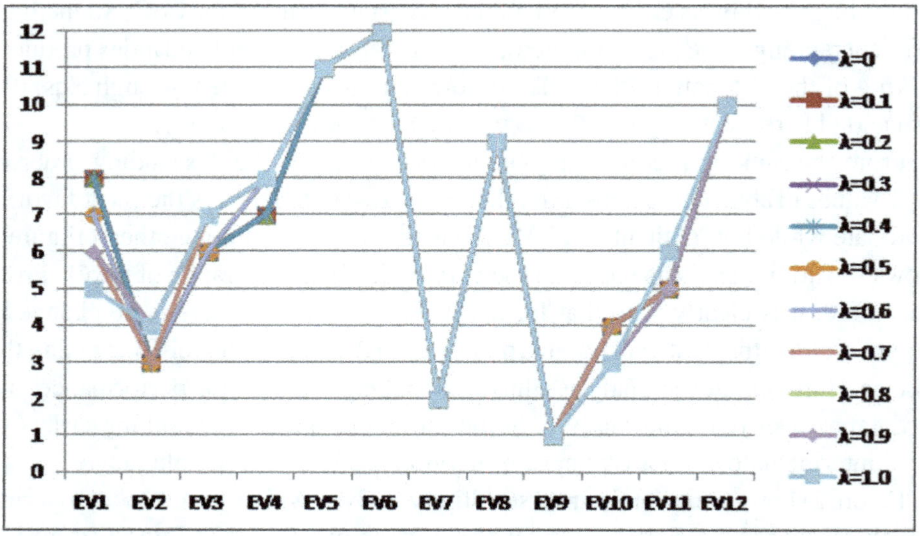

Figure 2. Comparison obtained results with other MCDM methods

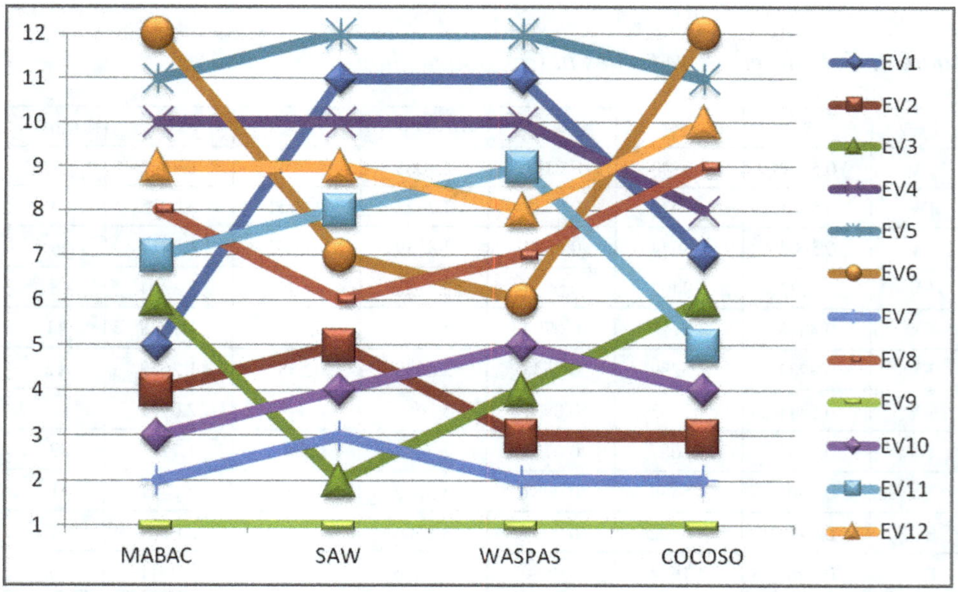

positions. Alternative Nissan leaf (EV_7) is on the second place using MABAC, WASPAS and COCOSO methods, while is in third place using SAW method. As alternatives EV_1, EV_3, EV_6, and EV_{11} have changed their ranking positions, thus, it is necessary to calculate the Spearman Correlation Coefficient (SCC) between the different ranking orders, as presented in Figure 3.

Results obtained using the proposed CRITIC-CoCoSo model is very similar with MABAC method with SCC value of 0.944. This may be attributed due to the use of same normalization technique in these two methods. SCC value between CoCoSo and WASPAS methods is observed as 0.773, while with same with SAW is 0.734. The overall average SCC value between all the considered methods is estimated as 0.817 which represents very high correlation according to following researchers (Stević et al. 2017; Stojić et al. 2018; Pamučar et al. 2019).

DISCUSSIONS AND CONCLUSION

An EV is one that utilizes a battery as the only power source. Their main advantage is the reduced environmental effect, as the tailpipe (direct) GHG emissions caused by

Figure 3. SCC for alternative ranking obtained using CoCoSo and other MCDM methods

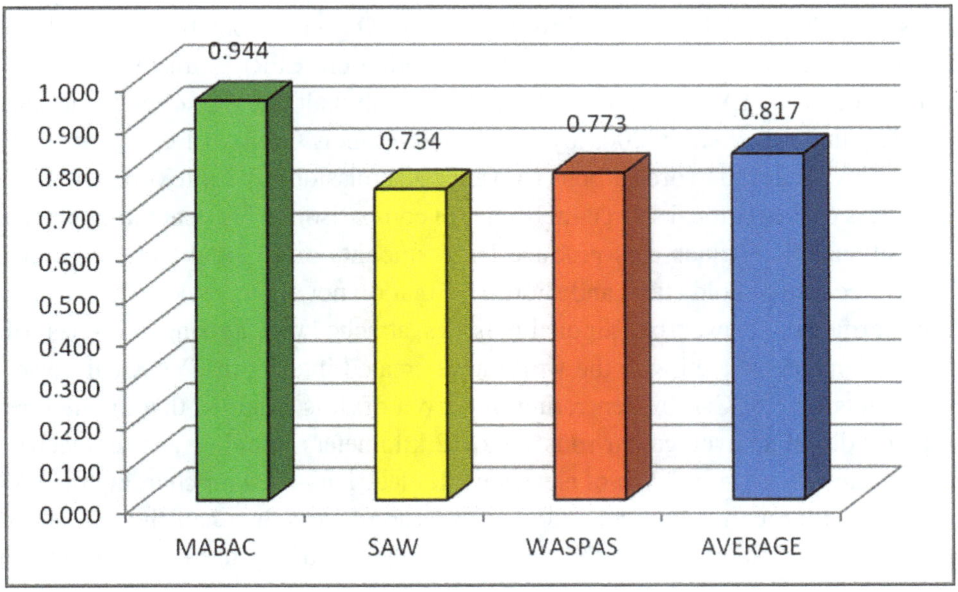

EVs are zero. Currently, the initial cost of EVs is relatively high, mainly because of the high costs of the batteries. EVs are also disadvantaged from a performance point of view when compared to IC engine because with the current battery technologies, the top speeds, driving ranges and charging facilities are limited. However, EVs have much lower operating costs than those of IC engines. Due to the limit in range of BEVs the cars that going to be introduced into the market most likely will be small city cars. However, several challenges including the high costs compared with conventional cars, the range anxiety and long charging time still impede the market penetration of EVs. However, depending on how the electricity which is used to power the motor is generated, a certain amount of indirect GHG emissions and/or air pollutant emissions would be unavoidable. But EVs pollute less than IC engine vehicles without considering where they are to be deployed and by what sources of electricity they are to be powered. If EV that is charged with energy from a clean source, like solar electric, hydroelectric power, will produce very little pollution, at the same time as one charged with energy from an unclean source, like coal or oil, may produce more pollution. Although a BEV produces zero vehicular emissions but emissions are produced at the generation site when the source fuel is converted into electrical power. The emissions of EVs therefore depend on the emissions profile of regional generating plants. Some researchers conclude that, in regions serviced by coal-fired plants, a switch to EVs may actually increase emissions of sulfur oxides (SOx) and particulate matter (PM), and perhaps increase emissions of CO_2. According to Electric Power Research Institute (EPRI), substituting EVs for conventional vehicles (CVs) would reduce urban emissions of non-methane organic gases (NMOG) by 98%, lower nitrogen oxide (NOx) emissions by 92%, and cut carbon monoxide (CO) emissions by 99%. In addition, EPRI estimates that, on a nationwide basis, EVs in the U.S. will produce only half the CO_2 of conventional vehicles. In another study 2005 by ©Taylor & Francis Group, LLC of four U.S. cities, BEVs reduced hydro carbon (HC) and CO emissions by approximately 97%, regardless of the regional source fuels mix. In comparison to big generating plants, conventional IC engines cars produce large amounts of HC and CO emissions, mainly because of cold starts and short trips that do not allow vehicles to become fully warmed up. Now, urban air quality issues, attached with a rising awareness of the problems associated with the world have created interest in EVs. Coal power plant emits 1 kg of CO_2 by generating of 1 kWh of. It is assumed that in one litre of petrol/diesel an average car runs 10 to 12 kilometers in cities; and an electric car can run 10 km with 1kWh of electricity. If electricity used for charging the EV is generated through fossil fuel, still CO_2 emission is less than half the emissions from petrol and diesel cars. In the case of EVs the electricity used for charging is produced in power plants located hundreds of kilometres away from the cities that are under attack with air pollution. A lot of strategies can be adopted from BEVs than combustion engine vehicles which reduces energy consumption. Driving with BEVs actually reduces the emission that contributes to climate change and smog, improving human health and reducing ecological damage. Charging the EVs on renewable energy like solar or wind or hydro power minimizes the emission even more. However, the batteries charging system which increase CO_2 emission, and proposes that minor use of battery could provide a potential way to promote EV market adoption and improve the sustainability performance of EVs through business model innovation like use of solar energy and EV battery reuse. In electric vehicle, regenerative braking system converts kinetic energy into electrical energy which is used to charge the battery. This system is particularly effective in recovering energy during city driving where driving patterns of repeated acceleration and deceleration are common. If electric vehicles are charged through renewable sources of energy like solar then emissions from EVs will be zero.

The proposed model for BEVs selection has dealt with finding of the best BEV from the available alternatives. After checking the various process parameters under different circumstances, it has observed that the proposed model is simple to use and easy to understand. Also, the computational time requirement is less and the stability of result obtained is high. Finally, a sensitivity analysis has shown to confirm the robustness of the ranking and further support the decision when selecting the final result. As the proposed CRITIC-CoCoSo model is robust, the number of alternatives and evaluation criteria in the decision matrix can be increased to get a more effective decision. This study can be extended for similar exercises in different vehicle segments, battery selection segments etc.. As the proposed model is generic in nature, it can be used for performance evaluation and ranking in other sectors of the society at large.

REFERENCES

Ajanovic, A., & Haas, R. (2016). Dissemination of electric vehicles in urban areas: Major factors for success. *Energy*, *115*(Part 2), 1451–1458. doi:10.1016/j.energy.2016.05.040

Behnam, V., Zandieh, M., & Moghaddam, R. T. (2011). Two novel FMCDM methods for alternative-fuel buses selection. *Applied Mathematical Modelling*, *35*(3), 1396–1412. doi:10.1016/j.apm.2010.09.018

Biswas, T., & Das, M. (2018). Selection of hybrid vehicle for green environment using multi-attributive border approximation area comparison method. *Management Science Letters*, *8*(2), 121–130. doi:10.5267/j.msl.2017.11.004

Biswas, T., & Das, M. (2018). Selection of hybrid vehicle for green environment using multi-attributive border approximation area comparison method. *Management Science Letters*, *8*(2), 121–130. doi:10.5267/j.msl.2017.11.004

Bühler, F., Cocron, P., Neumann, I., Franke, T., & Krems, J. F. (2014). Is EV experience related to EV acceptance? Results from a German field study. *Transportation Research Part F: Traffic Psychology*, *25*, 85–90. doi:10.1016/j.trf.2014.05.002

Cakır, S., & Percin, S. (2013). Çok kriterli karar verme teknikleriyle lojistik firmalarinda performans ölçümü. *EgeAkademik Bakış*, *13*(4), 449–459.

Chatterjee, P., Mondal, S., Boral, S., Banerjee, A., & Chakraborty, S. (2017). A novel hybrid method for non-traditional machining process selection using factor relationship and multi-attributive border approximation method. *Facta Universitatis Series: Mechanical Engineering*, *15*(3), 439–456. doi:10.22190/FUME170508024C

CNEV. (2017). *A survey of investment and capacity planning of new energy vehicles in China*. Retrieved from www.chinanev.net/news/newscontent/id/9582

Cobb, J. (2015b). *How Nissan and Renault are Dominating the Electric Car Game*. Retrieved from www.hybridcars.com/how-nissan-and-renault-are-dominating-the-electric-car-game/

Cocron, P., Bühler, F., Franke, T., Neumann, I., Dielmann, B., & Krems, J. F. (2013). Energy recapture through deceleration – regenerative braking in electric vehicles from a user perspective. *Ergonomics*, *56*(8), 1203–1215. doi:10.1080/00140139.2013.803160 PMID:23767823

Daimler. (2016a). *Interview Prof. Dr. Thomas Weber: Electric mobility at Daimler will be in the six figures by 2020*. Retrieved from http://media.daimler.com/marsMediaSite/en

Delucchi, M., Wang, Q., & Sperling, D. (1989). Electric vehicles: Performance, life-cycle costs, emissions and recharging requirements. Transpn. *Res A*, *23A*(3), 255–278.

Diakoulaki, D., Mavrotas, G., & Papayannakis, L. (1995). Determining objective weights in multiple criteria problems: The CRITIC method. *Computers Ops Res.*, *22*(7), 763–770. doi:10.1016/0305-0548(94)00059-H

Dijk, M., Orsato, R. J., & Kemp, R. (2013). The emergence of an electric mobility trajectory. *Energy Policy*, *52*, 135–145. doi:10.1016/j.enpol.2012.04.024

Ford. (2017). *MediaFord*. Retrieved from https://media.ford.com/content/fordmediamobile/fna/us/en/news/2017/01/03/ford-addingelectrified-f-150-mustang-transit-by-2020.html

Franke, T., Rauh, N., & Krems, J. F. (2015). Individual differences in BEV drivers' range stress during first encounter of a critical range situation. *Applied Ergonomics*. PMID:26456746

Guo, C., Wang, Y., & Jiang, W. (2013). An empirical study of evaluation index system and measure method on city's soft power: 17 cities in Shandong province. *Cross-Cultural Communication*, *9*(6), 27–31.

Hall, D., Moultak, M., & Lusey, N. (2017). *Electric Vehicle Capitals of the World: Demonstrating the Path to Electric Drive*. Washington, DC: International Council on Clean Transportation. Retrieved from www.theicct.org/sites/default/files/publications/Global-EV-Capitals_WhitePaper_06032017_vF.pdf

Helmbrecht, M., Bengler, K., & Vilimek, R. (2013). Strategies for Efficient Driving with Electric Vehicles. In E. Brandenburg, L. Doria, A. Gross, T. Günzler, & H. Smieszek (Eds.), Grundlagen and Anwendungen der Mensch-Maschine-Interaktion - Foundations and applications of human-machine interaction. 10. Berliner Werkstatt Mensch-Maschine-Systeme10-12. ZMMS, Technische Universität Berlin, prometeigraduiertenkolleg (pp. 83–91). Academic Press.

Honda. (2016). *Honda adds all-new dedicated hybrid model to growing electrified lineup*. Retrieved from http://news.honda.com/newsandviews/article.aspx?id=9463-en

Kazan, H., & Ozdemir, O. (2014). Financial performance assessment of large scale conglomerates via TOPSIS and CRITIC methods. *International Journal of Management and Sustainability*, *3*(4), 203–224.

Knittel, C. R. (2012). Reducing petroleum consumption from transportation. *The Journal of Economic Perspectives*, *26*(1), 93–118. doi:10.1257/jep.26.1.93

Lambert, F. (2017b). *BMW explains its electric vehicle plans in more details and it's going in the right direction*. Retrieved from https://electrek.co/2017/03/22/bmw-electric-vehicle-plans

Loveday, E. (2017). *LG Chem expects Chevrolet Bolt sales to exceed 30,000 in 2017*. Retrieved from http://insideevs.com/lg-chem-expects-chevrolet-bolt-sales-exceed-30000-2017/

Lutsey, N. (2015). *Global climate change mitigation potential from a transition to electric vehicles*. Academic Press.

Ma, J., Wang, N., & Kong, D. (2009). Market forecasting modeling study for new energy vehicle based on AHP and logit regression. *Journal of Tongji University*, *37*(8), 1079–1084.

MacCrimmon, K. R. (1968). *Decision Making among Multiple-Attribute Alternatives: A Survey and Consolidated Approach*. Santa Monica, CA: RAND Co.

Madic, M., & Radovanovic, M. (2015). Ranking of some most commonly used nontraditional machining processes using ROV and CRITIC methods. U.P.B. Sci. *Bull., Series D*, *77*(2), 193–204.

Mierlo, J. V., Maggetto, G., & Lataire, P. (2006). Which energy source for road transport in the future? A comparison of battery, hybrid and fuel cell vehicles. *Energy Conversion and Management*, *47*(17), 2748–2760. doi:10.1016/j.enconman.2006.02.004

Mouli, G. C., Bauer, P., & Zeman, M. (2016). System design for a solar powered electric vehicle charging station for workplaces. *Applied Energy*, *168*(Suppl C), 434–443. doi:10.1016/j.apenergy.2016.01.110

Pamučar, D., & Ćirović, G. (2015). The selection of transport and handling resources in logistics centers using Multi-Attributive Border Approximation area Comparison (MABAC). *Expert Systems with Applications*, *42*(6), 3016–3028. doi:10.1016/j.eswa.2014.11.057

Pamučar, D., Sremac, S., Stević, Ž., Ćirović, G., & Tomić, D. (2019). New multi-criteria LNN WASPAS model for evaluating the work of advisors in the transport of hazardous goods. *Neural Computing & Applications*, 1–24.

Rauh, N., Franke, T., & Krems, J. F. (2015). Understanding the impact of electric vehicle driving experience on range anxiety. *Human Factors*, *57*(1), 177–187. doi:10.1177/0018720814546372 PMID:25790577

Schalkwijk, W. V., & Scrosati, B. (2007). *Advances in lithium-ion batteries*. Springer Science & Business Media.

Stević, Ž., Pamučar, D., Vasiljević, M., Stojić, G., & Korica, S. (2017). Novel integrated multi-criteria model for supplier selection: Case study construction company. *Symmetry*, *9*(11), 279. doi:10.3390ym9110279

Stojić, G., Stević, Ž., Antuchevičienė, J., Pamučar, D., & Vasiljević, M. (2018). A Novel Rough WASPAS Approach for Supplier Selection in a Company Manufacturing PVC Carpentry Products. *Information*, *9*(5), 121. doi:10.3390/info9050121

Tesla. (2017a). *Tesla fourth quarter & full year 2016 update*. Retrieved from http://files.shareholder.com/downloads/ABEA-4CW8X0/4066363333x0x929284/22C29259-6C19-41AC-9CAB-899D148F323D/TSLA_Update_Letter_2016_4Q.pdf

UNFCCC. (2015b). *Adoption of the Paris Agreement*. United Nations Framework Convention on Climate Change. Retrieved from https://unfccc.int/resource/docs/2015/cop21/eng/l09r01.pdf

United Nations. (1992). *Report of the intergovernmental negotiating committee for a framework convention on climate change on the work of the second part of its fifth session*. United Nations Framework Convention on Climate Change 1992.

United Nations. (2014). *Report of the conference of the parties serving as meeting of the parties to the Kyoto Protocol on its tenth session*. United Nations Framework Convention on Climate Change 2015.

USEPA. (2014). *Inventory of US. Green house gas emissions and Sinks: 1990-2012*. EPA.

Volkswagen. (2016). *New Group strategy adopted: Volkswagen Group to become a world leading provider of sustainable mobility*. Retrieved from www.volkswagenag.com/en/news/2016/6/2025.html

Xuewu, X., & Yahui. (2012). Modeling and simulation of brushless dc motor control system for EPS applications. *2012 IEEE International Conference on Computer Science and Automation Engineering (CSAE).*

Yazdani, M., Zarate, P., Kazimieras Zavadskas, E., & Turskis, Z. (2018). A Combined Compromise Solution (CoCoSo) method for multi-criteria decision-making problems. *Management Decision*, MD-05-2017-0458. doi:10.1108/MD-05-2017-0458

Yılmaz, B., & Harmancıoglu, N. B. (2010). Multi-criteria decision making for water resource management: A case study of the Gediz River Basin, Turkey. *Water S.A.*, *36*(5), 563–576. doi:10.4314/wsa.v36i5.61990

Zavadskas, E. K., Turskis, Z., Antucheviciene, J., & Zakarevicius, A. (2012). Optimization of weighted aggregated sum product assessment. *Elektronika ir Elektrotechnika*, *122*(6), 3–6. doi:10.5755/j01.eee.122.6.1810

Chapter 4
A Hybrid AI-Based Conceptual Decision-Making Model for Sustainable Maintenance Strategy Selection

Soumava Boral
Indian Institute of Technology Kharagpur, India

Sanjay K. Chaturvedi
Indian Institute of Technology Kharagpur, India

V. N. A. Naikan
Indian Institute of Technology Kharagpur, India

Ian M. Howard
Curtin University, Australia

ABSTRACT

Selection of optimal maintenance strategy for critical systems/machinery is considered as a complex decision-making task that takes into account several available maintenance alternatives that are evaluated in terms of a set of different conflicting qualitative and quantitative factors. In the last few decades, progress has been made in different sustainable-based decision-making problems, where environmental, social, and economic factors played a pivotal role to arrive at the best decision. In this chapter, a hybrid artificial intelligence (AI)-based conceptual decision-making model is described by taking advantages of both expert system and case-based reasoning methodology to solve sustainable maintenance strategy selection problems. Adding to this, a flowchart of the model is suitably described by hypothetical examples of a sustainable maintenance strategy selection program.

DOI: 10.4018/978-1-5225-8579-4.ch004

INTRODUCTION

Classical theory of maintenance engineering has been based on the paradigm of fixing broken down assets to bring them back to the operative condition(s). However, even since the advent and advantages of several other improved paradigms of maintenance engineering, this ancient philosophy has been believed to be beneficial for non-critical or single shot items or where the repair is not an economically viable option. Therefore, for today's sophisticated and critical systems/machinery such as aerospace engines, boilers, high speed gearboxes, the earlier concept of maintenance engineering has transitioned towards a more advanced philosophy, where it is not only considered from a technical perspective (*i.e.,* vibration monitoring, temperature monitoring, visual inspections, etc.), but is extended towards managerial activities (*i.e.,* planning, coordination, personnel management, etc.) (Shafiee, 2015). Therefore, in the broader sense, it can be redefined as a set of activities comprising of technical, administrative, managerial routines during the life cycle of an asset and is carried out to preserve their values in terms of reliability, maintainability, availability and safety, which in turn directly influence the productivity and reputation of an organization. Adding to this context, it has been reported that depending on the type and size of organizations, a substantial percentage of the annual budget, varying from 15% to 60%, goes to deliver the maintenance program of an organization (Nezami & Yildirim, 2013; Mobley, 2002).

Now-a-days, most of the organizations demand that the working critical systems/machinery must have high availability and reliability, failing which might lead to a substantial amount of losses in terms of outgoing products' quality, generated revenue, missing targeted output, and even human casualties and environmental damages followed by litigations and lawsuits for the occurrence of certain types of failures. To avert such unwarranted failures of these systems and consequences thereof, they must be abetted with optimal maintenance strategies which not only reduces their annual maintenance cost but also alleviates them from unwanted operational inefficiencies and poor performance.

However, choosing the optimal maintenance strategy for critical systems/machinery is considered a complex decision-making task, where, at first, organizations have to identify a group of potential experts who have the required level of expertise and domain knowledge. Thereafter, a set of technical and managerial factors (*e.g.,* criticality of operation of the system, working condition, level of operators' involvement, environmental-social-economic impacts, etc.) are carefully defined by those expert panelists and in the very next step, based on those chosen factors, each of the maintenance strategies (*e.g.,* failure based maintenance, time-based maintenance, reliability centered maintenance, condition based maintenance, total productive maintenance, etc.) are rated accordingly, either in a subjective or objective manner. Upon completion of these, mathematical decision-making tool(s) are employed to arrive at the decision. It is worth to mention that the chosen factors are always conflicting in nature (*i.e.,* if the value of a factor upsurges then another one may rise or decline), which makes the situation more complex to arrive at the optimal decision.

The *Maintenance Strategy Selection Problem* (MSSP) can be stated, just like any other optimization problem with certain objective(s) satisfying a number of conflicting constraints, as the selection of the optimal alternatives with a balancing act of a set of conflicting criteria, thus, falling in the domain of multi-criteria decision making (MCDM) problems. Traditionally, most of the MSS problems have been considered from technical and economical perspectives; but, increasing concerns by scientific societies and /or governments' statutory regulations (*e.g.,* Clean Air Act (1970), Resource Conservation and Recovery Act (1976) and Toxic Substance Control Act (1976)) on the environmental impacts of hazardous wastes produced by such systems during the operational or maintenance phases have forced organizations to devise and adopt more viable and sustainable solutions.

Generally, the concept of sustainable development not only considers the problem domain from environmental perspectives but also includes the impacts of economic and social factors on any decision undertaken. It can be stated as the degree to which the present decisions of any organization will be impacting the natural environment, societies and business viability. These considerations benefit the organizations in formulating their long-term goals with associated pros and cons. Considering such multifaceted factors in any decision-making problem, however, is believed to make the problem more entwined and complex necessitating it to be addressed in a more systematic and scientific way.

Obviously, when the number of factors involved in a decision-making process increases, it becomes a difficult task for human decision makers to assess the relative importance among them by performing pairwise comparisons without resorting to the use of computational methods. In such circumstances, to reduce the complexity, the number of primarily considered factors are abridged by applying statistical analysis tool (*e.g.,* principal component analysis, a factor analysis), which is also considered as a time-consuming process. Until now, it can be observed that the existing literature has dealt with MSSP (conventional/sustainable approach) by employing a number of different MCDM approaches needing the participation of cross functional experts to arrive at an optimal decision. However, it is noteworthy that in a decision-making problem, each factor has a contribution to the final outcome, which may be nominal but must be included to arrive at the best decision. In this era of automation, our scientific society is considering how to eliminate the requirement of human experts in many routine day-to-day operations, besides, manual decision-making processes are not foolproof every time, and may lead to erroneous conclusions. Hence, such complex decision-making problems can be addressed by employing suitable *Artificial Intelligence* (AI) based approaches.

In the year of 1956, - a renowned scientist, John McCarthy, first coined the term AI and invited a group of researchers from versatile domains, *viz.,* mathematics, computer science, cognitive science, etc. to a summer workshop called *Dartmouth Summer Research Project on Artificial Intelligence* for further development of this field. Later, it proved its efficiency in the Turing test, Jupiter space project, Deep Blue Chess game (1997), etc. Now, in recent days, it has eliminated the requirement of human intervention in multiple domains, *e.g.,* medical diagnosis, law, rapid image processing, big-data analysis, etc. Although there are several definitions of AI, but the main theme of them is *"a computer program that simulates human intelligence or rationale thinking processes."*

The main objective of AI based approaches has been to solve extremely complex problems, which are beyond the ability of human cognitive reasoning and to reduce human's interventions in complex scenarios to avoid errors. *Expert System (ES)* and *Case-Based Reasoning (CBR)* are two such techniques from the AI paradigm, which have found their applications to solve several complex decision-making problems in engineering, *e.g.,* supplier selection, machine tool selection, fault diagnosis of complex machineries, etc.

Motivated by their potentiality, this chapter describes an AI-based hybrid decision-making model for selecting an optimal sustainable maintenance strategy. This model exploits prior information associated with several sustainable factors. Basically, these data are stored in the central database of any organization or can be made available as experts' opinions; however, it is observed that in certain situations, data on some factors might not be recorded or available that forces decision makers to arrive at an optimal decision with the available but incomplete information. Moreover, each of the considered factors have different influence on the final outcome and hence it is also necessary to consider the relative importance between them. In this chapter a hybrid ES and CBR technique based model is described. When a situation is delineated to the system then firstly it searches for the most suitable rule in the ES based module. If the system is not able to find an appropriate solution, it is next delegated to the CBR module that has an ability to provide an approximate solution. In both of the modules, it is assumed that the relative importance between the criteria is available, however, if it is not then it can be computed by any existing weight calculation method, *viz.,* entropy, analytical hierarchy process (AHP), analytical network process (ANP), decision making trial and evaluation laboratory (DEMATEL), etc.

RECENT CONTRIBUTIONS

The MCDM approaches employed in the MSSP domain can be classified into two broad categories, *viz.,* classical MCDM and fuzzy MCDM approaches. Table 1 summarizes some recent developments on conventional MSSP during the last three years, however, interested readers can refer to Shafiee (2015) covering the details on earlier research in this domain.

Table 1. Some recent contributions in MSSP research (2015-2018)

Sl. No.	Contributors	MCDM approach(es)	Factors chosen	Considered maintenance strategies	Area of application
1.	Kirubakaran & Ilangkumaran (2015)	ANP, GRA-TOPSIS	Safety, cost, added value, feasibility	CM, PdM, TBPM, CBM	Pumps of paper mill
2.	Joshua et al. (2016)	ANP	Cost (repair cost, annual maintenance cost, spares inventory), safety (personal safety, facility, machine security), strategic workers and union's acceptance, dispatch plan and training for employees) and time requirement (production shift, spare parts availability and workforce).	Scheduled maintenance, CM, PdM, Reactive maintenance.	Casting plant
3.	Lazakis & Ölçer (2016)	FAHP and TOPSIS	Maintenance cost, maintenance type efficiency, system reliability, management commitment, crew training, company investment, spare parts inventories, minimization of operation loss	CM, PM, and PdM	Diesel engine generator of a vessel
4.	Kirubakaran & Ilangkumaran (2016)	FAHP, GRA-TOPSIS	Safety, cost, added value, feasibility	CM, PdM, TBPM, CBM	Pumps of paper mill
5.	Özcan et al. (2017)	AHP, TOPSIS, and GP	Warehouse backup, maintenance pre-conditions, failure period, possible consequences, availability of measuring instrument, static, dynamic or electrical property of the equipment, troubleshooting time, detectability of failure, additional work requirement	CM, PdM, PM, Revision maintenance strategy	Machinery of hydroelectric power plant

continued on following page

Table 1. Continued

Sl. No.	Contributors	MCDM approach(es)	Factors chosen	Considered maintenance strategies	Area of application
6.	Shafiee et al. (2017)	ANP	Maintenance implementation costs and failure intensities	Failure based, risk-based, TBPM, CBM	Wind turbine
7.	Seiti et al. (2017)	Rough AHP	Added values, safety, cost, reliability and feasibility, time, efficiency and damage	CM, TBM, CBM, BM, TPM.	Rolling mill company
8.	Hemmati et al. (2018)	FANP	Cost, risk, added value	CM, TBPM, CBM and shutdown maintenance.	Sulphuric acid production plant.
9.	Emovon et al. (2018)	Delphi-AHP and Delphi-AHP-PROMETHEE	Cost (spare parts inventories, maintenance cost, crew training cost, equipment damage cost), safety (personnel, equipment, and environment), added value (minimization of operation loss, equipment reliability) and applicability (system failure characteristics, available monetary resources and equipment risk level).	CM, PM, and CBM	Ship machinery system
10.	Borjalilu & Ghambari (2018)	FANP	Organization, safety, administration, staff and technical requirements.	TBPM, CM, CBM, RCM, PdM.	5-MW powerhouse.
11.	Seiti et al. (2018)	FAD	Added value, safety, cost, feasibility, damage, efficiency, reliability, time.	CBM, CM, BM, TPM, and TBPM.	Rolling mill

[AHP = Analytic Hierarchy Process, FAHP = Fuzzy Analytic Hierarchy Process, ANP = Analytic Network Process, FANP = Fuzzy Analytic Network Process, GRA = Grey Relational Analysis, TOPSIS = Technique for Order Preferences by Similarity to Ideal Solution, GP = Goal Programming, PROMETHEE = Preference Ranking Organization METHod for Enrichment Evaluations, FAD = Fuzzy Axiomatic Design, CM = Corrective Maintenance, PdM = Predictive Maintenance, TBPM = Time Based Preventive Maintenance, CBM = Condition Based Maintenance, PM = Preventive Maintenance, TBM = Time Based Maintenance, TPM = Total Productive Maintenance, RCM = Reliability Centered Maintenance]

In the context of sustainable MSSP, Nezami & Yildirin (2013) first reduced the number of contributory factors by using the *factor analysis method* and applied fuzzy VIsekriterijumska optimizacija i KOm-promisno Resenje (VIKOR) approach. The approach was validated by taking an example from a manufacturing plant. Wang et al. (2015) established an evaluation index by considering six aspects, *viz.*, input cost, risk of failure, fault duration, maintenance cost, social factors, and environmental factors. The relative importance among the factors was calculated by AHP whereas the final decision was made by employing the VIKOR approach. Ighravwe & Ayoola (2017) solved a sustainable MSS problem by using the FAD principle and fuzzy TOPSIS approach. The entropy principle was utilized to calculate the relative importance between the sustainable factors. A cement producing plant in Africa was taken as a case study to show the efficacy of their approach.

It is needless to say that the preceding MCDM approaches have strong mathematical and statistical background and were developed by simulating the human decision-making processes. However, it is evident that the complexities in terms of mathematical rigor and involvement of decision-makers in each of the processes is quite burdensome. Further, it becomes a cumbersome task to solve these problems manually, when the number of factors involved are large in number in addition to the available alternatives.

ES, a subset of AI, has successfully been applied in the recent past in numerous application domains, such as truck selection (Chakraborty & Prasad, 2016), CNC turning center selection (Prasad & Chakraborty, 2015), material selection (Prasad et al., 2014), etc. It is proven to be a powerful tool for solving different types of decision-making problems and still now is an open field of research. The CBR approach is considered as an amalgamated domain of AI and the human cognitive process which traces its origin when Roger Schank and his students started the research at Yale University in the early 80's (Schank, 1982). Schank's dynamic memory model was the basis for the earlier CBR system, followed by the researches of Janet Kolodner's CYRUS (Kolodner, 1992). A rapid progress on CBR was noticed in the late 90's when several international conferences were held in Europe, Germany and many other parts of the world. It is considered as a powerful reasoning tool that mimics the human reasoning process by utilizing prior experiences stored in the database (or rather called as case-base) and not by extracting any knowledge from those stored cases.

This analysis shows that the application potential of the aforementioned AI-based techniques in selecting optimal sustainable MSS has either not been explored or been overlooked by earlier researchers over the ubiquitous MCDM approaches. Keeping in mind the potentialities of ES and CBR, this chapter intends to present a hybrid decision-making model for sustainable MSSP by utilizing the prior cases as a basis to arrive at the optimal decision. In the ensuing sections, a brief overview about these AI-based approaches are provided along with some ready references for further study.

BASIC COMPONENTS OF THE MODEL

Before suggesting the AI-based hybrid decision-making model intended to solve sustainable MSSP, a brief overview of the model components, viz., sustainability, ES and CBR are provided for the sake of completeness.

Sustainability

Sustainability is a complex issue, and an elusive one. It is very significant since it has to do with the chances of humankind surviving on this planet. At the rate at which the individuals/groups/nations are exploiting the scarce resources it seems that unless measures are taken now, and if there is still time, the upcoming civilization, at least as it is understandable now, is uncertain to say the least. It follows that such a complex subject has no simple and straightforward treatment, especially when one has to understand that sustainability is not a goal but an endless process. It leads to a better life for the present generation and survival for generations in years to come by enhancing their ability to cope with the world that they will inherit from the present.

The terms *sustainability* and *sustainable developments* are often used interchangeably and since 1980, it has gradually turned into a flourishing area of research in R&D groups of government and industrial sectors with the theme - *"sustainable development is a kind of development that fulfills the needs of the present generation without compromising the ability of future generations to meet their needs."* This theme consists of three prime terms, *viz.,* development, present and future.

After looking at the word '*development*', one instinctively thinks about economic development, however, from the sustainable perspective, it means the advancement in every area, *viz.,* economic development, social progress, and environmental protection. '*Present*' refers to acting in a structured way with a view of achieving growth not only in the economical context, but also in social and environmental contexts. The term '*future*' does not refer to the immediate future, but to the long term future which will be inhabited by future generations.

Now-a-days, sustainable development has become a major challenge for any organization, especially for manufacturing and process plant sectors, as several complex machineries or waste disposal actions are producing hazardous materials/gases on a regular basis that bear direct impacts on the surroundings. Moreover, catastrophic failures of these machineries due to poor decisions taken by managers reduces the overall profit generation (economic impact) and changes the workers' mindset (social impact) that directly affects the overall productivity of the organization. Adding to this context, organizations are now compelled to adhere to statutory rules and regulations enforced by government/international agencies for their sustainable development and to survive in the competitive business market.

Expert System

An ES differs from the conventional programs in several way, for instance, it is knowledge intensive, highly interactive, and divides the experts' knowledge into separate rules. In a broader sense, an ES consists of a knowledge base, a working memory, an inference engine, system analysis and graphical software and a user interface, which interacts with the end-user as shown in Figure 1. The knowledge contained in the knowledge base can be either prior knowledge or posterior knowledge. This knowledge can be represented as rules, semantic nets, frames, scripts, object oriented structures, conceptual graphs and so on. An inference engine examines the knowledge base and reasons the answer (how and why) to the end-user, which is also known as pruning. This task is carried out by following any of the methods, *viz.*, production rules, structured objects and predicate logic. Production rules consists of a rule set, a rule interpreter that specifies when and how to apply the rules and a working memory which holds the data, goals and intermediate results. Structured objects use vector representation of essential and accidental properties; whereas, predicate logic uses propositional and predicate calculi.

The process of building an ES/KBS is known as knowledge engineering wherein knowledge engineers embed the knowledge of human experts in it. This embedding process is carried out by decoding the linguistic terms given by the experts' into suitable programing codes, and in the absence of such experts, the ES provides decisions to a particular problem domain with which it has been constructed. Generally, ES provides solutions, which are derived from its knowledge base and contains declarative facts, as well as procedural (or heuristic rules) knowledge about the problem domain by using a reasoning process embedded in its inference engine, the '*thinking part*' of the system. It uses any of these three methods, *viz.*, backward chaining (top-down reasoning), forward chaining (bottom-up reasoning) or abduction as the basis of inference. At first, it looks for the 'most likely' hypothesis and then searches for the evidence for the hypothesis. If after receiving all the relevant information from the end-user, the initial hypothesis cannot be supported then it looks for the 'next most likely hypothesis' and so on, (Lucas & van der Garg, 1991).

Figure 1. Architecture of an Expert System

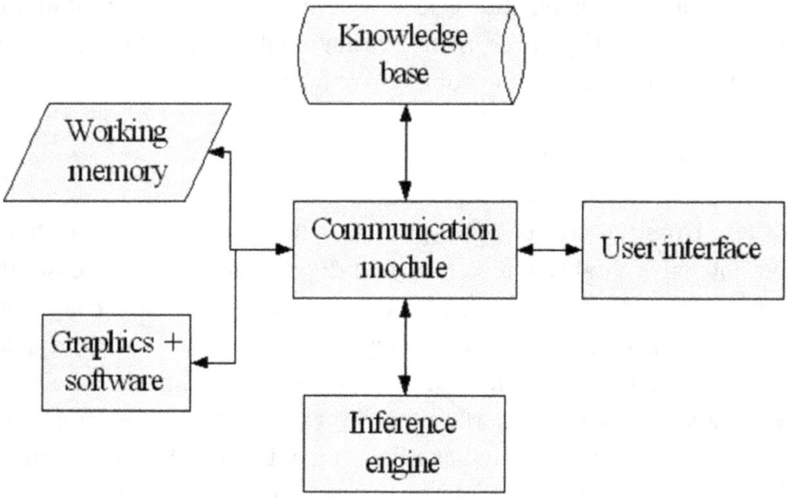

Usually, to develop a GUI based ES in an industry following steps are adopted (Lucas & van der Garg, 1991):

1. Define the deliverable or the outputs which are expected from an ES.
2. Several interviews are conducted with the active participation of knowledge engineers.
3. Store the Acquired knowledge in the knowledge-base.
4. Several rules are derived by exploiting these knowledge.
5. Store the problems and their associated solutions in the inference engine.
6. Develop a GUI by using any graphics enabled software or alike (e.g., Microsoft Visual Basic 6.0) to exploit the knowledge base and to interact with end-user.

ES is a useful tool in a decision-making process when the following bottlenecks are observed:

1. When there is a scarcity of experts' knowledge in an organization for a decision making purpose.
2. When sufficient amount of prior cases is available, but they are unexploited and simply stored in the database.

In other words, the concept of ES is useful when there are sufficient cases available. In most of the instances, an ES deduces solutions from these stored cases and from an exhaustive 'IF-THEN' rule-base. Despite its several advantages, the major problems associated with ES are (Boral & Chakraborty, 2016; Chakraborty & Boral, 2017):

1. Knowledge acquisition barrier, linguistic barrier, cognitive barrier, representation barrier.
2. It works on the principle of direct matching.
3. A rule-base model usually becomes unwieldy due to the exponential rise in the number of rules due to the number of factors and their respective levels. This not only creates the problem of assessment of rules by experts but also for many of the combinations, it becomes difficult to assess the consequent part of certain rule(s) even by the best expert available. This necessitates incorporation of an appropriate rule-reduction technique (Gragama & Chaturvedi, 2011).

Alternatively, the CBR approach is also a preferable option, where only a limited number of prior cases are available.

Case-Based Reasoning

Case-based reasoning (CBR) is an amalgamated domain of the human cognitive process and AI. It is a well-known concept and methodology that touches upon some fundamental issues related to knowledge representation, reasoning and incremental learning from experience. Its applications have recently been shown in several domains such as, non-traditional machining process selection (Boral & Chakraborty, 2016), machine tool selection (Chakraborty & Boral, 2017), fault detection and diagnosis of machinery (Xu et al., 2018). The term CBR is composed of three basic words, *viz.,* 'case' – which is nothing but the preceding contextualized experiences, 'based' – dependent on the prior cases, and finally, 'reasoning' – deducing some inference from the previous cases. It is nothing but a '*similarity-search algorithm*'. In CBR, previous cases play the pivotal role in the decision making process. It is based on the concept of how people learn new skills or how humans generate hypotheses about new situations based on their past experiences. It has a process model and a memory model (Pal & Shiu, 2004).

CBR is sometimes referred as '4-R' cycle, illustrated schematically in Figure 2, as it is composed of the following four procedural steps:

- Retrieve: When a new problem is delineated to the system by the end-user or program, the CBR system searches its case-base, to find a case that has the same problem specifications as the case under analysis. In this step, the most similar cases from the case-base are retrieved based on the richness of the attributes of the new situation. At first, the distance between each attribute of the prior and new case is calculated by different distance measuring formulas. Some popular distances used in the CBR literature are Euclidian distance, hamming distance (Boral & Chakraborty, 2016), Minkowski distance, etc. After that, the similarities among the new case and the prior cases (Chakraborty & Boral, 2017) are computed gradually and the case with the highest similarity measure is presented primarily to the end-user by using the *k-nearest neighbor* (k-NN) algorithm. In some CBR systems, to simplify the retrieval mechanism, n-step retrievals are carried out. In some situations, to consider the relative importance of the attributes, weighted similarity measures are calculated for better precision in decision making. To calculate the relative importance among the considered criteria, different available approaches could be used, *viz.,* Delphi method (Wang et al., 2014), AHP (Satty, 1990; Banerjee & Majumder, 2016), ANP (Rai & Bolia, 2014), DEMATEL (Ranjan et al., 2015), Entropy (Shemshadi et al., 2011).

- Reuse: In some situations, where the present and prior problem are identical, then it is assumed that the solution attached to the best previous case would be the solution of the present one. However, in most of the situations, prior cases do not match exactly with the present case. Hence some adjustments are needed which requires the phase of adaption, like human cognitive processes. During this adaption process, differences between the present case and the best prior case are taken into consideration, which are further utilized to modify the retrieved case for better solution accuracy.
- Revise: The retrieved case is revised as per the experts' opinion/by developing an expert system /some machine learning algorithms to provide the optimal solution.
- Retain: The case with a successful solution is further retained in the case-base for its future use.

To develop a CBR system in an industry, following steps are adopted (Pal & Shiu, 2004):

- Define the purpose and expected output of a CBR system.
- Define the problems, their associated features, values (either qualitative or quantitative).
- Define the solutions for each problems.
- Create an exhaustive case-base in any database management software (*e.g.,* Microsoft Access, Excel etc.). Each cases should have problems and their associated solutions.

Figure 2. CBR cycle or '4-R' cycle

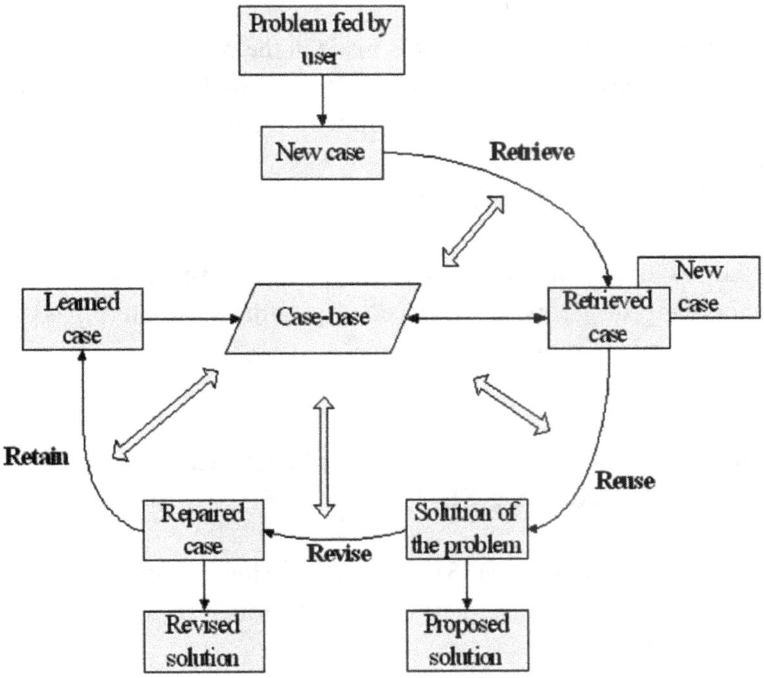

- Develop a GUI by using any graphics enabled software or alike (e.g., Microsoft Visual Basic 6.0) to interact with end-user. When an end-user provides an input query to the developed CBR based software module, it exploits the case-base to provide solutions.

Decision-makers can resort to build a CBR system under the following situations:

- When the previous cases repeat themselves.
- When limited prior cases are available in the central database of the organization and decisions are to be made by using them.
- When the decision-making problem does not have an underlying model or a model that is hard to interpret.
- When novel cases are encountered frequently.

There are several benefits of using CBR based systems for decision making, such as:

- The task of knowledge acquisition from the previous cases is reduced to a great extent.
- Avoiding the repetition of mistakes made in the past.
- Incremental learning is made in the CBR system. In contrast to the artificial neural network, there is no need to train the network each time, when a new case arrives.
- Sometimes, it can explain the problem's solution.

Since the sustainable maintenance strategy selection is greatly influenced by several factors and criteria, both qualitative and quantitative in nature, it is worthwhile to describe them before providing the model.

INFLUENCING FACTORS FOR SUSTAINABLE MAINTENANCE PROBLEMS

Factors involved in a sustainable MSSP can be divided into qualitative and quantitative components, which in turn have their own sub-factors, viz., economic, technical, social and environmental.

Qualitative Factors

Most of the time, these factors are evaluated by the experts in a linguistic manner on a nominal scale (*viz.,* high, very high, low, very low, etc.), which can later be decoded by following some already well-accepted and standardized scales (*e.g.,* Likert scale) within the organization for further processing to arrive at a decision, *e.g.,* very high = 9, high =7, medium = 5, low =3, very low =1). Some of the possible criteria under this heading could be:

Economic Factors

- Quality of output product after maintenance: Maintenance of a complex system/machinery is carried out by repairing, replacing or adjusting the components making that system (Elsayed, 2012). The repair actions may be perfect (as good as new/renewed), minimal (as bad as old, implying the failure rate of the system remains the same as it was just before the failure) or in-between these two extremities (better than minimal repair but less than perfect) (Doyen & Gaudoin, 2004). After carrying out maintenance tasks of a critical system/machinery by following the maintenance strategy, it is usually assessed in terms of the rejection rate, which should not be worse than the previous one.

- Ease of maintenance: Ease of maintenance is the simple way of maintaining a system/machinery and getting rid of faults and/or failures. It depends on a number of factors, such as clearance to carry out maintenance tasks, design of a machinery, time on the maintenance platform, and force magnitude needed on the maintenance platform. Ease of carrying out specific maintenance tasks could be considered as one of the major criteria for selecting the optimal one.

Technical Factors

- Technical feasibility: Sometimes, it may be possible that the particular critical machinery could not be brought under the specific maintenance strategy due to, say, cost or technical infeasibility. This infeasibility might be due to space impediment, climatic conditions and/or maintenance platform. So, it depends on the knowledge domain of the experts to suggest the next best technical alternative(s) of a particular maintenance strategy on that particular or group of machinery.
- Technical complexity: Each maintenance strategy has its own technical complexity in terms of knowledge requirements, use of software, use of prior knowledge, etc. Hence, this becomes an important criterion for selecting the best maintenance practice to optimize the resources needed. Such resources are mainly maintenance supply support, maintenance test and support equipment, maintenance personnel, maintenance facility, maintenance technical data and maintenance computer resources (Knezevic, 1993; Knezevic, 1997).
- Flexibility of maintenance: It measures the readiness of response of a maintenance strategy to unwarranted incidents of a critical system/machinery. Each of the maintenance strategies must be flexible enough (adaptive, responsive and agile) for an organization to have more options, with faster changing mechanisms to the changing situations with minimum time and efforts (Garg & Deshmukh, 2009). The elapsed time can be influenced by factors, viz., the personnel factors (e.g., motivation, skill, physical, attitude) of personnel involved, conditional factors influencing the operating environment and consequence of failure on the component, and environmental factors (humidity, noise, vibrations, time of the day…). It can also be measured in terms of factors such as maintenance capacity, maintenance facility, vertical integration, maintenance organizations, managerial flexibility, etc. There are also different techniques for flexible maintenance, such as distribution integration, risk pooling, multifunctional staffs, maintenance outsourcing, etc.

Social Factors

- Workers safety: It should be one of the major concerns while selecting a maintenance strategy for a critical system/machinery. For instance, a pressurized boiler cannot be brought under corrective or breakdown maintenance as a failure of such critical system may lead to severe injury or even death of the personnel. Hence, for such critical system/machinery other sophisticated maintenance policy like CBM or PdM should be adopted.
- Acceptance by workers: Workers are the primary drivers of the observation of any abnormalities in the system/machinery. The chosen maintenance strategy should therefore be accepted by the workers or the operators operating the system/machinery as well. Sometimes, it might be possible that a complicated maintenance strategy is not well-accepted by the workers. In this situation, management must train workers to help them understand, through training programs/simulators/brainstorming events from time to time, its advantages.
- Compliance with government regulations: Apart from the above, the organization has to follow the statutory regulations set out by government regulatory bodies to avoid unwarranted litigations.

Environmental Factors

- Compliance with environmental standards: The sustainability needs that an organization's liability and responsibility must have to protect the environment through recycling or safe disposal of the worn-out item. There are standards, e.g., ISO 14001, that specify the requirement of an environmental management system for small and large scale organizations. For each of the considered maintenance strategies, this standard should be followed for a viable and sustainable MSS program.
- Toxicity of generated waste: During the maintenance activities, (of e.g., nuclear reactors, boilers, etc.) several types of hazardous/non-hazardous wastes may be generated. This factor also has a bearing on having a sustainable MSS program.

Quantitative Factors

In industrial practice, it is hard to obtain the precise and quantified value of the many factors which play a pivotal role in the decision making process. For those, an assessment can be made either through measurement or it can be provided by a domain expert. For some factors, assessments are made either per repair basis (e.g., cost of production loss, mean time between failures, etc.) or annual basis (e.g., training cost, software cost, etc.). Some possible criteria/factors can be:

Economic Factors

This factor can be further decoupled into different sub-factors as follows:

- Hardware cost: Usually, it includes cost of electrical/electronic/computer hardware components, e.g., sensors used to detect the condition of the system to determine the overall health.
- Spare parts cost: Cost of spares, tools, special supplies and related inventories needed to support the maintenance process used during the maintenance procedure.
- Software cost: This includes the cost of different computerized systems or diagnostic software used to detect the faults and/or determine the health conditions to take a maintenance decision.
- Manpower cost: Costs of labor/workers, technicians/engineers to bring the system back to an operative condition.
- Training cost: Cost for training the operators to make them acquainted with the tools and their handling techniques (*e.g.,* signal processing in CBM) used in monitoring and maintenance procedures.
- Cost of production loss: Due to failure of a critical equipment/machinery in a system or production line, the total system/production line may be shut down, causing significant revenue/production loss.
- Return on investment: It is the ratio of net profit gained from that critical system/machinery and cost of investment made on them.

Technical Factors

It can be classified into the following categories:

- Mean time between failures: This is the arithmetic mean time between successive failures of a repairable system. Usually, failures of a repairable system are measured in global time, if the failure times are recorded as time since the initial start-up of the system. Whereas, the same are measured in terms of local time if the failure times are recorded as time since previous failure. For a chosen maintenance strategy, the mean time between failures should tend to increase rather than getting worse (Rigdon & Basu, 2000), with the higher the mean time the better.

- Mean time to repair: It measures the average time required to bring back a failed system/machinery to the working condition. It is a basic technical measure of the maintainability of equipment and repairable parts. Repair time of a system/machinery is the quantification of time that it is out of production due to some occurred faults and/or failures. For a good maintenance practice, this criterion should be less, i.e., the lower the better.
- Availability of spare machinery: Availability is considered as an important metric for repairable system performance and combines both reliability and maintainability. It is defined as the probability that the system is available for use when demanded. In the worst situations of maintenance (e.g., breakdown maintenance), it is required to replace the system/machinery with the redundant one, which should be in good working condition or available on demand.
- The risk level of the system/machinery: ISO 9000:2015 defines risk as the "effects of uncertainty on an expected result." For critical system/machinery, risks are considered from different points of view, such as financial, technical, operational, environmental, health, safety, and impact on business and social objectives. While selecting a maintenance strategy (either conventional or sustainable) for critical system/machinery, the level of risk of them must be included in the decision making process at the system level and/or component level.

Social Factor

Level of the performance of employees: Operators and employees are primarily responsible to carry out maintenance tasks. Their level of performance would be a major concern for selecting the optimal maintenance strategy for the system/ machinery. Several scales are available to measure their performance such as Global Vigor and Affect (GVA) scale, NASA TLX scale and the Subjective Workload Assessment Technique (SWAT) scale.

Environmental Factors

- Amount of toxic substance emissions: It should be a major concern for selecting the optimal maintenance strategy. Toxic substances (i.e., lubricating oils and their fumes, different harmful gases, like carbon monoxide, sulfur dioxides, etc.) are produced when the system/machinery is run at poor operating conditions or at faulty conditions. For instance, this toxicity can be produced during the task of preventively changing the lubricating oils from gearboxes, or through the generation of harmful gases due to oil-evaporation during a bearing failure of a gearbox.

- Amount of waste materials generated in a particular maintenance strategy: In each maintenance practice, various waste materials are generated, such as in PM, though the component is in a degraded state, it is replaced with a new one to mitigate the associated risks. Whereas, in CBM, components are allowed to run just prior to failure.
- Cost of cleaning the waste: In each maintenance practice, toxic or non-toxic wastes (e.g., scrap materials, used lubricating oils, metal burs, etc.) are generated. The careful extraction, storage and recycling of this waste entails cost, which might be substantial. Now-a-days, in most of the world-class organizations, to comply with the industrial regulations, toxic wastes are cleaned by robots or automated machines.

Table 2 provided the summary of the desired nature of factors discussed above for a sustainable MSS program.

Two simple examples will be considered further to illustrate the process by taking a small number of the above criteria before putting forward the optimum maintenance decision model. In these examples, the various factors in the groups are deliberately kept generic to assist in the illustration and to make the process easier to comprehend.

EXAMPLE

Consider that there are several multistage gearboxes functioning in a steel processing plant and each of them are managed with a sustainable maintenance program. Experts selected six sustainable factors, namely A, B, C, D, E and F, among which A, B and F are beneficial factors (*i.e.,* higher the better) and C, D and E are non-beneficial factors (*i.e.,* lower the better), as per their requirements. A, B, and C are assumed to be quantitative factors. For instance, in this example A is the *mean time between failures* of the gearbox, B is *availability of a spare gearbox,* and C is the *cost of production loss* due to the shutdown of the gearbox; whereas, D, E and F are assumed to be qualitative factors. Here, D represents *technical complexity*, E is *toxicity of generated waste*, and F is *ease of maintenance.* For better understanding of the readers as well as to reduce the complexity, linguistic assessment of each of the qualitative factors are converted into numerical value by following the scale given in Table 2. It is also assumed that all these data are already stored in the central database of the organization as shown in Table 3. Here, the *Case ID* represents the *'primary key'* associated with each case.

Table 2. Sustainable factors, sub-factors and their desired nature

	Factors	Sub-factors	Desired nature
Qualitative factors	Economic	Quality of output product after maintenance	↑
		Ease of maintenance	↑
	Technical	Technical feasibility	↑
		Technical complexity	↓
		Flexibility of maintenance program	↑
	Social	Worker's safety	↑
		Acceptance by workers	↑
		Compliance with government regulations	↑
	Environment	Compliance with environmental standards	↑
		Toxicity of generated wastes	↓
Quantitative factors	Economic	Hardware cost	↓
		Spare parts cost	↓
		Software cost	↓
		Manpower cost	↓
		Training cost	↓
		Cost of production loss	↓
		Return on investment	↑
	Technical	Mean time between failures	↑
		Mean time to repair	↓
		Availability of spare machinery	↑
		Risk level of system/machinery	↓
	Social	Level of performance of employees	↑
	Environment	Amount of toxic substance emission	↓
		Amount of waste material generated	↓
		Cost of cleaning the waste	↓

[↑ = higher the better, ↓ = lower the better]

Case-Input-1

Now suppose, a newly installed gearbox is to be supported with this sustainable maintenance strategy and the decision-maker/engineer has requirements of the following values for each selected factor as shown in Table 4.

This example can be solved from the inputs by using a hypothetical rule-based ES-1:

Table 3. Stored cases in database

	Antecedent factors							Consequent output
Case ID	**Machine ID**	**A (Hrs.)**	**B (factor)**	**C (Dollars)**	**D (scale)**	**E (scale)**	**F (scale)**	**Maintenance strategy**
1	A03	130-140	0.80-0.82	100-112	2-3	5-6	2-3	TBPM
2	A05	97-102	0.90-0.92	45-50	8-9	2-3	7-8	CBM
3	A04	170-175	0.75-0.78	198-206	3-4	7-8	5-6	CM
4	A02	150-154	0.84-0.86	112-118	4-5	5-6	2-3	Age based PM
:	:	:	:	:	:	:	:	:
:	:	:	:	:	:	:	:	:
N-1	A01	110-115	0.94-0.96	78-88	7-8	3-4	8-9	CBM
N	A07	98-102	0.81-0.83	96-104	1-2	8-9	4-5	CM

[Absolutely high = 9, very very high = 8, very high = 7, high = 6, medium = 5, low = 4, very low = 3, very very low = 2, absolutely low = 1]

Table 4. Case-input-1

Case-input						
Factors	**A**	**B**	**C**	**D**	**E**	**F**
Input values	132	0.81	112	3	5	2

1. The ES-1 builds an initial rule in the following manner:

$$"IF\ A\ is\ 132\ AND\ B\ is\ 0.81\ AND\ C\ is\ 112\ AND\ D\ is\ 3\ AND\ E\ is\ 5\ AND\ F\ is\ 2"$$

2. The built up rule is then forwarded to the knowledge base and then the knowledge base finds the most similar rule in it, which is stored in the following manner:

$$"IF\ A\ is\ BETWEEN\left(130\ and\ 140\right) AND\ B\ is\ BETWEEN\left(0.80\ and\ 0.82\right)$$

$$AND\ C\ is\ BETWEEN\left(110\ and\ 112\right) AND\ D\ is\ BETWEEN\left(2\ and\ 3\right) AND$$

$$E\ is\ BETWEEN\left(5\ and\ 6\right) AND\ F\ is\ BETWEEN\ \left(2\ and\ 3\right)"$$

3. Thereafter, the knowledge base triggers the inference engine for getting the attached solution with the above rule. As shown in Table 2, the solution is attached in the following manner:

$$"IF\ A\ is\ BETWEEN\left(130\ and\ 140\right)AND\ B\ is\ BETWEEN\left(0.80\ and\ 0.82\right)$$

$$AND\ C\ is\ BETWEEN\left(110\ and\ 112\right)AND\ D\ is\ BETWEEN\left(2\ and\ 3\right)AND$$

$$E\ is\ BETWEEN\left(5\ and\ 6\right)$$
$$AND\ F\ is\ BETWEEN\ \left(2\ and\ 3\right)"THEN\ Solution\ is\ "TBPM".$$

Clearly, the ES-1 advice is to opt for TBPM.

Case-Input-2

Now suppose the decision maker provides the values given in Table 5. It can be observed by comparing the values in Table 3 and Table 5 that none of the stored cases are matching with the given input, *e.g.*, for factor A, the input value is 145, but there is no value range in the stored cases between which 145 can fall. Similarly, for criterion B, 0.94 matches with Case-ID (N-1), C matches with none, D matches with Case-ID 1 & 3, etc. The ES-1 module will then fail to provide any feasible solution. Note that these types of situations are usually encountered in the real world. To solve this problem, an AI based hybrid decision-making model is described in the next section.

AI-BASED HYBRID DECISION-MAKING MODEL

A generic flowchart of the suggested model is shown in Figure 3. The model consists of mainly two modules, viz., ES-1 plus a CBR module. Within CBR, there

Table 5. Data for Case-Input 2

Case-input-2						
Factors	A	B	C	D	E	F
Factor weight	0.18	0.22	0.1	0.2	0.16	0.14
Input values	**145**	**0.94**	**62**	**3**	**4**	**2**

Figure 3. Workflow diagram of the suggested model

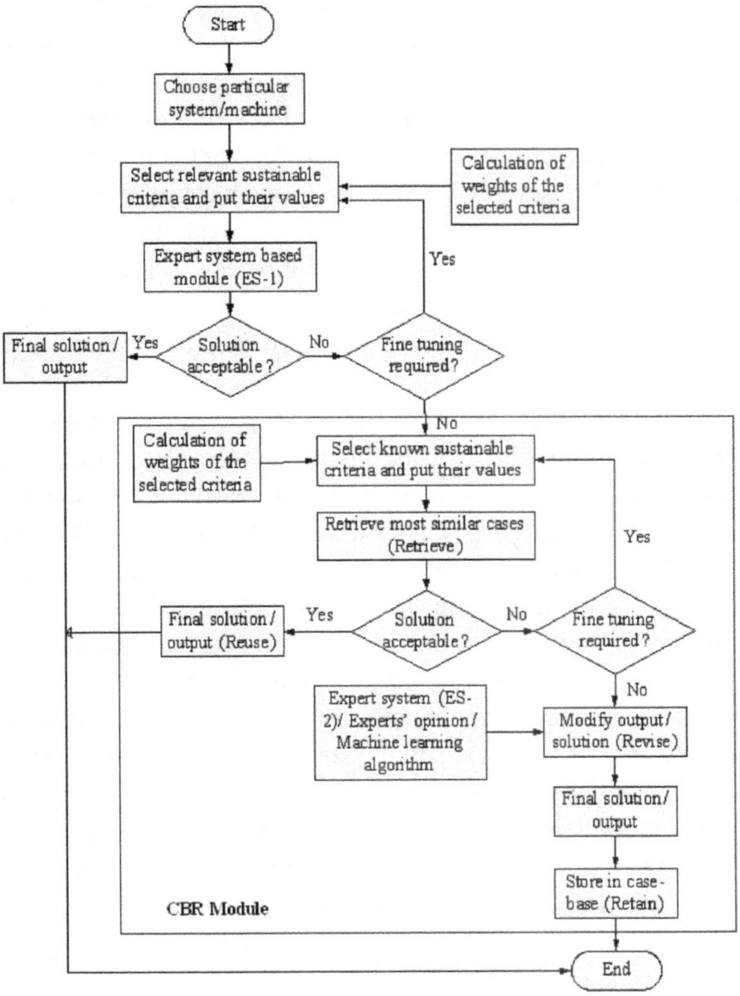

is another ES-2 whose purpose is to carry out the refinement on the outcome of the retrieve stage of the CBR module. In this model when a new and unseen input is encountered, first the ES-1 module tries to solve it by using its own rule-base. If the required rule(s) is not available in the knowledge base, the ES-1 module fails to provide any solution to the end-user. Next, it delegates the task to the CBR module, which provides an approximate solution of the current input. If necessary, the ES-2 could be used for a revised phase of the CBR. It is believed that by following the steps provide next for a given unseen input, one can easily build up software-based modules intended to solve much complex sustainable MSSP. The steps are:

Step 1: Referring to the same example of the gearbox, when a new sustainable MSS case arrives, the end user initially provides inputs like number of gears in a gearbox, their diameters, module of gears, etc. to seek for the machine which has almost the same technical specifications with the new one and is already available within this model as global input data. After that the end-user selects the relevant criteria to be used in the decision making process.

Step 2: Since there will be several factors/criteria inputted by the end-use, it becomes necessary to compute their relative importance due the varied impact they have on the final outcome. The relative importance of the selected criteria can be computed by either any of the well-known methods as mentioned in the introductory section or can be directly provided by the end-user by further processing.

Step 3: The model now triggers the ES-1 module that searches its knowledge base for the closest similarity with the input. It is worth mentioning that ES-1 makes the final decision based on a backward reasoning (top-down approach) approach. Thereafter, the best matched rule is forwarded to the inference engine for finding its associated solution. In the worst situation, when the ES-1 is unable to give a solution to the end-user as confronted in case-input-2, it may prompt for fine-tuning of the factors, or it will enter into the CBR module for providing an approximate solution to the complex problem.

Step 4: The CBR module of the model searches for the best matched case from the case-base (flat memory, serial search process), without extracting any knowledge from them. A set of prior similar cases are presented to the end-user along with their similarity score with the input case. In this context, it should also be mentioned that the module may not provide solutions exactly matching with the stored cases, rather it may provide solutions relevant to other installed machines in that organization having similar technical specifications.

Step 5: If the solution provided by the retrieval phase, in step 4, is accurate then it can be accepted by the end-user. However, there can be a situation when the end-user disagrees on the solution and it becomes necessary to fine-tune the criteria for a better similarity match. This fine tuning process may be carried out by selecting more number of factors and providing values of them. In some circumstances, based on the output provided by the developed model, it is sometimes necessary to adjust the values of input parameters to arrive at optimal decision. If after fine-tuning the system is still not able to provide a level of similarity (*e.g.,* a threshold similarity score) among the present problem and prior cases, it is required to modify the output, which is known as the 'revision' stage in CBR terminology.

Step 6: The revision task can be carried out in different ways, *viz.*, by ES-2, experts' opinion or by machine learning algorithms, as the case may be. The most popular approach to make a revision is by building an ES-2. Mainly, in ES-2 knowledge from knowledge engineers are incorporated to carry out the revision process of a retrieved case. Based on the output, provided by retrieve phase of a CBR system, different 'IF-THEN' rules are formed to carry out the revision task and to arrive at the optimal decision.

Step 7: After this revision phase, the final outcome is suggested to the end-user and is also stored in the case-base for its future use for new cases, providing the retain phase.

The solution to the case-inputs presented in the example can now be provided. In this model, the decision-making process is carried out by exploiting a single database which is available with the organization. This model takes the advantages of both ES and CBR to achieve an improved reasoning process.

SOLUTION BY THE MODEL

1. **Case-Input-1:** When the case-input-1 is provided to the model, it solves the problem by utilizing the ES-1 module only. Hence, there is no requirement to enter in the CBR module, and the final outcome is the same as previously obtained.
2. **Case-Input-2:** When the input query is provided to the model, the model works as follows:
   a. **Step 1:** ES-1 of the model will fail to offer a solution as already explained earlier and therefore, the input is passed to the CBR module. It is assumed here that each of the criterion has a relative importance score as A = 0.18, B = 0.22, C = 0.1, D = 0.2, E = 0.16 and F = 0.14.
   b. **Step 2:** The CBR system calculates the hamming distance for each of the selected factors between the prior cases and the input case (refer Table 6). It now calculates the weighted similarity score for all the stored cases (refer Table 7). In this context, it is to be noted that for calculating the weighted distance, higher/lower values are taken from the prior cases for 'higher the better/lower the better' factors.

Table 6. Hamming distances between input and stored cases

Case ID	A	B	C	D	E	F
1	0.064	0.571	0.236	0.125	0.143	0.143
2	0.551	0.095	0.106	0.625	0.286	0.857
3	0.385	0.762	0.845	0.000	0.429	0.571
4	0.115	0.381	0.311	0.125	0.143	0.143
5	0.385	0.095	0.099	0.500	0.143	1.000
6	0.551	0.524	0.211	0.250	0.571	0.429

Table 7. Similarity scores after considering criteria weights

Case ID	A	B	C	D	E	F	Weighted sum	Similarity score = (1- weighted sum)
1	0.012	0.126	0.024	0.025	0.023	0.020	0.229	0.771
2	0.099	0.021	0.011	0.125	0.046	0.120	0.421	0.579
3	0.069	0.168	0.084	0.000	0.069	0.080	0.470	0.530
4	**0.021**	**0.084**	**0.031**	**0.025**	**0.023**	**0.020**	**0.203**	**0.797**
5	0.069	0.021	0.010	0.100	0.023	0.140	0.363	0.637
6	0.099	0.115	0.021	0.050	0.091	0.060	0.437	0.563

In this example, let the threshold value of similarity score set in the model be 75%. The best similarity score provided by the CBR module belongs to case-ID 4. The solution attached to this score outputted by the model is Age based PM (similarity score = 79.7%). If the similarity score was less than 75%, then there exists a need for case-revision that can be carried out either by experts' opinion or by ES-2.

CONCLUSION AND FUTURE SCOPE

It is well known that ES are useful for those organizations where exhaustive prior cases and/or experts' are available. However, when the number of criteria and their levels are increased, the ES might become quite unwieldy. Alternatively, CBR plays a pivotal role for those organizations where limited cases are available, and the decision-maker has to provide approximate decisions, based on those prior cases, without extracting any knowledge from them. Moreover, it has the potentiality of incremental learning (without repeated training) and adaptation capability, which make it superior than other AI based approaches, viz., artificial neural network, support vector machine, etc.

In this chapter, a hybrid AI-based conceptual decision-making model has been described for a sustainable maintenance strategy selection. The model utilized the benefits of ES and CBR techniques. Several influencing factors (economic, social, technical and environmental), though the list may not be very exhaustive, have also been provided. The model is relevant in any industrial decision-making problem wherein the end-user/decision makers are forced to arrive at an optimal decision with the available information/data- structured or unstructured, complete or incomplete. Moreover, keeping in mind the recent trend of moving towards sustainable-based approaches, this hybrid MSS model would alleviate the gap and aid the decision-maker to choose an optimal sustainable maintenance strategy with ease.

The conceptual workflow diagram provided in the narrative would facilitate the readers to extend the idea of this decision-making model and develop a software for practical application. The model can be further extended to capture the vagueness in data/information through the applications of several theories such as Fuzzy, Grey, Rough, Belief or a combination of these. Adding to these, some new methods such as factor relationship (FARE) or full consistency method (FUCOM) for calculating the relative importance of considered criteria may also be explored for their efficacy in the model.

REFERENCES

Banerjee, S., & Majumder, M. (2016, April). Application of AHP for selection of source for Microgrid system. In *Energy Efficient Technologies for Sustainability (ICEETS), 2016 International Conference on* (pp. 112-115). IEEE.

Boral, S., & Chakraborty, S. (2016). A case-based reasoning approach for non-traditional machining processes selection. *Advances in Production Engineering & Management*, *11*(4), 311–323. doi:10.14743/apem2016.4.229

Borjalilu, N., & Ghambari, M. (2018). Optimal maintenance strategy selection based on a fuzzy analytical network process: A case study on a 5-MW powerhouse. *International Journal of Engineering Business Management*, *10*, 1–10. doi:10.1177/1847979018776172

Chakraborty, S., & Boral, S. (2017). A developed case-based reasoning system for machine tool selection. *Benchmarking: An International Journal*, *24*(5), 1364–1385. doi:10.1108/BIJ-07-2016-0103

Chakraborty, S., & Prasad, K. (2016). A QFD-based expert system for industrial truck selection in manufacturing organizations. *Journal of Manufacturing Technology Management*, *27*(6), 800–817. doi:10.1108/JMTM-02-2016-0020

Doyen, L., & Gaudoin, O. (2004). Classes of imperfect repair models based on reduction of failure intensity or virtual age. *Reliability Engineering & System Safety*, *84*(1), 45–56. doi:10.1016/S0951-8320(03)00173-X

Elsayed, E. A. (2012). *Reliability engineering* (Vol. 88). Hoboken, NJ: John Wiley & Sons.

Emovon, I., Norman, R. A., & Murphy, A. J. (2018). Hybrid MCDM based methodology for selecting the optimum maintenance strategy for ship machinery systems. *Journal of Intelligent Manufacturing*, *29*(3), 519–531. doi:10.100710845-015-1133-6

Garg, A., & Deshmukh, S. G. (2009). Flexibility in maintenance: A Framework. *Global Journal of Flexible Systems Management*, *10*(2), 21–33. doi:10.1007/BF03396559

Gargama, H., & Chaturvedi, S. K. (2011). Criticality assessment models for failure mode effects and criticality analysis using fuzzy logic. *IEEE Transactions on Reliability*, *60*(1), 102–110. doi:10.1109/TR.2010.2103672

Hemmati, N., Rahiminezhad Galankashi, M., Imani, D. M., & Farughi, H. (2018). Maintenance policy selection: A fuzzy-ANP approach. *Journal of Manufacturing Technology Management*, *29*(7), 1253–1268. doi:10.1108/JMTM-06-2017-0109

Ighravwe, D. E., & Ayoola Oke, S. (2017). Ranking maintenance strategies for sustainable maintenance plan in manufacturing systems using fuzzy axiomatic design principle and fuzzy-TOPSIS. *Journal of Manufacturing Technology Management*, *28*(7), 961–992. doi:10.1108/JMTM-01-2017-0007

Joshua, J., Mathew, S. G., & Harikrishnan, A. R. (2016). Selection of an optimum maintenance strategy for improving the production efficiency in a casting unit. *International Journal of Science Technology & Engineering*, *3*(2), 138–141.

Kirubakaran, B., & Ilangkumaran, M. (2015). The selection of optimum maintenance strategy based on ANP integrated with GRA-TOPSIS. *Journal for Global Business Advancement*, *8*(2), 190–215. doi:10.1504/JGBA.2015.069530

Kirubakaran, B., & Ilangkumaran, M. (2016). Selection of optimum maintenance strategy based on FAHP integrated with GRA–TOPSIS. *Annals of Operations Research*, *245*(1-2), 285–313. doi:10.100710479-014-1775-3

Knezevic, J. (1993). *Reliability, maintainability, and supportability: a probabilistic approach*. McGraw-Hill Companies.

Knezevic, J. (1997). Systems Maintainability Analysis, Engineering and Management. London: Chapman & Hall.

Kolodner, J. L. (1992). An introduction to case-based reasoning. *Artificial Intelligence Review*, *6*(1), 3–34. doi:10.1007/BF00155578

Lazakis, I., & Ölçer, A. (2016). Selection of the best maintenance approach in the maritime industry under fuzzy multiple attributive group decision-making environment. *Proceedings of the Institution of Mechanical Engineers, Part M: Journal of Engineering for the Maritime Environment*, *230*(2), 297–309.

Lucas, P. F. J., & van der Gaag, L. C. (1991). *Principles of Expert Systems*. Amsterdam: Centre for Mathematics and Computer Science.

Mobley, R. K. (2002). An introduction to predictive maintenance (2nd ed.). Elsevier Science.

Nezami, F. G., & Yildirim, M. B. (2013). A sustainability approach for selecting maintenance strategy. *International Journal of Sustainable Engineering*, *6*(4), 332–343. doi:10.1080/19397038.2013.765928

Özcan, E. C., Ünlüsoy, S., & Eren, T. (2017). A combined goal programming–AHP approach supported with TOPSIS for maintenance strategy selection in hydroelectric power plants. *Renewable & Sustainable Energy Reviews*, *78*, 1410–1423. doi:10.1016/j.rser.2017.04.039

Pal, S. K., & Shiu, S. C. (2004). *Foundations of soft case-based reasoning* (Vol. 8). Hoboken, NJ: John Wiley & Sons. doi:10.1002/0471644676

Prasad, K., & Chakraborty, S. (2015). Development of a QFD-based expert system for CNC turning centre selection. *Journal of Industrial Engineering International*, *11*(4), 575–594. doi:10.100740092-015-0122-x

Prasad, K., Mahanty, S., Maity, S. R., & Chakraborty, S. (2014). Development of an expert system for materials selection. *Journal of Materials Education*, *36*(5-6), 117–138.

Rai, R. N., & Bolia, N. (2014). Optimal decision support for air power potential. *IEEE Transactions on Engineering Management*, *61*(2), 310–322. doi:10.1109/TEM.2013.2293420

Ranjan, R., Chatterjee, P., & Chakraborty, S. (2015). Evaluating performance of engineering departments in an Indian University using DEMATEL and compromise ranking methods. *Opsearch*, *52*(2), 307–328. doi:10.100712597-014-0186-1

Rigdon, S. E., & Basu, A. P. (2000). *Statistical methods for the reliability of repairable systems*. New York: Wiley.

Saaty, T. L. (1990). How to make a decision: The analytic hierarchy process. *European Journal of Operational Research*, *48*(1), 9–26. doi:10.1016/0377-2217(90)90057-I

Schank, R. C. (1982). *Dynamic memory: A theory of reminding and learning in computers and people* (Vol. 240). Cambridge, UK: Cambridge University Press.

Seiti, H., Hafezalkotob, A., & Fattahi, R. (2018). Extending a pessimistic–optimistic fuzzy information axiom based approach considering acceptable risk: Application in the selection of maintenance strategy. *Applied Soft Computing*, *67*, 895–909. doi:10.1016/j.asoc.2017.11.017

Seiti, H., Tagipour, R., Hafezalkotob, A., & Asgari, F. (2017). Maintenance strategy selection with risky evaluations using RAHP. *Journal of Multi-Criteria Decision Analysis*, *24*(5-6), 257–274. doi:10.1002/mcda.1618

Shafiee, M. (2015). Maintenance strategy selection problem: An MCDM overview. *Journal of Quality in Maintenance Engineering*, *21*(4), 378–402. doi:10.1108/JQME-09-2013-0063

Shafiee, M., Labib, A., Maiti, J., & Starr, A. (2017). Maintenance strategy selection for multi-component systems using a combined analytic network process and cost-risk criticality model. *Proceedings of the Institution of Mechanical Engineers, Part O: Journal of Risk and Reliability*, 1-35.

Shemshadi, A., Shirazi, H., Toreihi, M., & Tarokh, M. J. (2011). A fuzzy VIKOR method for supplier selection based on entropy measure for objective weighting. *Expert Systems with Applications*, *38*(10), 12160–12167. doi:10.1016/j.eswa.2011.03.027

Wang, F., Shi, Q., & Hu, Z. (2015). A sustainability approach to maintenance strategy selection based on fuzzy AHP and VIKOR algorithms. *Recent Developments on Reliability, Maintenance and Safety*, *717*. doi:10.2495/QR2MSE140841

Wang, Y., Yeo, G. T., & Ng, A. K. (2014). Choosing optimal bunkering ports for liner shipping companies: A hybrid Fuzzy-Delphi–TOPSIS approach. *Transport Policy*, *35*, 358–365. doi:10.1016/j.tranpol.2014.04.009

Xu, F., Liu, X., Chen, W., Zhou, C., & Cao, B. (2018). Ontology-Based Method for Fault Diagnosis of Loaders. *Sensors*, *18*(3), 1-22.

Chapter 5
Optimum Selection of Biodiesel for Sustainable Assessment:
A Prospect Theory-Based Approach

Chiranjib Bhowmik
https://orcid.org/0000-0001-9338-7715
National Institute of Technology Silchar, India

Sumit Bhowmik
https://orcid.org/0000-0002-7787-756X
National Institute of Technology Silchar, India

Amitava Ray
Jalpaiguri Government Engineering College, India

ABSTRACT

This chapter aims to select the best biodiesel for a diesel power generator considering sustainable criteria. The study proposes the application of an almost unexplored prospect theory-based multi-criteria decision-making (MCDM) method, popularly known as Tomada de Decisao Interativa Multicriterio (TODIM). Several conflicting criteria including calorific value, cetane number, density, viscosity, flash point, and pour point are considered as the most predominant criteria, while pongamia oil, jatropha oil, cotton seed oil, linseed oil, madhuca indica oil, olive oil, and sunflower oil are among the considered biodiesel alternatives. Results show that madhuca indica oil scored the highest value followed by others. The study is also complemented by an analysis of the sensitivity of the numerical results obtained to show the robustness of the proposed method.

DOI: 10.4018/978-1-5225-8579-4.ch005

INTRODUCTION

Energy is a crucial outlook for any developing and developed country (Durairaj et al., 2014). Day by day the utilization of energy is increasing order due to improved standard of living and population growth (Durairaj et al., 2016). Most of the developing as well as developed countries fulfill their needs through relic fuels (Bhowmik et al., 2017; 2018). Conventional resources are the main sources of energy but leads to environmental problems due to high carbon dioxide and other emissions (Standford, 1998). Environmental snags can be prevented by the use of green energy sources such as solar, hydro, biogas, geothermal etc. with almost zero emissions (Arkesteijn and Oerlemans, 2005). In this regard biodiesel is also an ideal choice to meet the energy requirement of everyday life (Vilela, 2010). The biodiesel is extracted from both edible and non-edible oils and can be considered as one of the environmentally friendly resources (Demirbas, 2007). The extracted biodiesel from green resources can be used as alternative fuels for I.C. engines, boilers and diesel generators etc. (Demirbas, 2007).

In energy-mix market, it is very difficult to select the optimum alternative for sustainable assessment, because diversity of conflicting parameters. Today, energy market is facing daunting pressure to incorporate environmentally friendly key resources for sustainable future due to increased utilization of relic fuels. The extensive use of relic fuels in commercial and domestic sectors causes numerous environmental problems such as SOx, NOx, CO_2 emissions, resources depletion and global warming, forcing the governments as well as research activists to think about judicious use of green resources in a sustainable way. Therefore, green energy sources are considered as crucial elements for humanity to continue the economic empowerment and social development. In today's energy dilemma biodiesel is considered as one of the outmost green sources for pastoral power generation. Effectiveness of biodiesel depends on the quality of fuel input. It is quite difficult to select the optimum biodiesel for an application point of view because of various conflicting parameters present in the selection strategy (Sakthivel et al., 2015).

Most of the past researchers highlighted that the existing work mainly focuses on operating fuel condition with reference to NOx, smoke, BTE performance. Based on the factors the operating fuel is recommended as the optimum blend without considering other influencing factors (Godiganur et al., 2010; Behçet, 2011; Vedaraman et al., 2011; Shanmugam et al., 2011a, 2011b; Sakthivel & Nagarajan, 2011; Sakthivel et al., 2015). Biodiesel selection under suitable multi-criteria decision-making (MCDM) milieu were attempted previously by various researchers. MCDM is considered one of the general paradigms of operation research application mostly dealing with the problems under the presence of glut of factors and criteria.

MCDM approaches provides cultured phenomenon that are concerned with the support of decision makers in solving complex real-world decisions. The utilization of MCDM in automobile engineering has been progressively snowballing in the past few eras (Sakthivel et al., 2015). An application of hybrid MCDM technique based on analytical network process (ANP) integrated with Technique for Order Performance by Similarity to Ideal Solution (TOPSIS) and VlseKriterijumska Optimizacija I Kompromisno Resenje (VIKOR) is utilized (Sakthivel et al., 2015) for the selection of optimum fuel blend in fish oil biodiesel for internal combustion engine. Biomass to energy conversion system and the selection of bioenergy using different MCDM methods were investigated by various researchers (Scott et al., 2012; Yilmaz & Selim, 2013). Azadeh et al., (2014) proposed a stochastic linear programming model under multi-period planning framework to maximize the supply-chain vulnerability and shipping of biofuels according to customer demand. Optimum fuel alternative for land transportation in Singapore was investigated using analytical hierarchy process (Poh & Ang, 1999). One MCDM model was developed (Yedla et al., 2003) to investigate the selection of environmentally friendly suitable transportations system in Delhi. Tzeng et al., (2005) utilized TOPSIS and VIKOR to identify the alternative fuel buses for public transportation. Another MCDM model was used for selecting the alternative fuels for transportation sector (Rassafi, 2006). A decision support system for the evaluation of biofuels were anticipated by Perimenis et al., (2011). Evaluation of alternative fuels for road transport sector in Greek was investigated (Tsita & Pilavachi, 2012) using AHP method. Sivaraja & Sakthivel (2017) addressed a comparison study of hybrid MCDM methods, FAHP-TOPSIS, FAHP-VIKOR and FAHP-ELECTRE, to select the optimum biofuel blend at different injection timings. An integrated AHP methodology based on VIKOR and PROMETHEE II methods were used (Debbarma et al., 2017) to select the optimum fuel combination among the alternatives in relation to various conflicting factors. From the literature there is no trace of research that deals with selection of optimum biodiesel for sustainable planning based on MCDM technique. In this work, one example of biodiesel alternatives and their corresponding properties is solved using Stepwise weight assessment ratio analysis (SWARA) based TOmada de Decisao Interativa Multicriterio (TODIM) method which has not successfully explored in the field of biodiesel selection. The example clarifies the robustness of the adopted MCDM method followed by sensitivity analysis.

The rest of the study is organized as follows: in the second section, materials and methods are summarized. In the third section a case study analysis is presented. Fourth section shows the comparative analysis of the selected problem. In fifth section sensitivity analysis is explained. Finally, sixth section concludes the study with future research suggestion.

MATERIALS AND METHODS

In this study seven biodiesel alternative namely pongamia, jatropha, cotton seed, linseed, madhuca indica, olive oil and sunflower and their corresponding seven criteria viz. calorific value (kJ/kg), cetane number, density (kg/m^3), viscosity (mm^2/sec), flash point (°C), and pour point (°C) are selected from various published literature and their sourcing references are shown in Table 1 and Table 2 respectively. Firstly, SWARA method is utilized to calculate the weight of each criteria. Secondly, TODIM a discrete MCDM method based on prospect theory is applied to select

Table 1. Literature sources of biodiesel alternatives

Alternatives	Respective Symbol	Literature Source
Pongamia oil	A1	Ozcanli et al., 2013; Durairaj et al., 2014, 2016.
Jatropha oil	A2	Ozcanli et al., 2013; Durairaj et al., 2014, 2016.
Cotton seed oil	A3	Ozcanli et al., 2013; Durairaj et al., 2014, 2016.
Linseed oil	A4	Ozcanli et al., 2013; Durairaj et al., 2014, 2016.
Madhuca indica oil	A5	Ozcanli et al., 2013; Durairaj et al., 2014, 2016.
Olive oil	A6	Ozcanli et al., 2013.
Sunflower oil	A7	Ozcanli et al., 2013.

Table 2. Literature sources of corresponding criteria for biodiesel selection

Criterion	Respective Symbol	Literature Sources
Calorific value	C1	Sureshkumar et al., 2008; Sahoo & Das, 2009; Hazar, 2010; Puhan et al., 2005, 2009; Kalligeros et al., 2003; Rashid et al., 2008; Ozcanli et al., 2013.
Cetane number	C2	Sarin et al., 2009; Ahmad et al., 2009; Hazar, 2010; Puhan et al., 2009; Puhan et al., 2005; Ramos et al., 2009; Srivastava & Prasad, 2000; Ozcanli et al., 2013.
Density	C3	Sureshkumar et al., 2008; Banapurmath et al., 2008; Hazar, 2010; Puhan et al., 2005, 2009; Kalligeros et al., 2003; Srivastava & Prasad, 2000; Ozcanli et al. 2013.
Viscosity	C4	Sureshkumar et al., 2008; Sahoo & Das, 2009; Hazar, 2010; Puhan et al., 2005, 2009; Ramos et al., 2009; Srivastava & Prasad, 2000; Ozcanli et al., 2013.
Flash point	C5	Sarin et al., 2007; Sahoo & Das, 2009; Hazar, 2010; Puhan et al., 2005, 2009 Ramos et al., 2009; Srivastava & Prasad, 2000; Ozcanli et al., 2013.
Pour point	C6	Ahmad et al., 2009; Sarin et al., 2009; Sahoo & Das, 2009; Alptekin & Canakci, 2009; Puhan et al., 2009; Ghadge & Raheman, 2005; Kalligeros et al., 2003; Rashid et al., 2008; Ozcanli et al., 2013.

the optimum alternative. The next sub-section discusses the details about various criteria and utilized methodology in this work.

Brief Description of Selected Criteria for Biodiesel

The evaluating criteria for the selection of optimum biodiesel is identified from a published work (Ozcanli et al., 2013). These are discussed in brief as follows:

Calorific value: To measure the energy content i.e., the heat produced by the complete combustion of fuel caloric value acts as an indicator (Ozcanli et al., 2013; Durairaj et al., 2014).

Cetane number: Auto ignition after injection of fuel is measured by cetane number. If cetane number is more for any fuel then it shortens its combustion process (Ozcanli et al., 2013; Durairaj et al., 2014).

Density: The fuel's energy content can be corelate with density. For higher density the fuel gets its potential on volume basis (Ozcanli et al., 2013; Durairaj et al., 2014).

Viscosity: The quality of atomization, combustion and engine wear effected by viscosity, high viscous fuels leads to poor atomization and fuel injection respectively wherein low viscosity lead to power loss (Ozcanli et al., 2013; Durairaj et al., 2014).

Flash point: Flash point is the lowest temperature at which a liquid can form an ignitable mixture when an ignition source is introduced (Ozcanli et al., 2013; Durairaj et al., 2014).

Pour point: The pour point of a fluid is the temperature below which the liquid losses its flow characteristics (Ozcanli et al., 2013; Durairaj et al., 2014).

SWARA Method

There are numerous methods for evaluating the weights of choice criterion in the literature like entropy (Shannon, 1948), analytical hierarchy process (AHP) (Saaty, 1980), analytical network process (ANP) (Saaty & Vargas, 2001), factor relationship (FARE) (Ginevičius, 2011), SWARA (Keršuliene et al., 2010) and KEMIRA (Krylovas et al., 2014) etc. Among the aforementioned methods SWARA method is considered one of the novel tools compared to others. Accordingly, as per the methodology the most significant criterion is given the highest priority whereas, the most trivial criterion is given the least rank. The overall ranks are obtained according to the average value of ranks based on the expert judgment (Zolfani et al., 2018). The main benefit of this method is that each of the priority importance of each criterion are determined merely. Based on the overall outcome experts are able to sort the criteria from higher to lower. Further information related to SWARA method can be found in a scholarly published paper by Mardani et al., 2017. Dehnavi et al. (2015) proposed a novel hybrid model based on SWARA method and adaptive neuro-fuzzy inference system (ANFIS) to evaluate landslide susceptible areas using geographical information system (GIS). Zolfani et al., (2018) addressed the restriction of SWARA method and offered an extended version for criteria prioritizing process. Zolfani & Saparauskas (2013) utilized SWARA method as a new outline for evaluating and prioritizing sustainability assessment indicators of energy system. Keršuliene et al., (2010) applied SWARA method to select the dispute from economic, social and other viewpoints. Zarbakhshnia et al., (2018) proposed a multiple-attribute decision-making (MADM) model to rank and select the third-party reverse logistics providers (3PRLPs) based on fuzzy-SWARA to weight the evaluation criteria in the presence of various risk factors. In this research SWARA method is selected as a pertinent tool in the process of decision making (Karabasevic et al., 2016). The procedure for determining the weights by SWARA is shown in Figure 1.

TODIM Method

TODID is a distinct MCDM method grounded on the thought of prospect theory. However, based on the prospect theory TOMID method helps to compute the global measurement of value. Empirically this method was proved how people reacts in situation of facing risk. The value function is similar as the gain/loss function of prospect theory. The feasibility of TODIM is based on the global multi-attribute value function. The global measure permits the decision maker to complete the

Figure 1. Procedure for SWARA weight calculation

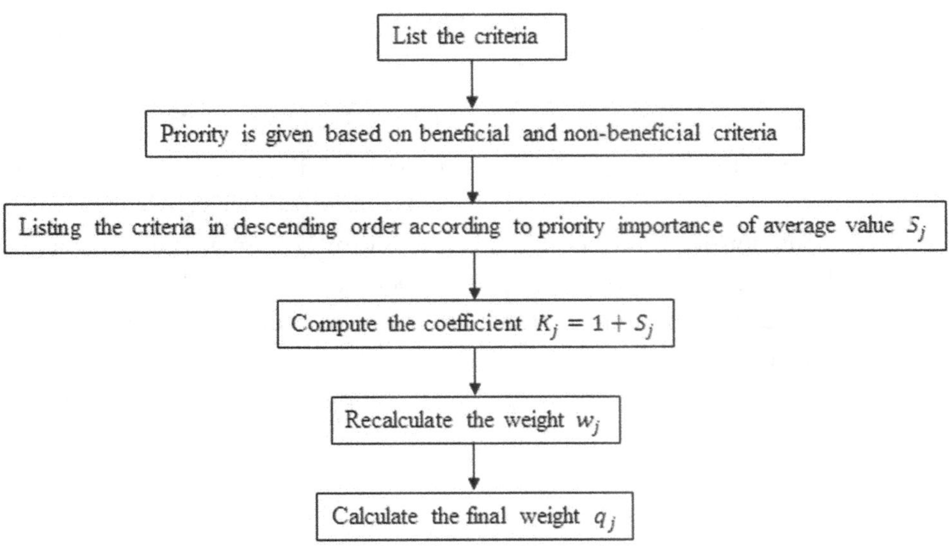

ranking order of all the alternatives. The problem of selection the best opinion for the destination of the natural gas reserve was investigated by Gomes et al., (2009) using TODIM method. As per the study, (Lourenzutti & Krohling, 2013) selection of any alternative that affect the performance of selection strategy can be handled using fuzzy-TODIM approach. Residential choice problem is a one-time decision for most of the people and investigated by (Uysal & Tosun, 2014) based on TODIM approach using objective and subjective factors. Araújo, 2015 utilized new fuzzy systematic evaluation approach based on TODIM method for the evaluation of sales channels for commercial products. Wang et al., (2016) developed a likelihood-based TODIM approach using multi-hesitant fuzzy linguistic information for the selection and evaluation of contractors in logistic outsourcing. Zindani et al., (2017) applied TODIM method for material selection. Wu et al., (2018) applied an extended TODIM method in conjunction with preference ranking organization method for enrichment evaluations (PROMETHEE-II) method for waste-to-energy plant site selection for sustainability. The brief idea about TODIM method can be found in many scholarly published articles (Tseng et al., 2014; Sen et al., 2016; Zindani et al., 2017; Llamazares, 2018; Li & Cao, 2018). The step-wise procedure for TODIM method is shown below:

Step 1: Development of decision matrix A based on n and m numbers of alternatives and evaluating criteria respectively.

$$A = \left[a_{ic_{n\times m}}\right] = \begin{bmatrix} a_{11} & \cdots & a_{1m} \\ \cdots & \cdots & \cdots \\ a_{n1} & \cdots & a_{nm} \end{bmatrix}, i = 1, 2, \ldots, n; c = 1, 2, \ldots, m \tag{1}$$

where a_{ic} denotes the performance of ith alternative with respect to j^{th} .criterion.

Step 2: Computation of normalized decision matrix to make the matrix dimensionless and all of its elements comparable. For beneficial criteria higher values using Equation (2) and for non-beneficial criteria lower values using Equation (3) are adopted.

$$K_{ij} = \frac{a_{ic}}{\sum_{i=1}^{n} a_{ic}} \text{. For beneficial criteria} \tag{2}$$

$$K_{ij} = \frac{\frac{1}{a_{ic}}}{\sum_{i=1}^{n} \frac{1}{a_{ic}}} \text{ For non-beneficial criteria} \tag{3}$$

where K_{ij} is the normalized value of a_{ic}.

Step 3: Using SWARA method, the priority weight q_j is calculated. The relative weight q_{cr} of criteria $q_c\left(c = 1, 2, \ldots, m\right)$ with respect to reference criteria C_r is computed using Equation (4).

$$q_{cr} = {q_c}/{q_r} \tag{4}$$

where q_r is the weight of the reference criteria having the highest weight value.

Step 4: Thereafter, dominance degree of alternative B_i over alternative B_j is calculated using following Equation (5).

$$\delta\left(B_i, B_j\right) = \sum_{c=1}^{m} \varnothing_c\left(B_i, B_j\right) \qquad \forall\left(i.j\right) \tag{5}$$

The dominance degree of alternative B_i over alternative B_j, i.e., $\varnothing_c\left(B_i.B_j\right)$ is computed using the following Equation (6, 7 & 8) respectively based on some conditions.

$$if\left(K_{ic} - K_{jc}\right) > 0 \ then \varnothing_c\left(B_i, B_j\right) = \left\{\sqrt{\frac{q_{cr}\left(K_{ic} - K_{jc}\right)}{\sum_{c=1}^{m} q_{cr}}}\right\} \tag{6}$$

$$if\left(K_{ic} - K_{jc}\right) = 0 then \varnothing_c\left(B_i, B_j\right) = 0 \tag{7}$$

$$if\left(K_{ic}-K_{jc}\right)<0\,then\,\varnothing_c\left(B_i,B_j\right)=\frac{-1}{\theta}\sqrt{\frac{\left(\sum_{c=1}^{m}q_{cr}\right)\left(K_{ic}-K_{jc}\right)}{q_{cr}}} \tag{8}$$

where $\left(K_{ic}-K_{jc}\right)>0$ and $\left(K_{ic}-K_{jc}\right)<0$ respectively stands for gain and loss of ith alternative over j^{th} alternative, and θ is the attenuation factor of losses.

Step 5: Calculation of overall dominance degree of alternative. Therefore, φ_i is determined using Equation (9).

$$\varphi_i=\frac{\sum_{j=1}^{n}\delta\left(B_i,B_j\right)-min\sum_{j=1}^{n}\delta\left(B_i,B_j\right)}{max\sum_{j=1}^{n}\delta\left(B_i,B_j\right)-min\sum_{j=1}^{n}\delta\left(B_i,B_j\right)} \tag{9}$$

Step 6: Alternatives are ranked according to descending order of their dominance scores and alternative having the maximum dominance score is selected as the best option.

CASE STUDY ANALYSIS

In order to show the application feasibility of the adopted SWARA based TODIM methodology, one illustrative case study of biodiesel selection is presented (Ozcanli et al., 2013). In this case study, the priority importance of seven biodiesel alternatives i.e., pongamia oil (A1), jatropha oil (A2), cotton seed oil (A3), linseed oil (A4), madhuca indica oil (A5), olive oil (A6) and sunflower oil (A7) are evaluated based on six corresponding criteria, i.e., calorific value, cetane number, density, viscosity, flash point and pour point. Among them calorific value and cetane number is considered as beneficial criteria and remaining are non-beneficial in nature. The corresponding decision matrix is shown in Table 3. In order to apply the TODIM methodology for solving this biodiesel selection problem, the decision matrix of Table 3 is first normalized, as shown in Table 4.

Thereafter the priority importance of all the criteria is determined by SWARA methodology as described in Figure 1 and shown in Table 5.

Study reveals that calorific value of the biodiesel alternative scored the highest priority importance and corresponding q_{cr} values are subsequently computed. Then

Table 3. Decision matrix for biodiesel selection (Ozcanli et al., 2013)

	Criteria					
	C1	**C2**	**C3**	**C4**	**C5**	**C6**
A1	35560	55.27	0.878	10.64	141	-6
A2	42673	57.4	0.87	4.23	148	4.2
A3	40430	48	0.87	3.7	75	6
A4	40759	48	0.865	4.2	161	-18
A5	36900	51	0.865	5.2	127	6
A6	32781	57	0.88	4.5	178	-3
A7	45300	49	0.88	4.6	183	-1

Table 4. Normalized decision matrix

	Criteria					
	C1	**C2**	**C3**	**C4**	**C5**	**C6**
A1	0.1295	0.1511	0.1415	0.0639	0.1354	0.1693
A2	0.1555	0.1569	0.1428	0.1607	0.1290	-0.2419
A3	0.1473	0.1312	0.1428	0.1837	0.2546	-0.1693
A4	0.1485	0.1312	0.1436	0.1618	0.1186	0.0564
A5	0.1344	0.1394	0.1436	0.1307	0.1504	-0.1693
A6	0.1194	0.1558	0.1411	0.1511	0.1073	0.3387
A7	0.1650	0.1340	0.1444	0.1478	0.1043	1.0161

Table 5. SWARA method in weighting of criterion

Criterion				
C1		1	1	0.2277
C2	0.25	1.25	0.8	0.1822
C3	0.1	1.1	0.7272	0.1656
C4	0.1	1.1	0.6611	0.1505
C5	0.05	1.05	0.6296	0.1434
C6	0.1	1.1	0.5724	0.1303

the dominance degree of each biodiesel alternative $\varnothing_c\left(B_i, B_j\right)$ over the remaining alternatives based on each considered criterion are computed employing Equation 6, 7 and 8 respectively. The value of attenuation factor of losses θ .is assumed as 1 which significantly contributes with their real value to the global value (Zindani et al., 2017). The dominance degree of alternative B_i over biodiesel alternative B_j. $\delta\left(B_i, B_j\right)$ is computed using Equation 5. The dominance degrees of first biodiesel alternative over the others considering each criterion are tabulated in Table 6. Similarly, for the remaining biodiesel alternatives the dominance degrees are computed, and overall dominance degrees for all the seven entrant biodiesels are determined using Equation 9, as shown in Table 7. From Table 7, it is observed that the biodiesel alternative A5 (madhuca indica oil) evolves as the best suited alternative

Table 6. Dominance degrees of biodiesel alternative A1 over others with respect to each criterion

Alternatives	C1	C2	C3	C4	C5	C6	
A2	-0.3373	-0.1787	-0.0952	-0.8618	0.0325	0.2488	-1.1917
A3	-0.2791	0.0601	-0.0952	-0.9588	-0.8483	0.2258	-1.8955
A4	-0.2884	0.0601	-0.1217	-0.8669	0.0527	0.1303	-1.0337
A5	-0.1464	0.0461	-0.1217	-0.7160	-0.3002	0.2258	-1.0125
A6	0.0480	-0.1611	0.0067	-0.8177	0.0682	-1.0604	-1.9163
A7	-0.3947	0.0558	-0.1437	-0.8022	0.0717	-2.3713	-3.584

Table 7. Overall dominance degrees of the biodiesel alternatives

Biodiesel Alternatives		φ_i	Rank
A1	-10.6343	0.8947	2
A2	-8.8700	0.7366	3
A3	-4.8305	0.3746	6
A4	-7.7916	0.6399	4
A5	-11.8089	1	1
A6	-6.7444	0.5461	5
A7	-0.6499	0	7

for biodiesel production. Alternative A7 (sunflower oil) scored the worst rank among all the alternatives in this study. Furthermore, a sensitivity analysis of this study is shown in the next section.

Sensitivity Analysis

Sensitivity analysis is the study of uncertainty in the output of an analysis which can be distributed to different sources of uncertainty as its inputs. In this research the uncertainty of biodiesel alternative rankings with various criteria is demonstrated using a mathematical model proposed by (Bhattacharjee et al., 2005). The generalized Equation of the proposed model are as follows:

$$SI_i = \left[\left(\alpha SFM_i\right) + \left(1-\alpha\right) OFM_i\right] \tag{10}$$

where

$$OFM_i = \frac{1}{\left[OFD_i \sum_{i=1}^{n} OFD^{-1}\right]} \tag{11}$$

where SI is the sensitivity index, α is the objective factor weight, SFM is the global priorities of each biodiesel alternatives and OFM is the objective factor measure. OFD is the objective factor dimension and n is the number of alternatives ($n = 7$). In this study the SFM values used in Equation (10) are the final scores found from Table 7. OFM values are taken from Table 3. The selection of the value of α is an important issue. It is the decision maker's perception depending upon the preference importance regarding objective and subjective factor measures. However, the evaluation practice may outline diverse sets of outcomes for various values of α for the same alternatives with different criteria. Therefore, a sensitivity plot shown in Figure 2, to analyze the effect of α in the biodiesel selection problem is strongly recommended (Bhowmik et al., 2018). Using Equation (10) and considering $\alpha = 0.051$, the biodiesel alternatives are ranked A5> A1> A2> A4> A6> A3> A7, which is similar to that found from Table 7.

The outcomes of the sensitivity plot are gathered in Table 8, which specifies that the appropriate value of the objective factor weight should be nominated judiciously. The motive is that the ascendency of the SFM values will be higher for a higher

Figure 2. Sensitivity analysis

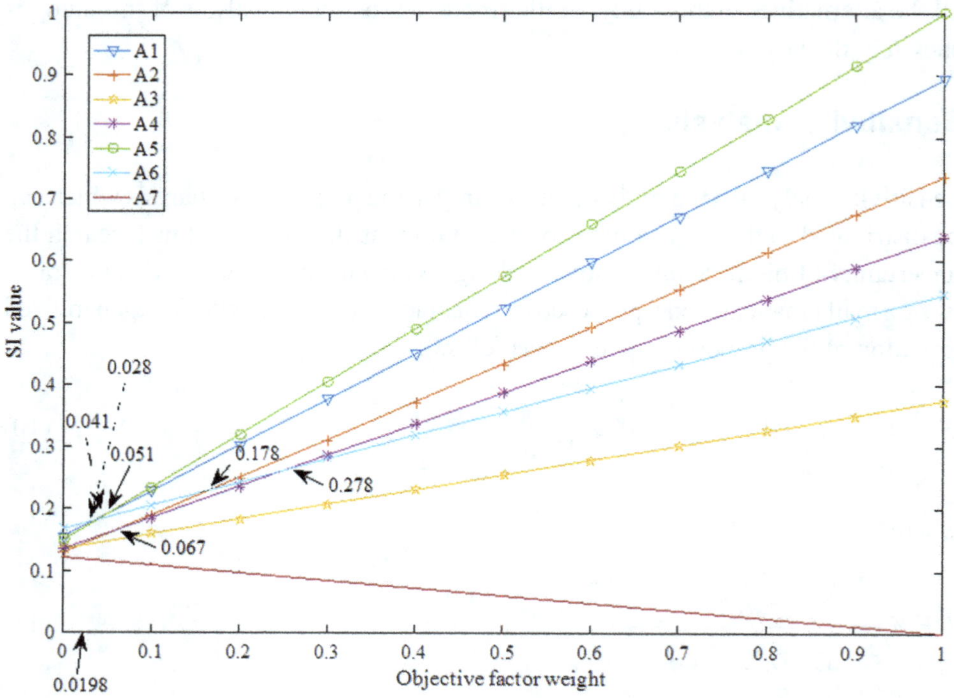

Table 8. Analysis of figure 2

Biodiesel Alternatives	Prime Range of α	Comparison Between
A1	$0.000 < \alpha = 0.041 < 0.1$	Between A5, A2 and A6
A6	$0.000 < \alpha = 0.028 < 0.1$	Between A5, A2 and A6
A2	$0.1 < \alpha = 0.178 < 0.2$	Between A2 and A6
A6	$0.2 < \alpha = 0.278 < 0.3$	Between A4 and A6
A2	$0.000 < \alpha = 0.0198 < 0.1$	Between A2 and A3
A2	$0.000 < \alpha = 0.067 < 0.1$	Between A2 and A4

value of α (Bhattacharjee et al., 2005). On the other hand, the ascendency of calorific values will be superior for an inferior value of α, and afterward, the SFM will have a lesser importance.

CONCLUSION

In this study the practicability of an almost unexplored MCDM technique in the form of TODIM method is vindicated while solving biodiesel selection problem. The SWARA based TODIM method identified madhuca indica oil is the optimum biodiesel alternative wherein, sunflower oil is the least alternative. In this study calorific value scored the highest performance value. Therefore, a sensitivity analysis of the SWARA-TODIM method with respect to calorific value shows the robustness of the analyzed results with similar rankings. Its effectiveness in solving other selection problems may be experienced and effect of different normalization tools on its rankings order may need to be studied as the future scope of this work.

REFERENCES

Ahmad, M., Zafar, M., Khan, M. A., & Sultana, S. (2009). Biodiesel from Pongamia pinnata L. oil: A promising alternative bioenergy source. Energy Sources. *Part A*, *31*(16), 1436–1442.

Alptekin, E., & Canakci, M. (2009). Characterization of the key fuel properties of methyl ester–diesel fuel blends. *Fuel*, *88*(1), 75–80. doi:10.1016/j.fuel.2008.05.023

Araújo, R. M. (2015). Multicriteria classification with TODIM-FSE. *Procedia Computer Science*, *55*, 559–565. doi:10.1016/j.procs.2015.07.043

Arkesteijn, K., & Oerlemans, L. (2005). The early adoption of green power by Dutch households: An empirical exploration of factors influencing the early adoption of green electricity for domestic purposes. *Energy Policy*, *33*(2), 183–196. doi:10.1016/S0301-4215(03)00209-X

Azadeh, A., Arani, H. V., & Dashti, H. (2014). A stochastic programming approach towards optimization of biofuel supply chain. *Energy*, *76*, 513–525. doi:10.1016/j.energy.2014.08.048

Banapurmath, N. R., Tewari, P. G., & Hosmath, R. S. (2008). Performance and emission characteristics of a DI compression ignition engine operated on Honge, Jatropha and sesame oil methyl esters. *Renewable Energy*, *33*(9), 1982–1988. doi:10.1016/j.renene.2007.11.012

Behçet, R. (2011). Performance and emission study of waste anchovy fish biodiesel in a diesel engine. *Fuel Processing Technology*, *92*(6), 1187–1194. doi:10.1016/j.fuproc.2011.01.012

Bhattacharya, A., Sarkar, B., & Mukherjee, S. K. (2005). Integrating AHP with QFD for robot selection under requirement perspective. *International Journal of Production Research*, *43*(17), 3671–3685. doi:10.1080/00207540500137217

Bhowmik, C., Baruah, A., Bhowmik, S., & Ray, A. (2018). Green energy sources selection for sustainable energy planning using multi-criteria decision-making approach. In IOP Conference Series: Materials Science and Engineering (Vol. 377, No. 1, p. 012029). IOP Publishing. doi:10.1088/1757-899X/377/1/012029

Bhowmik, C., Bhowmik, S., & Ray, A. (2018). Social acceptance of green energy determinants using principal component analysis. *Energy*, *160*, 1030–1046. doi:10.1016/j.energy.2018.07.093

Bhowmik, C., Bhowmik, S., Ray, A., & Pandey, K. M. (2017). Optimal green energy planning for sustainable development: A review. *Renewable & Sustainable Energy Reviews*, *71*, 796–813. doi:10.1016/j.rser.2016.12.105

Debbarma, B., Chakraborti, P., Bose, P. K., Deb, M., & Banerjee, R. (2017). Exploration of PROMETHEE II and VIKOR methodology in a MCDM approach for ascertaining the optimal performance-emission trade-off vantage in a hydrogen-biohol dual fuel endeavour. *Fuel*, *210*, 922–935. doi:10.1016/j.fuel.2017.08.016

Dehnavi, A., Aghdam, I. N., Pradhan, B., & Varzandeh, M. H. M. (2015). A new hybrid model using step-wise weight assessment ratio analysis (SWARA) technique and adaptive neuro-fuzzy inference system (ANFIS) for regional landslide hazard assessment in Iran. *Catena*, *135*, 122–148. doi:10.1016/j.catena.2015.07.020

Demirbas, A. (2007). Alternatives to petroleum diesel fuel. Energy Sources. *Part B*, *2*(4), 343–351.

Durairaj, S., Sathiyasekar, K., & Ilangkumaran, M. (2016). Selection of alternate fuel for electrical power generator using hybrid multi criteria decision making technique. *University Politehnica of Bucharest Scientific Bulletin Series C-Electrical Engineering and Computer Science, 78*(1), 247-258.

Durairaj, S., Sekar, K. S., Ilangkumaran, M., RamManohar, M., Thyalan, B., Yuvaraj, E., & Ramesh, S. (2014). Multi-Criteria Decision Model for Biodiesel Selection in an Electrical Power Generator Based on Fahp-Gra-Topsis. *Int. J. Res. Eng. Technol*, 2319-2363.

Ghadge, S. V., & Raheman, H. (2005). Biodiesel production from mahua (Madhuca indica) oil having high free fatty acids. *Biomass and Bioenergy*, *28*(6), 601–605. doi:10.1016/j.biombioe.2004.11.009

Ginevičius, R. (2011). A new determining method for the criteria weights in multicriteria evaluation. *International Journal of Information Technology & Decision Making*, *10*(06), 1067–1095. doi:10.1142/S0219622011004713

Godiganur, S., Murthy, C. S., & Reddy, R. P. (2010). Performance and emission characteristics of a Kirloskar HA394 diesel engine operated on fish oil methyl esters. *Renewable Energy*, *35*(2), 355–359. doi:10.1016/j.renene.2009.07.007

Gomes, L. F. A. M., Rangel, L. A. D., & Maranhão, F. J. C. (2009). Multicriteria analysis of natural gas destination in Brazil: An application of the TODIM method. *Mathematical and Computer Modelling*, *50*(1-2), 92–100. doi:10.1016/j.mcm.2009.02.013

Hazar, H. (2010). Cotton methyl ester usage in a diesel engine equipped with insulated combustion chamber. *Applied Energy*, *87*(1), 134–140. doi:10.1016/j.apenergy.2009.07.024

Kalligeros, S., Zannikos, F., Stournas, S., Lois, E., Anastopoulos, G., Teas, C., & Sakellaropoulos, F. (2003). An investigation of using biodiesel/marine diesel blends on the performance of a stationary diesel engine. *Biomass and Bioenergy*, *24*(2), 141–149. doi:10.1016/S0961-9534(02)00092-2

Karabasevic, D., Zavadskas, E. K., Turskis, Z., & Stanujkic, D. (2016). The framework for the selection of personnel based on the SWARA and ARAS methods under uncertainties. *Informatica*, *27*(1), 49–65. doi:10.15388/Informatica.2016.76

Keršuliene, V., Zavadskas, E. K., & Turskis, Z. (2010). Selection of rational dispute resolution method by applying new step-wise weight assessment ratio analysis (SWARA). *Journal of business economics and management*, *11*(2), 243–258. doi:10.3846/jbem.2010.12

Krylovas, A., Zavadskas, E. K., Kosareva, N., & Dadelo, S. (2014). New KEMIRA method for determining criteria priority and weights in solving MCDM problem. *International Journal of Information Technology & Decision Making*, *13*(06), 1119–1133. doi:10.1142/S0219622014500825

Kumar, R., & Ray, A. (2015). Optimal selection of material: an eclectic decision. *Journal of the Institution of Engineers (India): Series C, 96*(1), 29-33.

Li, M. Y., & Cao, P. P. (2018). Extended TODIM method for multi-attribute risk decision making problems in emergency response. *Computers & Industrial Engineering*. doi:10.1016/j.cie.2018.06.027

Llamazares, B. (2018). An analysis of the generalized TODIM method. *European Journal of Operational Research*, *269*(3), 1041–1049. doi:10.1016/j.ejor.2018.02.054

Lourenzutti, R., & Krohling, R. A. (2013). A study of TODIM in a intuitionistic fuzzy and random environment. *Expert Systems with Applications*, *40*(16), 6459–6468. doi:10.1016/j.eswa.2013.05.070

Mardani, A., Nilashi, M., Zakuan, N., Loganathan, N., Soheilirad, S., Saman, M. Z. M., & Ibrahim, O. (2017). A systematic review and meta-Analysis of SWARA and WASPAS methods: Theory and applications with recent fuzzy developments. *Applied Soft Computing*, *57*, 265–292. doi:10.1016/j.asoc.2017.03.045

Ozcanli, M., Gungor, C., & Aydin, K. (2013). Biodiesel fuel specifications: A review. *Energy Sources. Part A, Recovery, Utilization, and Environmental Effects*, *35*(7), 635–647. doi:10.1080/15567036.2010.503229

Perimenis, A., Walimwipi, H., Zinoviev, S., Müller-Langer, F., & Miertus, S. (2011). Development of a decision support tool for the assessment of biofuels. *Energy Policy*, *39*(3), 1782–1793. doi:10.1016/j.enpol.2011.01.011

Poh, K. L., & Ang, B. W. (1999). Transportation fuels and policy for Singapore: An AHP planning approach. *Computers & Industrial Engineering*, *37*(3), 507–525. doi:10.1016/S0360-8352(00)00020-6

Puhan, S., Jegan, R., Balasubbramanian, K., & Nagarajan, G. (2009). Effect of injection pressure on performance, emission and combustion characteristics of high linolenic linseed oil methyl ester in a DI diesel engine. *Renewable Energy*, *34*(5), 1227–1233. doi:10.1016/j.renene.2008.10.001

Puhan, S., Vedaraman, N., Ram, B. V., Sankarnarayanan, G., & Jeychandran, K. (2005). Mahua oil (Madhuca Indica seed oil) methyl ester as biodiesel-preparation and emission characterstics. *Biomass and Bioenergy*, *28*(1), 87–93. doi:10.1016/j.biombioe.2004.06.002

Rashid, U., Anwar, F., Moser, B. R., & Ashraf, S. (2008). Production of sunflower oil methyl esters by optimized alkali-catalyzed methanolysis. *Biomass and Bioenergy*, *32*(12), 1202–1205. doi:10.1016/j.biombioe.2008.03.001

Rassafi, A. A., Vaziri, M., & Azadani, A. N. (2006). Strategies for utilizing alternative fuels by Iranian passenger cars. *International Journal of Environmental Science and Technology*, *3*(1), 59–68. doi:10.1007/BF03325908

Saaty, L. T. (1980). *The Analytic Hierarchy Process*. New York: McGraw Hill Company.

Saaty, L. T., & Vargas, L. G. (2001). *Models, Methods, Concepts &Applications of the Analytical Hierarchy Process*. Boston: Kluwer Academic Publishers; doi:10.1007/978-1-4615-1665-1

Sahoo, P. K., & Das, L. M. (2009). Combustion analysis of Jatropha, Karanja and Polanga based biodiesel as fuel in a diesel engine. *Fuel*, *88*(6), 994–999. doi:10.1016/j.fuel.2008.11.012

Sakthivel, G., Ilangkumaran, M., & Gaikwad, A. (2015). A hybrid multi-criteria decision modeling approach for the best biodiesel blend selection based on ANP-TOPSIS analysis. *Ain Shams Engineering Journal*, *6*(1), 239–256. doi:10.1016/j.asej.2014.08.003

Sakthivel, G., & Nagarajan, G. (2011). *Experimental studies on the performance and emission characteristics of a diesel engine fuelled with ethyl ester of waste fish oil and its diesel blends* (No. 2011-01-2402). SAE Technical Paper.

Sarin, A., Arora, R., Singh, N. P., Sarin, R., Malhotra, R. K., & Kundu, K. (2009). Effect of blends of Palm-Jatropha-Pongamia biodiesels on cloud point and pour point. *Energy*, *34*(11), 2016–2021. doi:10.1016/j.energy.2009.08.017

Sarin, R., Sharma, M., Sinharay, S., & Malhotra, R. K. (2007). Jatropha–palm biodiesel blends: An optimum mix for Asia. *Fuel*, *86*(10-11), 1365–1371. doi:10.1016/j.fuel.2006.11.040

Scott, J. A., Ho, W., & Dey, P. K. (2012). A review of multi-criteria decision-making methods for bioenergy systems. *Energy*, *42*(1), 146–156. doi:10.1016/j.energy.2012.03.074

Sen, D. K., Datta, S., & Mahapatra, S. S. (2016). Application of TODIM (Tomada de Decisión Inerativa Multicritero) for industrial robot selection. Benchmarking. *International Journal (Toronto, Ont.)*, *23*(7), 1818–1833.

Shanmugam, P., Sivakumar, V., Murugesan, A., & Ilangkumaran, M. (2011a). Performance and exhaust emissions of a diesel engine using hybrid fuel with an artificial neural network. *Energy Sources. Part A, Recovery, Utilization, and Environmental Effects*, *33*(15), 1440–1450. doi:10.1080/15567036.2010.539085

Shanmugam, P., Sivakumar, V., Murugesan, A., & Umarani, C. (2011b). Experimental study on diesel engine using hybrid fuel blends. *International Journal of Green Energy*, *8*(6), 655–668. doi:10.1080/15435075.2011.588891

Shannon, C. E. (1948). The Mathematical Theory of Communication. *The Bell System Technical Journal*, *27*(3), 379–423. doi:10.1002/j.1538-7305.1948.tb01338.x

Sivaraja, C. M., & Sakthivel, G. (2017). Compression ignition engine performance modelling using hybrid MCDM techniques for the selection of optimum fish oil biodiesel blend at different injection timings. *Energy*, *139*, 118–141. doi:10.1016/j.energy.2017.07.134

Srivastava, A., & Prasad, R. (2000). Triglycerides-based diesel fuels. *Renewable & Sustainable Energy Reviews*, *4*(2), 111–133. doi:10.1016/S1364-0321(99)00013-1

Stanford, A. (1998). Liberalisation of the UK energy market: An opportunity for green energy. *Renewable Energy*, *15*(1-4), 215–217. doi:10.1016/S0960-1481(98)00160-8

Sureshkumar, K., Velraj, R., & Ganesan, R. (2008). Performance and exhaust emission characteristics of a CI engine fueled with Pongamia pinnata methyl ester (PPME) and its blends with diesel. *Renewable Energy*, *33*(10), 2294–2302. doi:10.1016/j.renene.2008.01.011

Tseng, M. L., Lin, Y. H., Tan, K., Chen, R. H., & Chen, Y. H. (2014). Using TODIM to evaluate green supply chain practices under uncertainty. *Applied Mathematical Modelling*, *38*(11-12), 2983–2995. doi:10.1016/j.apm.2013.11.018

Tsita, K. G., & Pilavachi, P. A. (2012). Evaluation of alternative fuels for the Greek road transport sector using the analytic hierarchy process. *Energy Policy*, *48*, 677–686. doi:10.1016/j.enpol.2012.05.079

Tzeng, G. H., Lin, C. W., & Opricovic, S. (2005). Multi-criteria analysis of alternative-fuel buses for public transportation. *Energy Policy*, *33*(11), 1373–1383. doi:10.1016/j.enpol.2003.12.014

Uysal, F., & Tosun, Ö. (2014). Multi criteria analysis of the residential properties in Antalya using TODIM method. *Procedia: Social and Behavioral Sciences*, *109*, 322–326. doi:10.1016/j.sbspro.2013.12.465

Vedaraman, N., Puhan, S., Nagarajan, G., & Velappan, K. C. (2011). Preparation of palm oil biodiesel and effect of various additives on NOx emission reduction in B20: An experimental study. *International Journal of Green Energy*, *8*(3), 383–397. doi:10.1080/15435075.2011.557847

Vilela, L., Mata, T. M., & Caetano, N. S. (2010, November). Biodiesel production from fish oil with high acidity. In *3rd international symposium on energy from biomass and waste* (pp. 1-14). Academic Press.

Wang, J., Wang, J. Q., & Zhang, H. Y. (2016). A likelihood-based TODIM approach based on multi-hesitant fuzzy linguistic information for evaluation in logistics outsourcing. *Computers & Industrial Engineering*, *99*, 287–299. doi:10.1016/j.cie.2016.07.023

Wu, Y., Wang, J., Hu, Y., Ke, Y., & Li, L. (2018). An extended TODIM-PROMETHEE method for waste-to-energy plant site selection based on sustainability perspective. *Energy*, *156*, 1–16. doi:10.1016/j.energy.2018.05.087

Yedla, S., & Shrestha, R. M. (2003). Multi-criteria approach for the selection of alternative options for environmentally sustainable transport system in Delhi. *Transportation Research Part A, Policy and Practice*, *37*(8), 717–729. doi:10.1016/S0965-8564(03)00027-2

Yılmaz, S., & Selim, H. (2013). A review on the methods for biomass to energy conversion systems design. *Renewable & Sustainable Energy Reviews*, *25*, 420–430. doi:10.1016/j.rser.2013.05.015

Zarbakhshnia, N., Soleimani, H., & Ghaderi, H. (2018). Sustainable third-party reverse logistics provider evaluation and selection using fuzzy SWARA and developed fuzzy COPRAS in the presence of risk criteria. *Applied Soft Computing*, *65*, 307–319. doi:10.1016/j.asoc.2018.01.023

Zindani, D., Maity, S. R., Bhowmik, S., & Chakraborty, S. (2017). A material selection approach using the TODIM (TOmada de Decisao Interativa Multicriterio) method and its analysis. *International Journal of Materials Research*, *108*(5), 345–354. doi:10.3139/146.111489

Zolfani, S. H., & Saparauskas, J. (2013). New application of SWARA method in prioritizing sustainability assessment indicators of energy system. *The Engineering Economist*, *24*(5), 408–414.

Zolfani, S. H., Yazdani, M., & Zavadskas, E. K. (2018). An extended stepwise weight assessment ratio analysis (SWARA) method for improving criteria prioritization process. *Soft Computing*, 1–7.

Chapter 6
Optimization of Performance and Emissions Parameters of a Biodiesel-Run Diesel Engine:
An Integrated MCDM Approach

Sumita Debbarma
National Institute of Technology Silchar, India

Biplab Das
National Institute of Technology Silchar, India

Jagadish
https://orcid.org/0000-0001-5699-5674
National Institute of Technology Raipur, India

ABSTRACT

Biodiesel has been immersed as an immediate alternative of fossil fuels for diesel engines. However, choosing a good combination of biodiesel blends based on both performance and emission depend on various factors. The chapter presents the modeling and optimization of performance and emissions parameters of a biodiesel-run diesel engine using an integrated MCDM approach. The integrated MCDM approach consists of entropy with MCRA method. An experimental case study on performance and emission study of diesel engine is considered to show the modeling capability of the proposed method. The results show that trail no. 4 yields the optimal setting compare to the other combinations. The trail no. 4 gives optimum operating condition such as 85-90% load and PB10 which provides optimum performance parameters like higher brake thermal efficiency (BTE), lower brake-specific energy consumption (BSEC), lower carbon monoxide (CO), lower hydro carbon (HC), and lower oxides of nitrogen (NOx), respectively.

DOI: 10.4018/978-1-5225-8579-4.ch006

INTRODUCTION

Natural resources play a significant role in the field of science and technology. But with the increase in population and advancement in technology, the requirement for these natural resources is also expanded day by day. Some of the natural resources are fossil fuels, air, wind, water, sunlight etc. Unfortunately, it has been forecast by the experts that by 2050 the availability of fossil fuels will diminish due to their increase in consumption rate which is 10^5 faster than their natural production (Satyanarayana et al., 2011). Among the fossil fuels, diesel is in high demand in the transportation sector, agricultural sector, domestic sector and industrial sector, to meet the high demand for power over petrol or Spark Ignition (SI) engines. Since largest section of our transportation needs is met by diesel engines, substitution of even a small fraction of diesel fuel needs by alternative fuels is very important in the present context. The diesel engine release harmful emissions like carbon monoxide (CO), hydrocarbon (HC), carbon-di-oxide (CO_2), soot, particulate matter (PM), oxides of nitrogen (NOx), and formaldehyde etc., which causes serious health issue to the living beings and ecological imbalance. Though, diesel engines are mostly preferred over SI engine because of higher thermal efficiencies, lower consumption of fuel, high compression ratio, utilization of better air-fuel mixtures, reliability, low operating and maintenance cost, lesser CO and HC emissions, but, NOx and PM are greater for diesel engines, respectively. In order to control these pollutants and to compensate for the increasing demand of diesel fuel presently the researchers focus on clean alternative fuels (i.e., biodiesel) those are locally available, environmentally acceptable and technically feasible have become a topic on the global agenda. Generally, biodiesel produced through transesterification process from various vegetable oil and animal fats as feedstocks in presence of homogeneous catalysts (NaOH and KOH). Biodiesel is the better selection as an alternative fuel because it is biodegradable, nontoxic, good lubricant properties, similar properties to that of diesel, reduces the environmental pollution with no sulphur compounds during combustion and so forth (Lin et al., 2011; Kannan &Anand, 2012; Arbab et al., 2013; Rahman et al., 2013; Agarwal & Das, 2010).

The global market of biodiesel by 2020 is expected to reach 45,291 million litres (Varatharajan & Cheralathan, 2013). Many studies have been performed to find the suitability of different edible oil as alternative oil for existing diesel engine (Nwafor, 2004; Ghadge & Raheman, 2005; Sahoo et al., 2007; Gui et al., 2008; Zhang et al., 2013; Plata et al., 2015; Debbarma & Misra, 2018). Ramadhas et al., (2005) studied the engine performance and emission of a diesel engine operated with rubber seed oil biodiesel and its blend (B10, B20, B50, B75 and B100). It was found maximum BTE obtained for B10 and B20 about 12% and 3% respectively higher than that of neat diesel. Moreover, it was observed that using B100, the BSFC was about 12% higher compared to neat diesel, whereas for lower percentage of biodiesel blend showed lower fuel consumption. For all loading conditions, it have been observed that CO, CO_2, NOx and Smoke emission were lower than the neat diesel in case of B20 blend. Qi et al., (2010) compared the properties, performance, emissions and combustion characteristics of diesel fuel with biodiesel and its blends from soybean oil in a direct injection compression ignition engine. They highlited a significant reduction of CO and smoke emissions and no significant differences for HC emissions at high engine loads. Biodiesel and its blends exhibited a minor increase in BSFC, decrease in BTE and slightly higher NOx compared to diesel fuel. Ong et al., (2013) produced biodiesels from Jatropha curcas oil, Sterculia foetida oil and Ceiba pentandra oil. They showed similar in fuel properties compared to diesel fuel. Debbarma et al., (2011) used jatropha biodiesel and its blend as a fuel to run a single cylinder diesel engine. They observed improvement in engine performance and lower emissions. Dey and Misra, (2016) investigated the performance and emission characteristics of palm biodiesel and its blends with palm straight vegetable oil (PSVO) as lubricant in a conventional diesel engine compared with diesel fuel in presence of mineral oil as lubricant operated for 60 h. They noted that the B30 palm biodiesel fuel blend with PSVO as lubricant was considered as a suitable blend among all the fuel blends combination. The result showed that suitable blend provided significant reductions in emissions NOx by 31.93%, CO by 50.69%, and HC 137.5% compared to diesel fuel with mineral lubricant. Debbarma and Misra, (2017) analyzed the combustion and emissions of asingle cylinder diesel engine fueled with a mixture of palm biodiesel blend (B30) and iron nannoparticles (50 and 75 ppm). In presence of nanoparticles B30 blend produced significant reductions in the CO and NOx emissions by 56% and 4%, improvement in BTE by 3% and breakspecific energy consumption (BSEC) by 3.3%, compared to diesel fuel.

Jena and Misra, (2014) used palm biodiesel and karanja biodiesel to study the exergetic performance of a CI engine. Among two biodiesels they pointed that palm biodiesel is better in terms of both energetic and exergetic performance than karanja biodiesel. Further presence of more oxygen content in the fuel blend the combustion was observed better and irreversibilities also get reduced. Aydin and Bayındir, (2010) investigated the performance and emission characteristics of cottonseed oil biodiesel and its blends (B5 to B75) in a single -cylinder diesel engine at various engine speed. It was observed that among all the blends B5 obtained higher torque including diesel (by 1.4%) and lower power around 1.3%. At higher speeds (2000 rpm) minimum BSFC was observed for neat diesel, B5 and B20. Reduction of CO and NOx emission for all the fuel blends was observed. Senthil and Silambarasan, (2015) investigated the performance, emission and combustion characteristics of

Annona biodiesel (B100) and its blends (B20 to B80) in a diesel engine. They observed, brake thermal efficiency (BTE) of B20 showed slightly nearer to neat diesel by 3%. The reduction of emission of CO by 39%, HC by 38% than that of neat diesel but increased in NOx by 7%. Finally concluded that B20 blend of Annona biodiesel can be used as fuel in diesel engine without any modification. Jindal and Goyal, (2012) found that Ambadi biodiesel blend produces 8% higher BTE compared to neat diesel. Liaquat et al., (2013) studied the effect of coconut biodiesel blends (B5 and B15) on the performance and emission characteristics using of a single-cylinder diesel engine. They found that reduces average torque for B5 by 0.69% and for B15 by 2.58%, average power over the entire speed range by 0.66% for B5 and by 2.61% for B15 with increase in average brake specific fuel consumption (BSFC) by 0.53% for B5 and by 2.11% for B15 compared to neat diesel. Emissions result indicated that reduction of HC and CO as 13.89% and 13.38% respectively for B5 and 22.88% and 21.51% for B15. Increased in CO_2 and NOx as 2.54% and 1.42% respectively for B5, and 4.64% and 3.19% for B15 compared to neat diesel were observed. Ozcanli et al., (2011) evaluated the engine performance and emission on a three-cylinder, four-stroke, direct-injection CI engine operated with Terebinth oil biodiesel and its blend (B10-B100). It was found reducing power for all the blends compared to neat diesel, the average rate of increase in BSFC by 2.59% with B25, 3.86% with B50, 5.62% with B100 and slight decrease in case of B5. According to the emission analysis, maximum reduction ratio of CO emission was 16.75% for B25, 27.27% for B50, and 34.54% for B100 at different speed. However, increasing trend of NOx emission was observed. The maximum increment ratio of NOx emission with B25, B50, and B100 were 19.57%, 22.09% and 32.97%, respectively. Overall, it was highlighted that B25 blend can be used in diesel engines in terms of performance and environmental aspects. Perumal and Ilangkumaran (2017) investigated the performance and emission characteristics using biodiesel derived from Pongamia oil in single -cylinder, four-stroke, direct injection, water cooled diesel engine at 1500 rpm. Different proportion of biodiesel blend from B20 to B100 were used and tested at various loading condition. They found lower BTE, HC and CO and increase in BSFC and NOx than diesel fuel. Ng et al. (2012) evaluated the engine performance of palm biodiesel and its blend running at 3500 rpm. It was observed the reduction of tailpipe NO, HC, and smoke opacity with maximum decrease of 5.0%, 26.2%, and 66.7%, respectively, compared to neat diesel. It was concluded that biodiesel has the potential to reduce emission in a light-duty diesel engine, both in neat and blended forms. Salvi and Jindal (2013) investigated performance of a single-cylinder, four-stroke, water cooled at 1500 rpm at different load with linseed biodiesel blend as fuel (B10-B40). It was found that improvement in BTE by 3.7% and lower BSFC by 5.5% with B10 as compared to neat diesel at full load. The emission of HC and smoke opacity were reduced by 80% and 54%, respectively but with slight increase in NOx by 20% as compared to neat diesel. It was highlighted that the linseed biodiesel can be used as renewable fuel for the vehicles of transportation purposes. Dinesh et al. (2016) used the *Calophyllum inophyllum* (Tamanu) biodiesel blends in a CI engine and significant increase in BTE and reduced in emissions

In the present study, Palm oil *(Elaeisguineensis)* is chosen as a source for biodiesel. Many studies on performance analysis of Palm biodiesel (PB) as an alternative fuel for diesel engine at different operating conditions are available in literatures (Madiwale et al. 2018; Kalam and Masjuki, 2008; Reijnders and Huijbregts, 2008; Ng and Gan, 2010a; Ng and Gan, 2010b; Ndayishimiye and Tazerout, 2011; Fattah et al. 2014; Rao and Reddy, 2016; Ali et al. 2016; Yazdani et al. 2018). Also, several techniques like MOORA (Sharma et al. 2018), ANP-TOPSIS (Sakthivel et al. 2015), Taguchi-Fuzzy set theory (Sakthivel and Ilangkumaran, 2017), Fuzzy-AHP-PROMETHEE (Ramesh et al. 2018), NSGA-II and TOPSIS (Etghani et al. 2013) for the selection of a suitable PB as alternative are available in literature.

From the literature review, it has been observed that, many investigators have been worked on optimization using different optimization techniques. However, consideration of precise priority weights and the degree of importance different operating conditions of the diesel engine have been ignored. In addition, existing optimization techniques have failed to consider the effect of both beneficial and non-beneficial operating conditions separately while determining the optimal performance and emission parameters for the diesel engine. Hence, the existing optimization methods could not give the optimal results for the diesel engine. Besides, none of the researchers in the literature have been worked on optimization of performance and emissions parameters of a palm biodiesel run diesel engine with consideration of different blends and loading condition.

Furthermore, performance and emission study of diesel engine is considered to be multi-response or multi-criteria decision making (MCDM) optimization problem. Hence, need of MCDM method is essential for modelling and optimal selection of performance and emission parameter for diesel engine. In this context, in the present book chapter, an attempt is made to study the engine performance and exhaust emissions parameters of the prepared palm biodiesel blends, and optimum range of operating conditions using a novel integrated MCDM method. The novel integrated method consists of entropy coupled with multi criteria ranking analysis (MCRA). Here, entropy method is used for determination of precise priority weights of the different operating conditions while optimum operating conditions to get higher performance characteristics like maximize brake thermal efficiency while to minimize brake specific fuel consumption, carbon monoxide, hydro carbon and oxides of nitrogen using MCRA.

INTEGRATED METHOD

In this section, an integrated MCDM method is discussed. The integrated method consist of entropy (Caliskan et al. 2013, Emma et al. 2013) combined with multi criteria ranking analysis (MCRA) (Brauers and Zavadskas 2006, Prasad and Chakraborty, 2012). The detailed steps of entropy-MCRA based MCDM method are as follows,

Step 1: Development of evaluation matrix

First, the evaluation matrix is developed which consists of criteria as different performance parameters and corresponding alternatives as different operating parameters settings. The evaluation matrix with *n* number of criteria and *m* number of alternatives is defined by following equation.

$$A_{ij} = \begin{bmatrix} & C1 & C2 & \ldots & Cn \\ A1 & X_{11} & X_{12} & \ldots & X_{1n} \\ A2 & X_{21} & X_{22} & \ldots. & X_{2n} \\ \ldots & \ldots & \ldots & \ldots. & \ldots. \\ Am & X_{m1} & X_{m2} & \ldots.. & X_{mn} \end{bmatrix} \quad (1)$$

where, A_{ij} is the evaluation matrix, C_1, $C2\ldots..C_n$ denotes number of criteria A_1, $A2\ldots..$ A_m are number of alternatives.

Step 2: Normalization of decision matrix

In this step, data normalization of the evaluation matrix is done using the Eqs. (2-3). Data normalization is carried out to convert the original sequence of data in to comparable sequence in the range of 0-1. As the data is combined with beneficial and non-beneficial data, the data normalization is to be done separately.

$$\left[P_{ij}\right]_{m\times n} = \left[\frac{X_{ij}}{Max\sum_{i=1}^{m} X_{ij}^{2}}\right]_{m\times n} \quad \text{for beneficial criterion} \quad (2)$$

$$\left[P_{ij}\right]_{m\times n} = \left[\frac{X_{ij}}{Min\sum_{i=1}^{m} X_{ij}^{2}}\right]_{m\times n} \quad \text{for non-beneficial criterion} \tag{3}$$

where, P_{ij} represents the normalized evaluation matrix and *max* (X_{ij}) indicates the maximum performance value of the criterion for beneficial criterion and *min* (X_{ij}) is the minimum performance value criterion for non-beneficial criterion.

Step 3: Determination of precise priority weights

After the normalization, precise priority weights for each of the criteria are determined using Eqs. (4-6). First, degree of divergence (d_j) for each of the criteria is calculated using Eqs. (4-5), which measures the distance of each criteria with respect to the comparable sequence. After that, precise priority weights are determined using Equation (6).

$$Y_j = -k\sum_{i=1}^{m} P_{ij} \ln\left(P_{ij}\right) \quad j = 1, 2, \ldots, n. \tag{4}$$

$$d_j = \left|1 - Y_j\right|. \tag{5}$$

$$\beta_j = \frac{d_j}{\sum_{j=1}^{n} d_j}. \tag{6}$$

where k is a constant term and calculated as [$k = 1 / \ln m, 0 \leq d_j \leq 1$], β_j are the precise priority weights for each of the criterion.

Step 4: Formulation of weighted evaluation matrix

After determination of precise priority weights for each of the criteria, the formulation of weighted evaluation matrix is done. The weighted evaluation matrix is formulated by multiplying the each priority weights to the each of the performance values of the evaluation matrix. The weighted evaluation matrix is calculated using following equation.

$$WD_{ij} = P_{ij} \times \beta_j \tag{7}$$

where, *WD* is the weighted evaluation matrix, β_j denotes the priority weights for each of the criteria and P_{ij} is the normalized evaluation matrix.

Step 5: Determination of closeness index and Ranking

The closeness index (*CI*) is determined for each of the beneficiary criteria (BC) and non-beneficiary criteria (NBC). *CI* converts the multi-criteria into single criteria with consideration of both BC and NBC of the parameters (Brauers and Zavadskas 2006, Prasad and Chakraborty, 2012). The *CI* for each of the alternatives calculated using the following expression.

$$BC = \sum_{i=1}^{b} WDi_j \tag{8}$$

$$NBC = \sum_{i=b+1}^{n} WD_{ij} \tag{9}$$

$$CI = BC - NBC \tag{10}$$

where, *BC* and *NBC* are the sum of beneficial and non-beneficial criteria values of weighted decision matrix respectively, *b* denotes the number of beneficiary criteria, *n* represents the total number of criteria and *CI* is the closeness index. At last, ranking of alternatives is done based on the *CI* values. The alternative with higher *CI* gives the optimal operating conditions to get higher performance for the system.

EXPERIMENTAL STUDY

The experimental investigations are conducted to examine the performance of a single-cylinder, four-stroke diesel engine fuelled with PB and diesel fuel blends at different loads. The blends are made by mixing 10%, 20% and 30% volume of biodiesel with neat diesel and named as PB10, PB20 and PB30 respectively. The schematic diagram of the experimental setup is shown in Figure 1 and the engine specifications are given in Table 1.

MODELING OF DIESEL ENGINE USING INTEGRATED METHOD

In this section, the modeling of diesel engine is done using Entropy-MCRA method. Here, first, development of evaluation of decision matrix is done using Equation (1) and tabulated in Table 2. In which trail numbers are considered as number of alternatives while output parameters as number of criteria. There are five criteria like BTE, BSFC, CO, HC, and NOx, and 18 trails that include operating conditions like different loading and PB blending. Second, normalization of the evaluation of decision matrix is carriedout using Eqs. (2-3). It is done to convert original sequence data in to comparable sequence in the rage of 0-1. Third, precise priority weights of the criteria such as BTE, BSFC, CO, HC and NOx are determined using entropy method (Caliskan et al. 2013, Emma et al. 2013). In this method, first, degrees of divergence (dj) for each of the criteria / output parameters are calculated using Eqs. (4-5). The degree of divergence measures the distance of each criteria with respect to the comparable sequence as 1. If the criteria closer to the comparable sequence value that criteria have more degree of importance than the other. After that, precise priority weights for each of the criteria are determined using Equation (6) and results are tabulated in Table 3.

The precise priority weights (βj) for the diesel engine such as BTE, BSFC, CO, HC and NOx are calculated to be 0.2032, 0.2016, 0.1932, 0.2015 and 0.2005 respectively. The results show that, criteria BTE is the most influencing performance parameter followed by BSFC, CO, HC and NOx. Thereafter, formulation of weighted normalized evaluation matrix is carried out using Equation (7), in which priority weights for each performance parameters obtained by entropy method are multiplied to the each of the performance values of the normalized evaluation matrix and the results of weighted normalized evaluation matrix is shown in Table 4.

At last, closeness index for each of the criteria's that includes both beneficiary criteria and non-beneficiary criteria are determined using Eqs. (8-10). This converts multi-criteria/response in to single criteria/response subjected to the certain constraints with consideration of both beneficiary criteria and non-beneficiary criteria (Brauers and Zavadskas 2006, Prasad and Chakraborty, 2012). The closeness indexes (CI) for each of the alternatives are tabulated in Table 5.

Table 1. Specification of the test engine

Parameters:	Setup Specification:
Make	Kirloskar
Type	Single cylinder, Four stroke, Diesel engine
Power	3.5 kW@ 1500 rpm
Cooling type	Water cooling
Injection variation	0-25° BTDC
Compression ratio range	12:1-18:1
Stroke × Bore	110 mm × 87.5 mm
Displacement volume	661 cc
Connecting rod length	234 mm
Dynamometer arm length	185 mm
Fuel measuring intervals	60 s
Pulse per revolution	360°
Loading System	Eddy current dynamometer
Ignition System	Compression ignition
Combustion chamber	Hemispherical
Air box	M.S. fabricated with orifice meter and manometer
Fuel tanks	Two tanks 7.5 liters capacity with measuring tube
Calorimeter	Pipe in pipe type
Rotameter	Engine cooling 40-400 LPH; Calorimeter 25-250 LPH
Pump	Monoblock type
Lubricating oil pump	Gear type
Data acquisition Software	"EnginesoftLV" Engine performance analysis software
Engine weight	160 Kg
Overall dimensions	617 L x 504 W x 877 H
Pressure sensor	Piezo type, range 5000 PSI, with low noise cable
Crank angle sensor	Resolution 1°, speed 5500 rpm with TDC pulse.
Temperature sensor	PT100 (RTD)type, and K type Thermocouple
Load sensor and indicator	Strain gauge type load cell with digital indicator, range 0-50 kg
Data acquisition device	NI USB-6210,16-bit, 250 kS/s.
Temperature transmitter	Type two wire, Input RTD PT100, Range 0–100 °C, Output 4–20 mA Type two wire, Input Thermocouple, Range 0–1200 °C, Output 4–20 mA
Fuel flow transmitter	DP transmitter, range 0-500 mm WC
Air flow transmitter	Pressure transmitter, range (-) 250 mm WC

Figure 1. Experimental setup

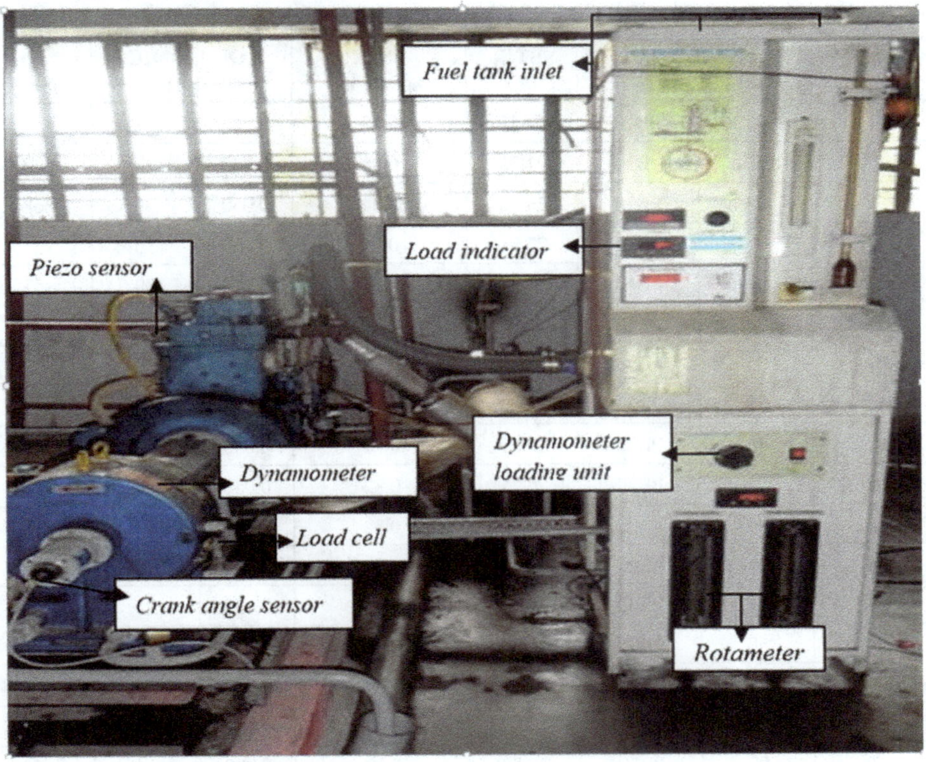

RESULTS AND DISCUSSION

Parametric Analysis of Output Parameters

Variations of output parameters with load are highlighted in this section. Variation of BTE with load is highlighted in Figure 2. Results reveal that for all the fuel blends BTE keeps on increasing with the increase in load, until 85% to 90%, beyond which it starts decreasing. It indicates the optimum loading conditions of the engine. An initial increase in efficiency with the load may be because of the comparatively higher percentage of frictional losses at lower load, i.e., most of the produced power is absorbed to overcome this loss. Quantity wise the magnitude of frictional losses remains almost constant and ratio wise it decreases at higher loading conditions. In addition, at full load BTE decreases compare to 90% loading condition. This is due to comparatively poor combustion. Further, it can be seen that with the

Table 2. Experimental results for different PB blends and loading condition

Trail No	Input Parameters		Output Parameters				
	Load (%)	Blend	BTE (%)	BSFC (kg/kW-hr)	CO (%)	HC (ppm)	NOx (ppm)
1	25	PB10	14.22	0.58	0.060	26	342
2	50	PB10	21.75	0.38	0.040	25	640
3	75	PB10	24.68	0.34	0.050	30	868
4	85	PB10	27.91	0.30	0.060	35	922
5	90	PB10	26.39	0.31	0.110	39	953
6	100	PB10	26.87	0.31	0.160	50	1001
7	25	PB20	15.10	0.55	0.060	26	361
8	50	PB20	22.00	0.38	0.050	28	675
9	75	PB20	25.56	0.33	0.050	30	782
10	85	PB20	26.21	0.32	0.055	38	960
11	90	PB20	26.55	0.32	0.090	43	976
12	100	PB20	26.66	0.32	0.140	48	1010
13	25	PB30	14.17	0.60	0.050	21	372
14	50	PB30	21.53	0.40	0.050	29	639
15	75	PB30	25.00	0.34	0.040	24	880
16	85	PB30	26.09	0.33	0.050	29	939
17	90	PB30	26.58	0.32	0.080	34	1021
18	100	PB30	26.46	0.32	0.120	45	1025

Table 3. Precise priority weights of criteria

Weights of criterias	BTE (%)	BSFC (kg/kW-hr)	CO (%)	HC (ppm)	NOx (ppm)
β_j	0.2032	0.2016	0.1932	0.2015	0.2005

increase in blend ratio, PB10 and PB20 tends to vary insignificantly except at 85% loading. However, PB30 blend always shows an inferior performance. Detritions in performance may be due to the reduction of calorific value with increase in the percentage of biodiesel in the blend. Thus, manually it can be said that either PB10 or PB20 may be chosen as the optimum blend. But energy security point view PB20 will be better since in PB20 blend more amount of conventional diesel is being replaced, that is depleting rapidly.

Table 4. Weighted and normalized decision matrix

Trail No	BTE (%)	BSFC (kg/kW-hr)	CO (%)	HC (ppm)	NOx (ppm)
1	0.0284	0.0713	0.0337	0.0359	0.0194
2	0.0435	0.0467	0.0225	0.0345	0.0364
3	0.0493	0.0418	0.0281	0.0414	0.0494
4	0.0558	0.0369	0.0337	0.0483	0.0524
5	0.0528	0.0381	0.0618	0.0539	0.0542
6	0.0537	0.0381	0.0899	0.0691	0.0569
7	0.0302	0.0676	0.0337	0.0359	0.0205
8	0.0440	0.0467	0.0281	0.0387	0.0384
9	0.0511	0.0406	0.0281	0.0414	0.0445
10	0.0524	0.0393	0.0309	0.0525	0.0546
11	0.0531	0.0393	0.0506	0.0594	0.0555
12	0.0533	0.0393	0.0787	0.0663	0.0574
13	0.0283	0.0737	0.0281	0.0290	0.0212
14	0.0430	0.0492	0.0281	0.0401	0.0363
15	0.0500	0.0418	0.0225	0.0332	0.0500
16	0.0522	0.0406	0.0281	0.0401	0.0534
17	0.0531	0.0393	0.0450	0.0470	0.0581
18	0.0529	0.0393	0.0674	0.0622	0.0583

Variation of BSFC with load is shown in Figure 3. It is seen that within the increase in load quality of combustion improves and comparatively lower amount fuel is consumed, this is also in line with the trend of efficiency. At higher load contribution of frictional loss is lower and most of the power production is available as brake power. This also indicates the economic loading conditions. Further, among all the fuel blends PB10 shows the lower value of BSFC. At higher blends, density and viscosity of the biodiesel may play a crucial role. Higher value of density and viscosity cause in poor atomisation and ultimately results in improper combustion.

The variation of emission of CO with load is depicted in Figure 4. It is observed that CO emission increases with the increase in load for all the considered fuel samples. However, at lower loads ($\leq$ 75%), no significant CO emissions trend is observed. Also, it is observed that CO emission decreases with the increases in blend ratio. This may be due to the presence of oxygen (about 10-12%) in palm biodiesel helps in better oxidation during combustion of biodiesel blended fuel molecules and as a result, CO emissions are reduced.

Table 5. Closeness index value for alternatives

Trail No	BC	NBC	CI
1	0.0274	0.0369	0.0095
2	0.0123	0.0204	0.0081
3	0.0064	0.0332	0.0268
4	0.0043	0.0398	0.0032
5	0.0030	0.0581	0.0551
6	**0.0021**	**0.0869**	**0.0848**
7	0.0256	0.0335	0.0079
8	0.0118	0.0241	0.0123
9	0.0047	0.0287	0.0240
10	0.0034	0.0432	0.0398
11	0.0027	0.0549	0.0522
12	0.0025	0.0774	0.0749
13	0.0275	0.0373	0.0098
14	0.0127	0.0243	0.0116
15	0.0058	0.0313	0.0255
16	0.0036	0.0363	0.0327
17	0.0026	0.0482	0.0456
18	0.0029	0.0681	0.0652

Figure 2. Variation of brake thermal efficiency with load

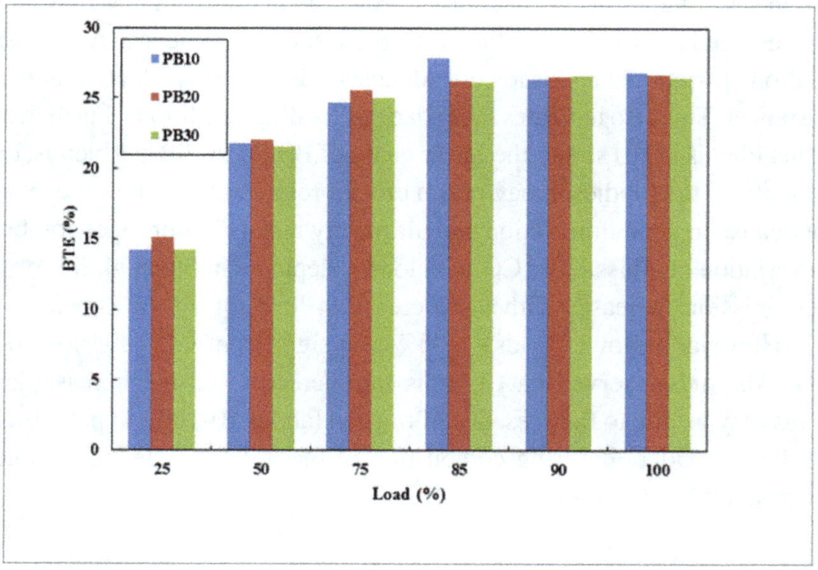

Figure 3. Variation of BSFC with load

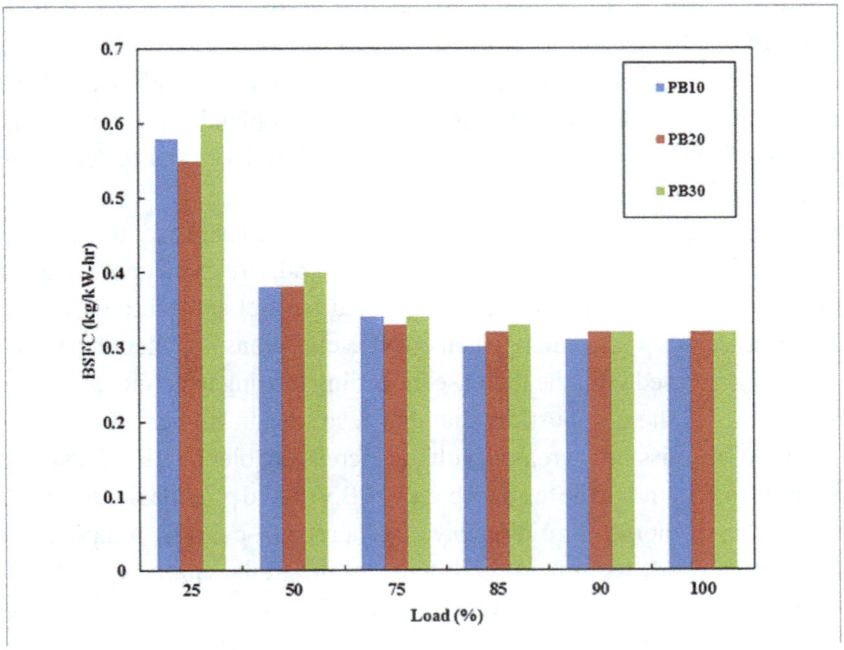

Figure 4. Variation of emission of CO with load

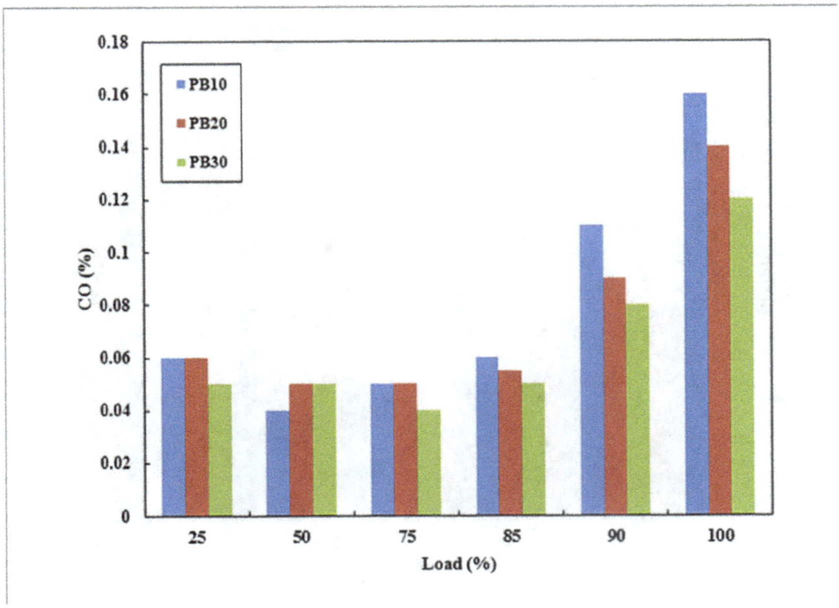

The variation of emission of HC with load is depicted in Figure 5. It is observed that the HC emission increases with the increase in load. From Figure 5 it clearly shows that the HC emission level decrease with the increase in blend ratio. The HC emission for PB30 blend is decreased compared to PB10 and PB20. This may be attributed to higher density and viscosity of PB20 blend with the presence of oxygen in it, which are responsible for shorter ignition delay and better pre-mixed combustion resulting in reduced HC emissions.

The variation of emission of NOx with load is depicted in Figure 6. It is observed that the level of NOx increases with the increase in load, irrespective of the fuel type. It is a well-known phenomenon that oxygenated biofuel substantially trim down carbonated emissions with a small forfeit on NOx emissions. Further, the overall air-fuel ratio gets increased with the increase in loading leading to an increased average gas temperature in the combustion chamber resulting in higher NOx formation. Further, the NOx emission increases with the increase in blend ratio. Because of the presence of oxygen, among the test fuel blends, PB30 blend provides better-premixed combustion due to shorter ignition delay and thereby in-cylinder temperature gets increased with a residence time of high temperature in the engine cylinder. These factors are responsible for higher thermal NOx and therefore, higher NOx emission is observed with PB30 blend compared to that PB10 and PB20.

Figure 5. Variations of HC with load

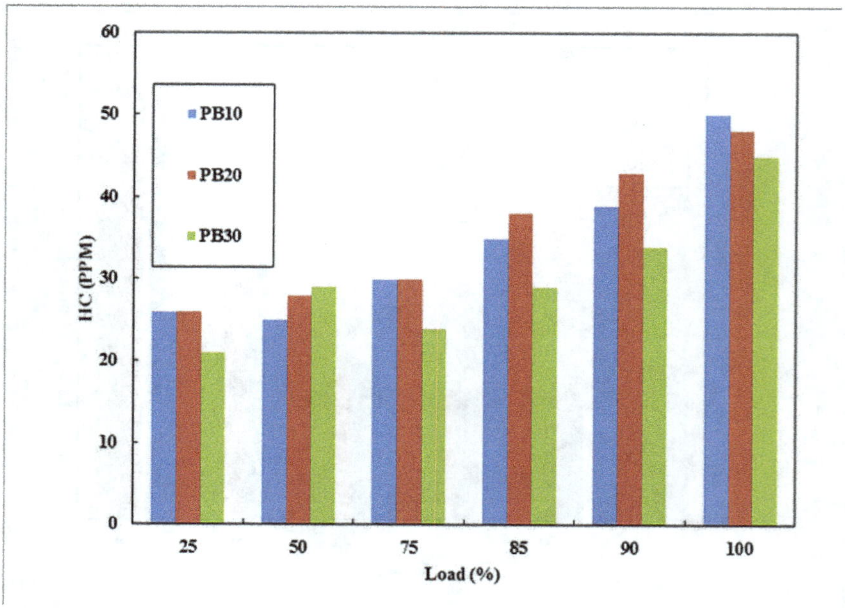

Figure 6. Variation of NOx with load

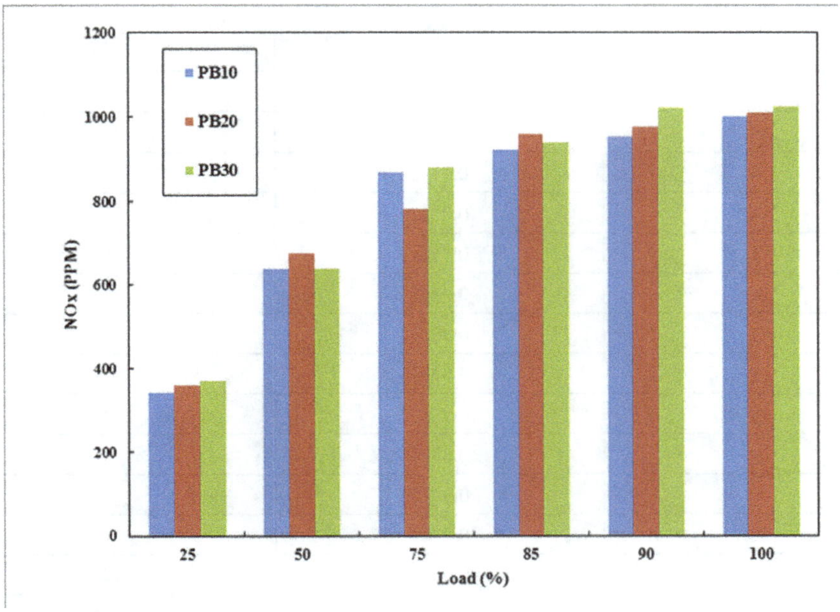

Optimization of Diesel Engine Parameters

The closeness index values for each of the alternatives are used for finding the optimum operating conditions. The results of the closeness index values for each of the alternatives are tabulated in Table 6. Thereafter, a ranking of the trails no is done based on the closeness index values. The trail no. 4 yields the highest closeness index value compare to the other trails no. The closeness index value for the optimal operating condition is 0.9812 and shows highest among the other.

The trail no. 6 gives optimum operating condition such as 100% load and PB10 which provides optimum performance parameters like higher brake thermal efficiency, while lower brake specific fuel consumption, lower carbon monoxide, lower hydro carbon and lower oxides of nitrogen respectively. Furthermore, a recommended optimal condition increases the efficiency and performance of the diesel engine during the 100% loading and PB10 blend conditions.

CONCLUSION

With the rapid depletion of diesel fuel, search for alternative fuel is an active field of research of present era. Biodiesel which shows as a partial replacement of diesel

Table 6. Ranking of alternatives

Trail No	Closeness Index (CI)	Rank
1	0.0095	15
2	0.0081	16
3	0.0268	9
4	0.0032	18
5	0.0551	4
6	**0.0848**	**1**
7	0.0079	17
8	0.0123	12
9	0.0240	11
10	0.0398	7
11	0.0522	5
12	0.0749	2
13	0.0098	14
14	0.0116	13
15	0.0255	10
16	0.0327	8
17	0.0456	6
18	0.0652	3

fuel for compression ignition engine. Further, choosing a good combination of replacement of diesel with biodiesel-diesel blends is depending on the various factors. In the present chapter, modeling and optimal selection of performance parameters of the diesel engine using entropy-MCRA method is discussed. Experimental case study on performance and emission study of diesel engine is illustrated via proposed method. The results show that trail no.6 yields the optimal setting (i.e. 100% load and PB10) compare to the other combination and provides optimal performance parameters like higher brake thermal efficiency, while lower brake specific fuel consumption, lower carbon monoxide, lower hydro carbon and lower oxides of nitrogen respectively. At last, optimal setting i.e. 100% loading and PB10 blend can be a recommended for diesel engine to get higher performance and efficiency.

REFERENCES

Agarwal, A. K., & Das, L. M. (2000). Biodiesel development and characterization for use as a fuel in compression ignition engines. *Journal of Engineering for Gas Turbines and Power*, *123*(2), 440–447. doi:10.1115/1.1364522

Ali, O. M., Mamat, R., Abdullah, N. R., & Abdullah, A. A. (2016). Analysis of blended fuel properties and engine performance with palm biodiesel-diesel blended fuel. *Renewable Energy*, *86*, 59–67. doi:10.1016/j.renene.2015.07.103

Arbab, M. I., Masjuki, H. H., Varman, M., Kalam, M. A., Imtenan, S., & Sajjad, H. (2013). Fuel properties, engine performance and emission characteristic of common biodiesels as a renewable and sustainable source of fuel. *Renewable & Sustainable Energy Reviews*, *22*, 133–147. doi:10.1016/j.rser.2013.01.046

Aydin, H., & Bayindir, H. (2010). Performance and emission analysis of cottonseed oil methyl ester in a diesel engine. *Renewable Energy*, *35*(3), 588–592. doi:10.1016/j.renene.2009.08.009

Brauers, W. K. M., & Zavadskas, E. K. (2006). The MOORA method and its application to privatization in a transition economy. *Control and Cybernetics*, *35*(2), 445–469.

Caliskan, H., Kursuncu, B., Kurbanoglu, C., & Guven, S. Y. (2013). Material selection for the tool holder working under hard milling conditions using different multi criteria decision making methods. *Materials & Design*, *45*, 473–479. doi:10.1016/j.matdes.2012.09.042

Debbarma, S., Das, B., Giri, A., Majumder, S., & Bose, P. K. (2011). Back propagation artificial neural network (BPANN) based performance analysis of diesel engine using biodiesel. *Journal of Renewable and Sustainable Energy*, *3*(1), 013101–013112. doi:10.1063/1.3517229

Debbarma, S., & Misra, R. D. (2017). Effects of Iron Nanoparticles Blended Biodiesel on the Performance and Emission Characteristics of a Diesel Engine. *Journal of Energy Resources Technology*, *139*(042212), 1–12.

Debbarma, S., & Misra, R. D. (2018). Effects of Iron NanoparticleFuel Additive on thePerformance and ExhaustEmissions of a CompressionIgnition Engine Fueled With Diesel and Biodiesel. *Journal of Thermal Science and Engineering Applications*, *10*(4), 041002-1–12. doi:10.1115/1.4038708

Dey, A. R., & Misra, R. D. (2016). Effect of infiltration of bio-lubricant on the performance of a compression ignition engine fuelled with biodiesel blends. *Clean Technologies and Environmental Policy*, 1–11.

Dinesh, K., Tamilvanan, A., Vaishnavi, S., Gopinath, M., & Rajmohan, K. S. (2016). () Biodiesel production using Calophyllum inophyllum (Tamanu) seed oil and its compatibility test in a CI engine. *Biofuels*, 1–7. doi:10.1080/17597269.2016.1187543

Emma, M., Kieran, S., & Vida, M. (2013). An assessment of sustainable housing affordability using a multiple criteria decision making method. *Omega*, *41*(2), 270–279. doi:10.1016/j.omega.2012.05.002

Etghani, M. M., Shojaeefard, M. H., Khalkhali, A., & Akbari, M. (2013). A hybrid method of modified NSGA-II and TOPSIS to optimize performance and emissions of a diesel engine using biodiesel. *Applied Thermal Engineering*, *59*(1–2), 309–315. doi:10.1016/j.applthermaleng.2013.05.041

Fattah, I. M. R., Masjuki, H. H., Kalam, M. A., Mofijur, M., & Abedin, M. J. (2014). Effect of antioxidant on the performance and emission characteristics of a diesel engine fueled with palm biodiesel blends. *Energy Conversion and Management*, *79*, 265–272. doi:10.1016/j.enconman.2013.12.024

Gan, S., & Ng, H. K. (2010). Effects of antioxidant additives on pollutant formation from the combustion of palm oil methyl ester blends with diesel in a non-pressurised burner. *Energy Conversion and Management*, *51*(7), 1536–1546. doi:10.1016/j.enconman.2010.02.012

Ghadge, S. V., & Raheman, H. (2005). Biodiesel production from mahua (Madhumica indica) oil having high free fatty acids. *Biomass and Bioenergy*, *28*(6), 601–605. doi:10.1016/j.biombioe.2004.11.009

Gui, M. M., Lee, K. T., & Bhatia, S. (2008). Feasibility of edible oil vs. non-edible oil vs. waste edible oil as biodiesel feedstock. *Energy*, *33*(11), 1646–1653. doi:10.1016/j.energy.2008.06.002

Jena, J., & Misra, R. D. (2014). Effect of fuel oxygen on the energetic and exergetic efficiency of a compression ignition engine fuelled separately with palm and karanja biodiesels. *Energy*, *68*, 411–419. doi:10.1016/j.energy.2014.02.079

Jindal, S., & Goyal, K. (2012). Evaluation of performance and emission of *Hibwascus cannabinus* (Ambadi) seed oil biodiesel. *Clean Technologies and Environmental Policy*, *14*(4), 633–639. doi:10.100710098-011-0426-5

Kalam, M. A., & Masjuki, H. H. (2008). Testing palm biodiesel and NPAA additives to control NOx and CO while improving efficiency in diesel engines. *Biomass and Bioenergy*, *32*(12), 1116–1122. doi:10.1016/j.biombioe.2008.02.009

Kannan, G. R., & Anand, R. (2012). Effect of injection pressure and injection timing on DI diesel engine fuelled with biodiesel from waste cooking oil. *Biomass and Bioenergy*, *46*, 343–352. doi:10.1016/j.biombioe.2012.08.006

Karande, P., & Chakraborty, S. (2012). Application of multi-objective optimization on the basis of ratio analysis (MOORA) method for materials selection. *Materials & Design*, *37*(1), 317–324. doi:10.1016/j.matdes.2012.01.013

Liaquat, A. M., Masjuki, H. H., Kalam, M. A., Fattah, I. M. R., Hazrat, M. A., Varman, M., ... Shahabuddin, M. (2013). Effect of coconut biodiesel blended fuels on engine performance and emission characteristics. *Procedia Engineering*, *56*, 583–590. doi:10.1016/j.proeng.2013.03.163

Lin, L., Cunshan, Z., Vittayapadung, S., Xiangqian, S., & Mingdong, D. (2011). Opportunities and challenges for biodiesel fuel. *Applied Energy*, *88*(4), 1020–1031. doi:10.1016/j.apenergy.2010.09.029

Madiwale, S., Karthikeyan, A., & Bhojwani, V. (2018). Properties investigation and performance analysis of a diesel engine fuelled with Jatropha, Soybean, Palm and Cottonseed biodiesel using Ethanol as an additive. *Materials Today: Proceedings*, *5*(1), 657–664. doi:10.1016/j.matpr.2017.11.130

Ndayishimiye, P., & Tazerout, M. (2011). Use of palm oil based biofuel in the internal combustion engines. *Energy*, *36*(3), 1790–1796. doi:10.1016/j.energy.2010.12.046

Ng, H. K., & Gan, S. (2010). Combustion performance and exhaust emissions from the non-pressurised combustion of palm oil biodiesel blends. *Applied Thermal Engineering*, *30*(16), 2476–2484. doi:10.1016/j.applthermaleng.2010.06.020

Ng, J. H., Ng, H. K., & Gan, S. (2012). Characterisation of engine-out responses from a light duty diesel engine fuelled with palm methyl ester (PME). *Applied Energy*, *90*(1), 58–67. doi:10.1016/j.apenergy.2011.01.028

Nwafor, O. M. I. (2004). Emission characteristics of diesel engine operating on rapeseed methyl ester. *Renewable Energy*, *29*(1), 119–129. doi:10.1016/S0960-1481(03)00133-2

Ong, H. C., Silitonga, A. S., Masjuki, H. H., Mahlia, T. M. I., Chong, W. T., & Boosroh, M. H. (2013). Production and comparative fuel properties of biodiesel from non-edible oils: Jatropha curcas, Sterculia foetida and Ceiba pentandra. *Energy Conversion and Management*, *73*, 245–255. doi:10.1016/j.enconman.2013.04.011

Özcanli, M., Keskin, A., & Aydın, K. (2011). Biodiesel production from terebinth (Pwastacia terebinthus) oil and its usage in diesel engine. *International Journal of Green Energy*, *8*(5), 518–528. doi:10.1080/15435075.2011.588766

Perumal, V., & Ilangkumaran, M. (2017). Experimental analysis of engine performance, combustion and emission using pongamia biodiesel as fuel in CI engine. *Energy*, *129*, 228–236. doi:10.1016/j.energy.2017.04.120

Plata, V., Gauthier-Maradei, P., Romero-Bohórquez, A. R., Kafarov, V., & Castillo, E. (2015). Characterization of insoluble material isolated from Colombian palm oil biodiesel. *Biomass and Bioenergy*, *74*, 6–14. doi:10.1016/j.biombioe.2014.12.024

Qi, D. H., Chen, H., Geng, L. M., & Bian, Y. Z. (2010). Experimental studies on the combustion characteristics and performance of a direct injection engine fueled with biodiesel/diesel blends. *Energy Conversion and Management*, *51*(12), 2985–2992. doi:10.1016/j.enconman.2010.06.042

Rahman, S. M. A., Masjuki, H. H., Kalam, M. A., Abedin, M. J., Sanjid, A., & Sajjad, H. (2013). Impact of idling on fuel consumption and exhaust emissions and available idle-reduction technologies for diesel vehicles – a review. *Energy Conversion and Management*, *74*, 171–182. doi:10.1016/j.enconman.2013.05.019

Ramadhas, A. S., Jayaraj, S., & Muraleedharan, C. (2005). Biodiesel production from high FFA rubber seed oil. *Fuel*, *84*(4), 335–340. doi:10.1016/j.fuel.2004.09.016

Ramesh, W. V., & Sakthivel, G. (2018). Selection Of Optimum Fuel Blend In An Ic Engine Using FAHP-PROMETHEE. *IOP Conf. Series: Materials Science and Engineering*, *390*(1), 1-12. 10.1088/1757-899X/390/1/012028

Rao, G. V., Reddy, C., & Chakravarthy, V. V. K. (2016). Parthasarthy. Performance and emissions of diesel engine fuelled with ethanol and palm oil methyl ester. *Intenational Journal of Engineering and Technology*, *3*, 108–113.

Rao, R. R., & Kalyankar, V. D. (2014). Optimization of Modern Machining Processes Using Advanced Optimization Techniques: A Review. *International Journal of Advanced Manufacturing Technology*, *47*(5-8), 1159–1188. doi:10.100700170-014-5894-4

Reijnders, L., & Huijbregts, M. A. J. (2008). Palm oil and the emission of carbon based greenhouse gases. *Journal of Cleaner Production*, *16*(4), 477–482. doi:10.1016/j.jclepro.2006.07.054

Sahoo, P. K., Naik, S. N., Babu, M. K. G., & Das, L. M. (2007). Biodiesel development from high acid value polanga seed oil and performance in a CI engine. *Fuel*, *86*(3), 448–454. doi:10.1016/j.fuel.2006.07.025

Sakthivel, G., & Ilangkumaran, M. (2017). Optimisation of compression ignition engine performance with fishoil biodiesel using Taguchi-Fuzzy approach. *Journal International Journal of Ambient Energy*, *38*(2), 146–160.

Sakthivel, G., Ilangkumaran, M., & Gaikwad, A. (2015). A hybrid multi-criteria decision modeling approach for the best biodiesel blend selection based on ANP-TOPSIS analysis. *Ain Shams Engineering Journal*, *6*(1), 239–256. doi:10.1016/j.asej.2014.08.003

Salvi, B.L., & Jindal, S. (2013). A Comparative Study of engine performance and exhaust emission characteristics of linseed oil biodiesel blends with diesel fuel in a direct injection diesel engine. *Journal of The Institution of Engineers (India): Series C, 94*(1), 1–8.

Satyanarayana, K., Mariano, A., & Vargas, J. (2011). A review on microalgae, a versatile source for sustainable energy and materials. *International Journal of Energy Research*, *35*(4), 291–31. doi:10.1002/er.1695

Senthil, R., & Silambarasan, R. (2015). Annona: A new biodiesel for diesel engine: A comparative experimental investigation. *Journal of the Energy Institute*, *88*(4), 459–469. doi:10.1016/j.joei.2014.09.011

Sharma, A., Muqeem, M., Sherwani, A., & Ahmad, M. (2018). Optimization of diesel engine input parameters running on Polanga biodiesel to improve performance and exhaust emission using MOORA technique with standard deviation. *Energy Sources. Part A, Recovery, Utilization, and Environmental Effects*, *40*(22), 1–18. doi:10.1080/15567036.2018.1511647

Varatharajan, K., & Cheralathan, M. (2013). Effect of aromatic amine antioxidants on NOx emissions from a soybean biodiesel powered DI diesel engine. *Fuel Processing Technology*, *106*, 526–532. doi:10.1016/j.fuproc.2012.09.023

Yazdani, M., Chatterjee, P., Zavadskas, E. K., & Streimikiene, D. (2018). A novel integrated decision-making approach for evaluation and selection of renewable energy technologies. *Clean Technologies and Environmental Policy*, *20*(2), 403–420. doi:10.100710098-018-1488-4

Zhang, Y., Dube, M. A., McLean, D. D., & Kates, M. (2003). Biodiesel production from waste cooking oil: 1. Process design and technological assessment. *Bioresource Technology*, *89*(1), 1–16. doi:10.1016/S0960-8524(03)00040-3 PMID:12676496

Chapter 7
Application of Modified Similarity-Based Method for Cotton Fiber Selection

Kanika Prasad
National Institute of Technology Jamshedpur, India

Rishi Dwivedi
Xavier Institute of Social Service, India

ABSTRACT

In today's business environment, sustainability is one of the most significant issues that organizations needs to deal with. Fiber choice is often the initial stride that designers and product developers will consider in minimizing the ecological impact of a garment while achieving sustainability. Cotton fiber, whether alone or blended with some other fiber, is being extensively used in textile industries. However, the selection of optimal cotton fiber depends on its several properties like length, strength, fineness, length uniformity, short fiber content, etc., thus making this selection process s multi-criteria decision-making problem. Thus, for the first time, this chapter examines the applicability and feasibility of a modified similarity-based approach for selection of cotton fiber which in turn assists in achieving sustainable design characteristics. Two illustrative examples from past researches and one existing cotton fiber selection problem of a small textile organization are solved applying the adopted method to validate its veracity and robustness.

DOI: 10.4018/978-1-5225-8579-4.ch007

INTRODUCTION

Global trade in textiles and garments has played a vital role in the progress of numerous nations and has also assisted their assimilation in to the worldwide financial system. In the industrialized countries, the procedure of development and ensuing opulence in a way started with the automation of textile manufacturing in the early 19th Century. However, in the developing nations, the said industry has occupied an imperative position among all other sectors based on its contribution to national output, employment and exports. On the basis of a report published by India Brand Equity Foundation in the year 2018, Indian textiles industry is currently computed to be at around US$ 150 billion, and anticipated to touch US$ 250 billion by 2019. It added 7% of the industry output of India in 2017-18. India's textiles industry contributed 2% to its national GDP and provided employment to more than 45 million people in 2017-18. Besides, in 2017-18, 15% of the export earnings of India are contributed through the said industry. With rapidly changing business ecosystem, the textile and clothing industry needs to undergo structural changes globally. Organisations in this sector need to adopt sustainable and flexible production systems for their production lines. Additionally, global sustainable structure is being characterized by higher potential for profit from innovation, marketing, and retailing rather than assembling, finishing and packaging alone. Moreover, with growing environmental consciousness in recent years, the application of renewable raw materials as well as eco-friendly and sustainable fibers has become very significant from the textile industry's perspective. Hence, selection of long lasting fiber provides a sustainable approach to said sector. In addition, the manufacturing of numerous synthetic fibers contaminates the eco system. Thus, for managers of textile enterprises, it is indispensable to bestow importance to selection of fiber while achieving sustainability. Amongst all different types of fibers used in the textile industries, cotton fiber is the most widely used natural fiber. Its natural schema and features yield functional and marketable benefits for consumer, institutional and industrial products. Cotton fiber almost constitutes half of the textile and clothing materials. Since, the textile industries in India are predominantly cotton-based and cotton fiber is their mainstay/backbone, it is very important to acquire the best quality of fiber for manufacturing a specified product. Not only textiles, but industries as varied as consumer products, healthcare and automotive appreciate the favorable outcomes of including cotton fiber in their diverse product segments because of its dependability and adaptability for assorted applications. In different regions of India several varieties of cotton fibers are cultivated. These zones differ from each other in various aspects, like topography, type of soil and irrigation facilities that ultimately affect the quality and yield/productivity of cotton fibers. The quality of cotton bale is determined by its various physical properties, such as length, length uniformity, strength, amount of trash, color grade etc. Besides, the cotton fibers utilized for weaving/knitting the yarns greatly affect the value and utility of textile/cloth produced. Therefore, identification of the most appropriate cotton fiber from a vast variety of available options is essential for attaining sustainability and advanced quality control in textile production. However, many physical and chemical properties influence the selection of cotton fiber and there may exist complex interrelationship among these selection criteria. For example, longer fibers allow higher spinning speed and also yield quality end product by producing stronger yarns, but a higher value of micronaire fiber becomes a limiting factor in spinning fine yarn. Therefore, evaluation of available cotton fiber alternatives based on influencing criteria/properties makes the selection process a multi-criteria decision-making (MCDM) problem. Several MCDM methods which can deal with this type of selection problem have been discussed in past. The objective of this paper is to apply a distance-based approach modified similarity-based method for selection of the most suitable cotton fiber from various alternatives available. Options in this method are evaluated with respect to positive and negative ideal solutions on a comprehensive concept. The overall performance index of all choices across each criterion is estimated based on the combination of the degree of similarity to positive and negative ideal solutions employing alternative gradient and magnitude. Lastly, cotton fiber alternative with highest performance index is identified as best option on the basis of selected criteria.

LITERATURE REVIEW

Numerous approaches have been developed and applied to solve MCDM problems. Although the literature covers a wide variety of such approaches, this review primarily focuses on two major refrains/themes which appear repetitively throughout the literature reviewed. The first part contains review on application of different MCDM techniques for cotton fiber selection and in other related fields, whereas the other part deals with implementation of similarity-based approach in different domains.

Majumdar et al. (2004) utilized analytic hierarchy process (AHP) technique to evaluate eight cotton fiber alternatives based on six properties to maximize strength of ring spun yarn. Majumdar et al. (2005) integrated AHP and technique for order preference by similarity to ideal solution (TOPSIS) methods for ranking 33 cotton fiber alternatives based on six evaluation criteria. The AHP and TOPSIS methods were applied to determine the criteria weights and rank the fiber alternatives respectively. A comparison between the said integrated approach with the other traditional methods was also presented. Majumdar et al. (2005) presented a multiplicative AHP model to determine the technological values of cotton fibers. The work also compared the performance of model with respect to three other indexing approaches, i.e. fiber quality index (FQI), spinning consistency index (SCI) and premium-discount index (PDI). Majumdar et al. (2006) applied TOPSIS method for grading, categorization and lay down of cotton fibers, and established its applicability as an effective MCDM tool that could be utilized to determine the overall quality of cotton fiber. Majumdar (2010) determined the most appropriate cotton fiber while integrating fuzzy logic with AHP technique to take the vague acuity of the decision makers and qualitative values of different fiber properties into consideration. Majumdar et al. (2010) designed and developed a decision support system (DSS) based on AHP and TOPSIS methods for cotton fiber ranking and selection. Majumdar and Singh (2014) presented a novel approach based on genetic algorithm (GA) for estimating the quality of cotton fibers and studied the fiber quality on the basis of two yarn properties, i.e. tenacity and unevenness. Banwet and Majumdar (2014) compared the relative performance of two integrated approaches, i.e. AHP-TOPSIS and GA-TOPSIS, and propounded that both the combined approaches would outperform the existing methods for cotton fiber selection and subsequent grading. Ghorabaee et al. (2015) introduced a new method evaluation based on distance from average solution (EDAS) for multi-criteria inventory classification problems. A comparative analysis was also performed to show the validity and stability of the proposed method in MCDM problems. Chakraborty et al. (2017) adopted a grey fuzzy logic-based approach for choosing the most apt cotton fiber from 17 alternatives based on six fiber properties. Chakraborty and Bandhopadyay (2017) combined preference ranking organization method for enrichment of evaluations (PROMETHEE) and geometrical analysis for interactive aid (GAIA) to decipher a cotton fiber selection problem. Chatterjee and Chakraborty (2017) combined design of experiments with TOPSIS method to develop a regression meta-model for determining the technological values of cotton fibers. Khandekar and Chakraborty (2017) developed a software prototype based on fuzzy axiomatic design principles to recognize the most suitable cotton fiber for ring spun yarn from a set of accessible alternatives. Stević et al. (2017) carried out supplier selection in a construction company, where weight coefficients were obtained using decision making trial and evaluation laboratory (DEMATEL) method, based on the rough numbers. Badi et al. (2018) proposed combinative distance-based assessment (CODAS) method to evaluate the suppliers of raw materials to the Libyan iron and steel company. Chakraborty and Prasad (2018) designed an expert system based on quality function deployment (QFD) technique to incorporate customers' perception in the selection process. Chakraborty et al. (2018) integrated two MCDM methods DEMATEL and VIKOR for selection and evaluation of cotton fibers. Mukhametzyanov and Pamucar (2018) presented a model for result consistency evaluation of MCDM methods and selection of the optimal one. The study also demonstrated the results of an analysis of the sensitivity of decision-making based on various rank methods with variations in the elements in the decision matrix within a given error/imprecision.

It is evident from the above literature review that selecting an appropriate cotton fiber is of great importance in textile industries, and numerous MCDM methods have already been described to solve this decision-making problem. Moreover, similarity-based approach has been successfully applied for ranking a fleet of jet fighters (Deng, 2007), ranking banks (Safari et al., 2013), personnel selection (Chaghooshi et al., 2014), ranking countries with respect to human development index (Safari and Ebrahimi, 2014), evaluating economic companies (Moradi and Ebrahimi, 2014), multi-objective optimization in drilling operation (Sonkar et al., 2014), cutting fluid selection (Prasad and Chakraborty, 2018) etc. Therefore, for the first time, this paper evaluates the applicability and potentiality of a modified similarity-based approach in textiles sector for selection of cotton fiber in consistency with the application requirements.

METHODOLOGY

Numerous MCDM approaches have been employed previously for solving cotton fiber selection problem, which requires simultaneous consideration of several factors instead of single criterion. TOPSIS is one of the most widely used MCDM method due to various advantages associated with it. It is a distance-based approach and considers Euclidean Distance for evaluating the performance of alternatives. However, a major drawback of TOPSIS method is its application of Euclidean Distance which does not consider correlation of attributes into consideration (Velasquez and Hester, 2013). In this paper a similarity based approach for solving cotton fiber selection problem is utilized. It is a modified form of TOPSIS methodology, based on ideal solution theory, where the most preferred alternative has the highest degree of similarity to the positive ideal solution (*PIS*) and lowest degree of likeness to the negative ideal solution (*NIS*). Alternative gradient and magnitude is incorporated in this modified similarity based technique to calculate the ultimate performance index of each alternative on diverse criteria. A multi-criteria selection problem encompasses an array of alternatives, and there may be existence of conflict between two options. Among several ways of reflecting this conflict, the concept of alternative gradient is the most frequent one. Therefore, the concept of variable's gradient is also employed to calculate the conflicting index as it can explain conflicts between alternatives in MCDM problems very clearly. The conflicting index amid two choices shows the degree of conflict between them (Safari et al., 2013).

If in a given MCDM problem, A_i and A_j denotes two alternatives, they can be identified as two vectors, and the angle flanked by them in the m-dimensional real space represents the conflict between them. Degree of conflict amongst alternatives A_i and A_j is calculated employing the following equation:

$$cos\,\theta_{ij} = \frac{\sum_{k=1}^{m} x_{ik} x_{jk}}{\sqrt{\left(\sum_{k=1}^{m} x_{ik}^{2}\right)}\sqrt{\left(\sum_{k=1}^{m} x_{jk}^{2}\right)}} \tag{1}$$

where θ_{ij} shows the angle flanked by the gradients of alternatives, and $(x_{i1},x_{i2},\ldots,x_{im})$ and $(x_{j1},x_{j2},\ldots,x_{jm})$ are the gradients of two options A_i and A_j respectively.

The conflict index represented by $\theta_{ij} = 0$ shows that the consequent gradient vectors lie in the same direction of advancement. In the similar way, the conflict index indicated by $\theta_{ij} = \pi/2$ represents that their gradient vectors have vertical association with each other. The steps involved in application of modified similarity-based approach for solving MCDM problems are provided as below:

Step 1: In the primary step, the decision matrix consisting of a set of options A_i $(i = 1,2,\ldots,n)$ to be evaluated on the basis of identified set of criteria C_j $(j = 1,2,\ldots,m)$ is developed.

$$X = \begin{bmatrix} x_{11} & x_{12} & \cdots & x_{1m} \\ x_{21} & x_{22} & \cdots & x_{2m} \\ \vdots & \vdots & \vdots & \vdots \\ x_{n1} & x_{n2} & \cdots & x_{nm} \end{bmatrix} \tag{2}$$

Step 2: Weight of each criterion needs to be assigned in this step. Let w_j denotes the relative significance of criterion C_j with respect to the ultimate objective of the selection problem.

$$W = (w_1, w_2, \ldots, w_m)\,(j = 1,2,\ldots,m) \tag{3}$$

Step 3: Next, the decision matrix is normalized to make all the values in it dimensionless so that they can be compared with each other. The ranking of alternatives begins with normalizing the decision matrix through the vector normalization technique.

$$x'_{ij} = \frac{x_{ij}}{\sqrt{\left(\sum_{j=1}^{m} x_{ij}^2\right)}} \tag{4}$$

Therefore, the normalized matrix can be derived as follows:

$$X' = \begin{bmatrix} x'_{11} & x'_{12} & \cdots & x'_{1m} \\ x'_{21} & x'_{22} & \cdots & x'_{2m} \\ \vdots & \vdots & \vdots & \vdots \\ x'_{n1} & x'_{n2} & \cdots & x'_{nm} \end{bmatrix} \tag{5}$$

Step 4: The performance of all options with respect to the evaluation criterion is estimated through multiplying the normalized decision matrix with the weights. Thus, the resultant matrix so obtained is termed as performance matrix.

$$X'_w = \begin{bmatrix} w_1 x'_{11} & w_2 x'_{12} & \cdots & w_3 x'_{1m} \\ w_1 x'_{21} & w_2 x'_{22} & \cdots & w_3 x'_{2m} \\ \vdots & \vdots & \vdots & \vdots \\ w_1 x'_{n1} & w_2 x'_{n2} & \cdots & w_3 x'_{nm} \end{bmatrix} = \begin{bmatrix} y_{11} & y_{12} & \cdots & y_{1m} \\ y_{21} & y_{22} & \cdots & y_{2m} \\ \vdots & \vdots & \vdots & \vdots \\ y_{n1} & y_{n2} & \cdots & y_{nm} \end{bmatrix} \tag{6}$$

Step 5: In this step, estimating the *PIS* and *NIS* is on the basis of the concept that they comprise of the best (or worst) criteria values computable from all the alternatives if each criterion takes monotonically rising or diminishing values. For solving real time decision making problems, this notion has been extensively employed in assorted multi-criteria analysis models. *PIS* and *NIS* is estimated utilizing the equation given below:

$$\begin{cases} A_j^+ = \max\ y_{ij} \\ A_j^- = \min\ y_{ij} \end{cases} \qquad i = 1, 2, \ldots, n;\ j = 1, 2, \ldots, m \tag{7}$$

Step 6: The degree of conflict between each alternative A_i and the *PIS* and *NIS* in the m-dimensional real space is calculated in the next step. θ_i^+ (θ_i^-) is the angle flanked by A_i and A_j^+ (A_j^-) in the *m*-dimensional real space, and an excellent measure of divergence amid the vectors. The degree of conflict amongst the options is estimated employing the following equations:

$$\begin{cases} \cos\theta_i^+ = \dfrac{\sum_{j=1}^m y_{ij} A_j^+}{\sqrt{(\sum_{j=1}^m y_{ij}^2)}\ \sqrt{(\sum_{j=1}^m A_j^{+2})}} \\ \cos\theta_i^- = \dfrac{\sum_{j=1}^m y_{ij} A_j^-}{\sqrt{(\sum_{j=1}^m y_{ij}^2)}\ \sqrt{(\sum_{j=1}^m A_j^{-2})}} \end{cases} \quad i = 1, 2, ..., n \tag{8}$$

Step 7: Computing the degree of similarity amid each pair of alternative, and the *PIS* and *NIS* is the succeeding step after the degree of conflict is estimated. S_i^+ and S_i^- represents the degrees of similarity of the alternative A_i to A_j^+ and alternative A_i to A_j^- respectively. Prior to estimating the similarity index, the following aspect concerning Deng's similarity technique required to be discussed. It is derived from Figure 2 that *p* is the projection of alternative vector A_i on the *PIS* vector A_j^+. The A_i vector is equivalent or smaller than *PIS*, as the *PIS* has the biggest value among the alternatives. Consequently, values for S_i^+ range in the midst of 0 and 1. In the same way, *q* is the projection of alternative vector A_i on the *NIS* vector A_j^-. However, *NIS* has the minimum value amongst the alternatives, and therefore *q* is equal or longer than A_j^-. This is the main area of concern in Deng's method. Therefore, additional vector is required with a view to resolve the said problem. *r* is the projection of the *NIS* vector A_j^- on the alternative vector A_i that is at all times smaller than the alternative vector, as shown in Figure 2 (Safari and Ebrahimi, 2014). Thus, the similarity index is estimated using following equation:

$$\begin{aligned} S_i^+ &= \frac{\cos\theta_i^+ \times \sqrt{(\sum_{j=1}^m y_{ij}^2)}}{\left|A_j^+\right|} \\ S_i^- &= \frac{\left|A_j^-\right|}{\cos\theta_i^- \times \sqrt{(\sum_{j=1}^m y_{ij}^2)}} \end{aligned} \tag{9}$$

Step 8: In the last step of this modified similarity-based method, the overall performance score for each alternative with respect to the identified criteria is computed. The alternative with the higher performance index would be preferred.

$$P_i = \frac{S_i^+}{S_i^+ + S_i^-} (i = 1,2,\ldots,n) \tag{10}$$

In this technique, as the angle flanked by alternative A_i and *NIS* (*PIS*) is not comprehensible when it is equivalent to *PIS* (*NIS*), the value for the overall performance score is different than one (zero). Lastly, the rank for each option is estimated in diminishing order of their performance scores.

ILLUSTRATIVE EXAMPLES

With a view to prove the applicability, sturdiness and solution accuracy of the modified similarity index based method, two real time cotton fiber selection problems from the literature are solved. Then, veracity of the developed model is demonstrated while selecting cotton fiber for a real time small scale textile enterprise.

Example 1

Chakraborty and Prasad (2018) considered a cotton fiber selection problem for producing ring spun yarns to be used for developing fabrics for household purposes and solved it while using a QFD-based expert system. This selection problem evaluates nine alternatives on four criteria, i.e. fiber strength (in CN/Tex), uniformity index, micronaire and yellowness. Among these four criteria; fiber strength and uniformity index are beneficial attributes for which higher value is desirable, whereas micronaire and yellowness are non-beneficial attributes for which lower values are preferred. The related decision matrix for the selection problem is shown in Table 1. In addition, the last row of Table 1 contains the priority weights of all the criteria considered for evaluation which are computed using QFD methodology. Chakraborty and Prasad (2018) identified Giza 87 (A_4) as the most preferred choice for the defined application and Sille1 (Stone Vile) (A_9) emerged as the least preferred choice. The ranking preorder of the considered alternative cotton fibers was derived as 4-6-7-1-2-8-5-3-9.

While solving the problem using modified similarity index-based method, the initial step is to obtain the normalized decision matrix. The normalized values within a scale of 0 to 1 are estimated for the decision matrix employing the vector normalization technique. The corresponding normalized values are displayed in Table 2.

Table 1. Decision matrix for example 1

Cotton	Fiber Strength (C_1)	Micronaire (C_2)	Uniformity Index (C_3)	Yellowness (C_4)
1007 (A_1)	26.02	3.28	83.60	8.93
Estamble (A_2)	24.02	3.73	85.05	8.78
G. Cot 13 (A_3)	22.53	3.25	82.33	8.80
Giza 87 (A_4)	46.30	3.14	87.15	9.40
Giza 92 (A_5)	46.56	3.91	86.73	8.78
Laxmi (A_6)	23.73	3.43	82.53	8.75
Mech1 (A_7)	24.05	3.28	83.48	8.83
Sankar 6 (A_8)	28.71	3.13	83.55	8.50
Sille1 (stone vile) (A_9)	23.31	3.90	84.00	8.35
Weight	0.3166	0.2555	0.3537	0.0742

Table 2. Normalized decision matrix for example 1

Cotton	C_1	C_2	C_3	C_4
A_1	0.2809	0.3157	0.3306	0.3384
A_2	0.2593	0.3591	0.3364	0.3328
A_3	0.2432	0.3129	0.3256	0.3335
A_4	0.4998	0.3023	0.3447	0.3562
A_5	0.5026	0.3764	0.3430	0.3328
A_6	0.2562	0.3302	0.3264	0.3316
A_7	0.2596	0.3157	0.3302	0.3346
A_8	0.3099	0.3013	0.3304	0.3221
A_9	0.2516	0.3754	0.3322	0.3165
Weight	0.3166	0.2555	0.3537	0.0742

Subsequently, following the steps of the modified similarity-based approach as discussed in the preceding section, the columns of the normalized decision matrix are multiplied by the related criteria weights to obtain the corresponding performance matrix. The criteria weights for fiber strength, micronaire, uniformity index, and yellowness are considered as 0.3166, 0.2555 0.3537 and 0.0742 respectively, as computed from the QFD-based approach (Chakraborty and Prasad, 2018). Now, the degree of conflict among each alternative and *PIS* ($\cos\theta^+$) and *NIS* ($\cos\theta^-$) is computed from the performance matrix. The degree of similarity between the alternatives and *PIS* (S^+) and *NIS* (S^-) is then estimated. At last, the overall performance score, P for all options is calculated employing Equation (10), and each and every value thus computed are arranged in Table 3.

In this example, A_5 is identified as the best material for developing sheets for household usages on the basis of customers' particular requirements and A_4 is the second-best choice, as exhibited in Table 3. However, there is a very small differentiation between them with respect to their performance scores (score for $A_5 = 0.5707$ and score for $A_4 = 0.5510$). The rankings of these two cotton fiber alternatives at the top two positions precisely match with those of Chakraborty and Prasad (2018). From this method, the ranking of the alternative cotton fibers is derived as 6-4-9-2-1-7-8-3-5. It can be observed that more or less comparable type of ranking patterns is obtained by the past researchers (Chakraborty and Prasad, 2018). Moreover, a Spearman's rank correlation coefficient (r_s) of 0.6667 between modified similarity-based method and QFD-based technique signifies a good corroboration between the past and present study. This further authenticates similarity-based method's flexibility and potentiality as an almost accurate and complete MCDM tool.

Example 2

For solving this selection problem employing modified similarity-based approach, normalized decision matrix is first obtained. Table 5 exhibits the normalized decision matrix. With respect to the performance scores of 17 alternatives as shown in Table 6, their ranking order is computed as 5-3-6-1-14-8-16-7-2-17-4-12-15-13-

Table 3. Weighted normalized matrix for example 1

Cotton	C_1	C_2	C_3	C_4	Cos θ⁺	Cos θ⁻	S⁺	S⁻	P	Rank
A_1	0.0889	0.0807	0.1169	0.0251	0.9706	0.9986	0.7347	0.9467	0.4370	6
A_2	0.0821	0.0917	0.1190	0.0247	0.9565	0.9983	0.7389	0.9279	0.4433	4
A_3	0.0770	0.0799	0.1152	0.0248	0.9558	0.9999	0.6909	0.9901	0.4110	9
A_4	0.1582	0.0772	0.1219	0.0264	0.9970	0.9484	0.9609	0.7829	0.5510	2
A_5	0.1591	0.0962	0.1213	0.0247	1.0000	0.9558	0.9977	0.7504	0.5707	1
A_6	0.0811	0.0843	0.1154	0.0246	0.9604	0.9993	0.7130	0.9646	0.4250	7
A_7	0.0822	0.0807	0.1168	0.0248	0.9620	0.9998	0.7128	0.9661	0.4246	8
A_8	0.0981	0.0770	0.1169	0.0239	0.9807	0.9946	0.7560	0.9334	0.4475	3
A_9	0.0797	0.0959	0.1175	0.0235	0.9516	0.9962	0.7350	0.9300	0.4414	5

Table 4. Decision matrix for example 2

Cotton	C_1	C_2	C_3	C_4	C_5	C_6	C_7	C_8	C_9	C_{10}	C_{11}	C_{12}	C_{13}
F_1	28.7	6.5	1.09	81.0	4.4	79.9	8.0	0.90	13.8	0.84	338	14	30
F_2	28.5	6.6	1.15	80.2	3.5	77.4	9.3	0.97	11.9	0.82	482	16	61
F_3	28.7	5.7	1.10	79.2	3.7	80.3	8.6	0.86	18.4	0.82	442	22	32
F_4	30.8	6.4	1.13	82.6	4.3	74.5	9.7	0.98	9.80	0.89	257	29	134
F_5	26.5	5.8	1.09	87.5	3.8	77.4	10	0.92	8.40	0.89	233	16	106
F_6	27.5	6.3	1.07	82.8	4.5	75.1	9.8	0.95	8.40	0.88	267	16	44
F_7	29.2	5.3	0.98	80.0	4.5	77.5	9.3	0.81	16.9	0.90	281	18	76
F_8	29.0	6.7	1.05	81.9	4.2	78.2	9.3	0.91	10.9	0.84	337	26	77
F_9	30.3	6.7	1.10	83.2	4.4	77.5	9.0	0.96	8.70	0.87	344	41	65
F_{10}	28.1	6.3	1.01	80.7	3.8	78.0	8.8	0.83	15.5	0.81	420	20	84
F_{11}	30.6	6.6	1.07	83.1	4.7	77.2	8.8	0.93	9.00	0.87	189	18	85
F_{12}	28.7	6.7	1.05	81.0	3.9	74.3	11.4	0.88	11.8	0.84	375	27	121
F_{13}	28.3	6.5	0.97	81.5	3.8	75.9	8.3	0.89	13.1	0.81	545	32	172
F_{14}	29.0	6.6	1.06	80.7	3.1	78.1	8.8	0.89	11.3	0.87	270	21	222
F_{15}	27.7	5.5	1.05	81.5	4.7	73	9.0	0.89	11.7	0.86	321	15	110
F_{16}	29.1	6.1	1.05	81.7	4.0	76	8.9	0.9	11.2	0.85	267	16	98
F_{17}	28.6	5.7	1.04	82.4	4.2	78.1	9.5	0.93	10.8	0.83	352	16	60
Weight	0.1723	0.0339	0.2190	0.0756	0.1022	0.0197	0.0207	0.1670	0.0619	0.0422	0.0283	0.0281	0.0289

Table 5. Normalized decision matrix for example 2

Cotton	C_1	C_2	C_3	C_4	C_5	C_6	C_7	C_8	C_9	C_{10}	C_{11}	C_{12}	C_{13}
F_1	0.2490	0.2586	0.2559	0.2475	0.2680	0.2597	0.2169	0.2485	0.2811	0.2458	0.2426	0.1530	0.0703
F_2	0.2473	0.2626	0.2700	0.2451	0.2132	0.2516	0.2522	0.2678	0.2424	0.2400	0.3459	0.1749	0.1429
F_3	0.2490	0.2268	0.2583	0.2420	0.2253	0.2610	0.2332	0.2374	0.3748	0.2400	0.3172	0.2405	0.0750
F_4	0.2672	0.2546	0.2653	0.2524	0.2619	0.2421	0.2630	0.2706	0.1996	0.2605	0.1844	0.3170	0.3139
F_5	0.2299	0.2307	0.2559	0.2674	0.2314	0.2516	0.2712	0.2540	0.1711	0.2605	0.1672	0.1749	0.2483
F_6	0.2386	0.2506	0.2512	0.2530	0.2741	0.2441	0.2657	0.2623	0.1711	0.2575	0.1916	0.1749	0.1031
F_7	0.2533	0.2108	0.2301	0.2445	0.2741	0.2519	0.2522	0.2236	0.3443	0.2634	0.2017	0.1968	0.1781
F_8	0.2516	0.2665	0.2465	0.2503	0.2558	0.2542	0.2522	0.2512	0.2220	0.2458	0.2419	0.2842	0.1804
F_9	0.2629	0.2665	0.2583	0.2543	0.2680	0.2519	0.2441	0.2650	0.1772	0.2546	0.2469	0.4482	0.1523
F_{10}	0.2438	0.2506	0.2371	0.2466	0.2314	0.2535	0.2386	0.2291	0.3157	0.2371	0.3014	0.2186	0.1968
F_{11}	0.2655	0.2626	0.2512	0.2539	0.2862	0.2509	0.2386	0.2568	0.1833	0.2546	0.1356	0.1968	0.1991
F_{12}	0.2490	0.2665	0.2465	0.2475	0.2375	0.2415	0.3091	0.2429	0.2404	0.2458	0.2691	0.2951	0.2835
F_{13}	0.2455	0.2586	0.2277	0.2491	0.2314	0.2467	0.2251	0.2457	0.2668	0.2371	0.3911	0.3498	0.4030
F_{14}	0.2516	0.2626	0.2489	0.2466	0.1888	0.2538	0.2386	0.2457	0.2302	0.2546	0.1938	0.2296	0.5201
F_{15}	0.2403	0.2188	0.2465	0.2491	0.2862	0.2373	0.2441	0.2457	0.2383	0.2517	0.2304	0.1640	0.2577
F_{16}	0.2525	0.2427	0.2465	0.2497	0.2436	0.2470	0.2413	0.2485	0.2281	0.2488	0.1916	0.1749	0.2296
F_{17}	0.2481	0.2268	0.2442	0.2518	0.2558	0.2538	0.2576	0.2568	0.2200	0.2429	0.2526	0.1749	0.1406

Table 6. Weighted normalized matrix for example 2

Cotton	C_1	C_2	C_3	C_4	C_5	C_6	C_7	C_8	C_9	C_{10}	C_{11}	C_{12}	C_{13}	$\cos\theta^+$	$\cos\theta^-$	S^+	S^-	P	Rank
F_1	0.0429	0.0088	0.0560	0.0187	0.0274	0.0051	0.0045	0.0415	0.0174	0.0104	0.0069	0.0043	0.0020	0.9870	0.9954	0.8943	0.8834	0.5031	5
F_2	0.0426	0.0089	0.0591	0.0185	0.0218	0.0050	0.0052	0.0447	0.0150	0.0101	0.0098	0.0049	0.0041	0.9863	0.9963	0.9104	0.8664	0.5124	3
F_3	0.0429	0.0077	0.0566	0.0183	0.0230	0.0051	0.0048	0.0397	0.0232	0.0101	0.0090	0.0068	0.0022	0.9890	0.9898	0.8935	0.8911	0.5007	6
F_4	0.0460	0.0086	0.0581	0.0191	0.0268	0.0048	0.0054	0.0452	0.0124	0.0110	0.0052	0.0089	0.0091	0.9906	0.9953	0.9393	0.8442	0.5266	1
F_5	0.0396	0.0078	0.0560	0.0202	0.0237	0.0050	0.0056	0.0424	0.0106	0.0110	0.0047	0.0049	0.0072	0.9854	0.9968	0.8669	0.9086	0.4883	14
F_6	0.0411	0.0085	0.0550	0.0191	0.0280	0.0048	0.0055	0.0438	0.0106	0.0109	0.0054	0.0049	0.0030	0.9820	0.9962	0.8790	0.8936	0.4959	8
F_7	0.0437	0.0071	0.0504	0.0185	0.0280	0.0050	0.0052	0.0373	0.0213	0.0111	0.0057	0.0055	0.0051	0.9911	0.9880	0.8614	0.9279	0.4814	16
F_8	0.0434	0.0090	0.0540	0.0189	0.0261	0.0050	0.0052	0.0420	0.0137	0.0104	0.0068	0.0080	0.0052	0.9910	0.9964	0.8848	0.8956	0.4970	7
F_9	0.0453	0.0090	0.0566	0.0192	0.0274	0.0050	0.0051	0.0443	0.0110	0.0107	0.0070	0.0126	0.0044	0.9870	0.9942	0.9214	0.8586	0.5176	2
F_{10}	0.0420	0.0085	0.0519	0.0186	0.0237	0.0050	0.0049	0.0383	0.0195	0.0100	0.0085	0.0061	0.0057	0.9942	0.9922	0.8553	0.9335	0.4782	17
F_{11}	0.0457	0.0089	0.0550	0.0192	0.0293	0.0049	0.0049	0.0429	0.0113	0.0107	0.0038	0.0055	0.0058	0.9848	0.9958	0.9042	0.8715	0.5092	4
F_{12}	0.0429	0.0090	0.0540	0.0187	0.0243	0.0048	0.0064	0.0406	0.0149	0.0104	0.0076	0.0083	0.0082	0.9948	0.9946	0.8798	0.9059	0.4927	12
F_{13}	0.0423	0.0088	0.0499	0.0188	0.0237	0.0049	0.0047	0.0410	0.0165	0.0100	0.0111	0.0098	0.0116	0.9977	0.9859	0.8665	0.9306	0.4822	15
F_{14}	0.0434	0.0089	0.0545	0.0186	0.0193	0.0050	0.0049	0.0410	0.0142	0.0107	0.0055	0.0065	0.0150	0.9911	0.9885	0.8759	0.9120	0.4899	13
F_{15}	0.0414	0.0074	0.0540	0.0188	0.0293	0.0047	0.0051	0.0410	0.0148	0.0106	0.0065	0.0046	0.0074	0.9914	0.9929	0.8804	0.9037	0.4935	9
F_{16}	0.0435	0.0082	0.0540	0.0189	0.0249	0.0049	0.0050	0.0415	0.0141	0.0105	0.0054	0.0049	0.0066	0.9901	0.9973	0.8767	0.9023	0.4928	11
F_{17}	0.0428	0.0077	0.0535	0.0190	0.0261	0.0050	0.0053	0.0429	0.0136	0.0103	0.0071	0.0049	0.0041	0.9879	0.9971	0.8768	0.9004	0.4934	10

9-11-10. It is observed from Table 6, that first two ranked cotton fibers estimated through implementation of modified similarity-based technique exactly match with the findings of Banwet and Majumdar (2014). Besides, the value for spearman's correlation coefficient between modified similarity-based method and combined AHP-TOPSIS is obtained as $r_s = 0.60$. Hence, it can be said that the outcomes of said technique are quite robust. By using this method, it is also feasible to review the comparative performance or appropriateness of the cotton fiber alternatives with respect to the selected criterion.

Example 3

Cotton is acknowledged for its flexibility, presentation and innate comfort. It is utilized for manufacturing diverse range of clothes and home wares in addition to industrial purposes, like canvases, tents, hotel sheets and military uniforms. However, more than half of the cotton fiber cultivated is used in the apparel industry to make clothes for wearing. This example illustrates selection of cotton fiber for producing yarns which would be used for making apparels in a small textile enterprise. Durability and comfort of the fabric are two of the important criteria which often influence the selection of clothes. Additionally, the fabric must be resistant to abrasion and free from any kind of imperfection. In this selection problem, 10 cotton fiber alternatives are evaluated based on five criteria. Based on the end requirements of the product, fiber strength (in CN/Tex), short fiber content (in %), seed coat neps, micronaire and uniformity index are chosen as the selection criteria. Here, fiber strength and uniformity index are identified as beneficial criteria, and seed coat neps, micronaire and short fiber content are categorized as non-beneficial attribute. Table 7 shows the decision matrix for this selection problem. In this example Shannon's entropy method (Rao, 2007) is applied for calculating priority weights of the considered criteria. Next, the normalized decision matrix is estimated using vector normalization technique as depicted in Table 8. Table 9 provides the final performance scores for chosen 10 cotton fiber options with respect to five recognized criteria as calculated using modified similarity-based method. It is apparent from the nearer view at the specifications of 'A_9' that it can effectively satisfy all the criteria value. This undoubtedly rationalizes its selection as the best cotton fiber option for producing different types of apparels based on recognized customers' needs.

FUTURE RESEARCH DIRECTIONS

The future scope of this current research work may include application of modified similarity model for selection of best supplier, machinery, dying agents etc. for a

Table 7. Decision matrix for example 3

Cotton	Fiber Strength (C_1)	Uniformity Index (C_2)	Micronaire (C_3)	Short Fiber% (C_4)	Seed Coat Neps (C_5)
A_1	28.7	81	4.4	13.8	14
A_2	28.5	80.2	3.5	11.9	16
A_3	28.7	79.2	3.7	18.4	22
A_4	30.8	82.6	4.3	9.8	29
A_5	26.5	81.5	3.8	8.4	16
A_6	27.5	82.8	4.5	8.4	16
A_7	29.2	80	4.5	16.6	18
A_8	29	81.9	4.2	10.9	26
A_9	30.3	83.2	4.4	8.7	41
A_{10}	28.1	80.7	3.8	15.5	20
Weight	0.2110	0.2115	0.2094	0.1894	0.1787

Table 8. Normalized decision matrix for example 3

Cotton	C_1	C_2	C_3	C_4	C_5
A_1	0.3156	0.3150	0.3373	0.3431	0.1910
A_2	0.3134	0.3119	0.2683	0.2958	0.2183
A_3	0.3156	0.3080	0.2836	0.4574	0.3002
A_4	0.3387	0.3212	0.3296	0.2436	0.3957
A_5	0.2914	0.3169	0.2913	0.2088	0.2183
A_6	0.3024	0.3220	0.3450	0.2088	0.2183
A_7	0.3211	0.3111	0.3450	0.4127	0.2456
A_8	0.3189	0.3185	0.3220	0.2710	0.3548
A_9	0.3332	0.3235	0.3373	0.2163	0.5595
A_{10}	0.3090	0.3138	0.2913	0.3853	0.2729

textile organization in order to identify the superior performing entities of entire value chain of the said enterprise. This in turn will further help mangers of the organization to achieve long term sustainability for them. Moreover, the modified similarity model is developed for selection of cotton fiber for a textile enterprise, but in future, there is enough scope for demonstrating its applicability in other agriculture-based organizations.

Table 9. Weighted normalized decision matrix for example 3

Cotton	C_1	C_2	C_3	C_4	C_5	Cos θ^+	Cos θ^-	S^+	S^-	P	Rank
A_1	0.0666	0.0666	0.0706	0.0650	0.0341	0.9379	0.9862	0.7217	0.8619	0.4557	7
A_2	0.0661	0.0659	0.0562	0.0560	0.0390	0.9510	0.9933	0.6784	0.9232	0.4236	9
A_3	0.0666	0.0651	0.0594	0.0867	0.0536	0.9733	0.9552	0.8114	0.8215	0.4969	2
A_4	0.0715	0.0679	0.0690	0.0462	0.0707	0.9736	0.9767	0.7939	0.8213	0.4915	4
A_5	0.0615	0.0670	0.0610	0.0396	0.0390	0.9355	0.9991	0.6372	0.9613	0.3986	10
A_6	0.0638	0.0681	0.0722	0.0396	0.0390	0.9255	0.9955	0.6695	0.9083	0.4243	8
A_7	0.0678	0.0658	0.0722	0.0782	0.0439	0.9581	0.9737	0.7913	0.8134	0.4931	3
A_8	0.0673	0.0673	0.0674	0.0513	0.0634	0.9769	0.9832	0.7713	0.8427	0.4779	5
A_9	0.0703	0.0684	0.0706	0.0410	0.1000	0.9699	0.9299	0.8721	0.7823	0.5271	1
A_{10}	0.0652	0.0664	0.0610	0.0730	0.0488	0.9709	0.9759	0.7631	0.8528	0.4722	6

DISCUSSION

Every enterprise as well as society progress organically. But, in today's modernized world, development is carried out at rapid speeds, thanks to progression in machineries. The only issue is that not all considers the snags that come along with uneven economic growth including impacts on people's health and surroundings. It's about time populace start to modify their perception on unequal economic development through viewing the world in an entirely diverse manner. Sustainable development is about enhancing the well being of everybody wherever they are and accomplishing this landmark together. It is the practice of utilizing guidelines for environmentally accountable and energy savings to construct innovative development projects and to manage and retrofit older ventures. It can include using green materials in manufacturing, designing projects that can collect their own energy to lessen load on a power grid, or that incorporate selection of best material in order to produce sustainable product while minimizing harmful ecological impact. Cotton is the most widely used natural fiber in the world, and has a variety of usage, like fiber for textiles, cottonseed oil as a food ingredient, and meal and hulls from crushed cotton seeds in feed for livestock, poultry and fish. Therefore, to manufacture textile products of long term sustenance, a framework for selecting more environmental and human friendly cotton fiber is ardently felt. This paper has applied a modified similarity-based model for selecting best suited cotton fiber for a range of scenarios while establishing its role in improving sustainable textile production. The results derived from the application of designed framework could help the mangers across the cotton industry to promote best practices while reducing environmental impact and augmenting sustainability. Selection of best cotton fiber for a variety of products in textile industry through implementation of modified similarity-based methodology could also assist organization in expanding its market access and increasing its incomes.

CONCLUSION

In the perspective of business practices, sustainable development may basically mean that the overall corporate strategies are not self-defeating or treacherous to the organization's long term health and repute. Implementing environmentally sustainable business practices can have diverse noteworthy benefits. Sustainable businesses are time and again the most ingenious for the reason that they are persistently reviewing processes to find new-fangled solutions. The focus on ensuring sustainable development also builds a tradition of answerability throughout the enterprise. Moreover, sustainable business methods are an important component of corporate risk management. For fabric and attire industry, sustainability has been a blazing subject for numerous associated fretful members. Therefore, designers need to be conscious of ecological impacts and incorporate environmental awareness into the design with a view to attain sustainable progress. Product sustainability is the easiest feature to change for an apparel brand, as this is where a corporation has the majority and unswerving control through design and product development. Transforming product sustainability may be achieved by means of a range of aspects, like fiber/textile selection, processing methods, use behaviours, and reuse/recycle strategies. Therefore, this paper emphasizes on examining applicability of modified similarity-based technique in responding the complex character of cotton fiber selection problem and facilitating judicious selection of cotton fiber from an inclusive list of available alternatives. In the first illustrative example, A_5 is identified as the most suitable cotton fiber for producing ring spun yarns to be used for developing fabrics for household purposes and in the second example, for a selection problem with 17 cotton fiber alternatives and 13 criteria, F4 emerges out as the best cotton fiber. The results obtained employing this MCDM technique almost corroborate with those as derived by the past researchers. It displays the viability, relevancy and straightforwardness of the modified similarity-based approach in solving cotton fiber selection decision-making problems. Besides, a real time cotton fiber selection problem of a small textile enterprise involved in manufacturing apparels is also solved using the considered technique. The results derived from application of this modified similarity-based model exactly matches with the practiced choices of the industrial experts. Thus it can be comprehended that implementation of this methodology would serve as an efficient approach for measurement of the aggregate functional performance of cotton fibers so as to achieve product sustainability.

REFERENCES

Badi, I., Abdulshahed, A., & Shetwan, A. (2018). A case study of supplier selection for steel making company in Libya by using combinative distance-based assessment (CODAS) model. *Decision Making: Applications in Management and Engineering*, *1*(1), 1–12.

Banwet, D. K., & Majumdar, A. (2014). Comparative analysis of AHP TOPSIS and GA-TOPSIS methods for selection of raw materials in textile industries. *Proceedings of the International Conference on Industrial Engineering and Operations Management*, 2071-2080.

Chaghooshi, A. J., Janatifar, H., & Dehghan, M. (2014). An application of AHP and similarity-based approach to personnel selection. *International Journal of Business Management and Economics*, *1*(1), 24–32.

Chakraborty, S., & Bandhopadyay, S. (2017). Cotton fiber selection and grading-A PROMETHEE-GAIA-based approach. *International Journal of Clothing Science and Technology*, *29*(5), 646–660. doi:10.1108/IJCST-12-2016-0137

Chakraborty, S., Chatterjee, P., & Prasad, K. (2018). An Integrated DEMATEL-VIKOR Method-based approach for cotton fiber selection and evaluation. *Journal of the Institution of Engineers (India): Series E, 99*(1), 63-73.

Chakraborty, S., Das, P. P., & Kumar, V. (2017). A grey fuzzy logic approach for cotton fiber selection. *Journal of the Institution of Engineers (India): Series E, 98*(1), 1-9.

Chakraborty, S., & Prasad, K. (2018). A quality function deployment-based expert system for cotton fiber selection. *Journal of the Institution of Engineers (India): Series E, 99*(1), 43-53.

Chatterjee, P., & Chakraborty, S. (2017). Development of a metamodel for the determination of technological value of cotton fiber using design of experiments and the TOPSIS method. *Journal of Natural Fibers, 15*(6), 882–895. doi:10.1080/15440478.2017.1376303

Deng, H. (2007). A similarity-based approach to ranking multicriteria alternatives. *International Conference on Intelligent Computing, Lecture Notes in Artificial Intelligence, 4682*, 253-262. 10.1007/978-3-540-74205-0_28

Keshavarz Ghorabaee, M., Zavadskas, E. K., Olfat, L., & Turskis, Z. (2015). Multi-criteria inventory classification using a new method of evaluation based on distance from average solution (EDAS). *Informatica, 26*(3), 435–451. doi:10.15388/Informatica.2015.57

Khandekar, A. V., & Chakraborty, S. (2017). Application of fuzzy axiomatic design principles for cotton fiber selection. *Journal of the Textile Institute, 109*(6), 730–739. doi:10.1080/00405000.2017.1366700

Majumdar, A. (2010). Selection of raw materials in textile spinning industry using fuzzy multi-criteria decision making approach. *Fibers and Polymers, 11*(1), 121–127. doi:10.100712221-010-0121-x

Majumdar, A., Majumdar, P. K., & Sarkar, B. (2005b). Determination of the technological value of cotton fiber: A comparative study of the traditional and multi-criteria decision-making approaches. *AUTEX Research Journal, 5*(2), 71–80.

Majumdar, A., Mangla, R., & Gupta, A. (2010). Developing a decision support system software for cotton fiber grading and selection. *Indian Journal of Fibre and Textile Research, 35*(3), 195–200.

Majumdar, A., Sarkar, B., & Majumdar, P. K. (2004). Application of analytic hierarchy process for the selection of cotton fibers. *Fibers and Polymers, 5*(4), 297–302. doi:10.1007/BF02875528

Majumdar, A., Sarkar, B., & Majumdar, P. K. (2005a). Determination of quality value of cotton fiber using hybrid AHP-TOPSIS method of multi-criteria decision-making. *Journal of the Textile Institute, 96*(5), 303–309. doi:10.1533/joti.2005.0013

Majumdar, A., Sarkar, B., & Majumdar, P. K. (2006). A new algorithm of cotton fiber selection and laydown using TOPSIS method of multi-criteria decision making. *Indian Journal of Fibre and Textile Research, 31*(2), 248–255.

Majumdar, A., & Singh, S. P. (2014). A new approach to determine the quality value of cotton fibers using multi-criteria decision making and genetic algorithm. *Fibers and Polymers*, *15*(12), 2658–2664. doi:10.100712221-014-2658-6

Moradi, M., & Ebrahimi, E. (2014). Applying fuzzy AHP and similarity-based approach for economic evaluating companies based on corporate governance measures. *Global Journal of Management Studies and Researches*, *1*(1), 10–20.

Mukhametzyanov, I., & Pamucar, D. (2018). A sensitivity analysis in MCDM problems: A statistical approach. *Decision Making: Applications in Management and Engineering*, *1*(2), 51–80.

Prasad, K., & Chakraborty, S. (2018). Application of the modified similarity-based method for cutting fluid selection. *Decision Science Letters*, *7*, 273–286. doi:10.5267/j.dsl.2017.8.002

Rao, R. V. (2007). *Decision making in the manufacturing environment using graph theory and fuzzy multiple attribute decision making methods*. London, UK: Springer-Verlag.

Safari, H., & Ebrahimi, E. (2014). Using modified similarity multiple criteria decision making technique to rank countries in terms of human development index. *Journal of Industrial Engineering and Management*, *7*(1), 254–275.

Safari, H., Khanmohammadi, E., Hafezamini, A., & Ahangari, S. S. (2013). A new technique for multi criteria decision making based on modified similarity method. *Middle East Journal of Scientific Research*, *14*(5), 712–719.

Sonkar, V., Abhishek, K., Datta, S., & Mahapatra, S. S. (2014). Multi-objective optimization in drilling of GFRP composites: A degree of similarity approach. *Procedia Materials Science*, *6*, 538–543. doi:10.1016/j.mspro.2014.07.068

Stević, Ž., Pamučar, D., Vasiljević, M., Stojić, G., & Korica, S. (2017). Novel integrated multi-criteria model for supplier selection: Case study construction company. *Symmetry*, *9*(11), 279. doi:10.3390ym9110279

Velasquez, M., & Hester, P. T. (2013). An analysis of multi-criteria decision making methods. *International Journal of Operations Research*, *10*(2), 56–66.

ADDITIONAL READING

Ayağ, Z., & Özdemir, R. G. (2012). Evaluating machine tool alternatives through modified TOPSIS and alpha-cut based fuzzy ANP. *International Journal of Production Economics*, *140*(2), 630–636. doi:10.1016/j.ijpe.2012.02.009

Bevilacqua, M., Ciarapica, F. E., & Giacchetta, G. (2006). A fuzzy-QFD approach to supplier selection. *Journal of Purchasing and Supply Management*, *12*(1), 14–27. doi:10.1016/j.pursup.2006.02.001

Bhattacharya, A., Geraghty, J., & Young, P. (2010). Supplier selection paradigm: An integrated hierarchical QFD methodology under multiple-criteria environment. *Applied Soft Computing*, *10*(4), 1013–1027. doi:10.1016/j.asoc.2010.05.025

Dawal, S. Z. M., Yusoff, N., Nguyen, H.-T., & Aoyama, H. (2013). Multiattribute decision-making for CNC machine tool selection in FMC based on the integration of the improved consistent fuzzy AHP and TOPSIS. *ASEAN Engineering Journal Part A*, *3*(2), 15–31.

Liu, H.-T. (2011). Product design and selection using fuzzy QFD and fuzzy MCDM approaches. *Applied Mathematical Modelling*, *35*(1), 482–496. doi:10.1016/j.apm.2010.07.014

Prasad, K., & Chakraborty, S. (2015). Development of a QFD-based expert system for CNC turning centre selection. *Journal of Industrial Engineering International*, *11*(4), 575–594. doi:10.100740092-015-0122-x

Roy, M. K., Ray, A., & Pradhan, B. B. (2014). Non-traditional machining process selection using integrated fuzzy AHP and QFD techniques: A customer perspective. *Production & Manufacturing Research*, *2*(1), 530–549.

Tho, N. H., Dawal, S. Z. M., Yusoff, N., Tahriri, F., & Aoyama, H. (2013). Selecting a CNC machine tool using the intuitionistic fuzzy TOPSIS approach for FMC. *Applied Mechanics and Materials*, *315*, 196–205. doi:10.4028/www.scientific.net/AMM.315.196

KEY TERMS AND DEFINITIONS

Attributes: It is trait that an object possesses.

Criterion: It is the yardstick on which an item is evaluated.

Multi-Criteria Decision Making: It is a valuable tool that is apt to solve problems that are characterized as a choice among alternatives based on conflicting criteria.

Negative Ideal Solution: It is the result that maximizes the cost criteria and minimizes the benefit criteria.

Normalization: It is a technique used to minimize data redundancy and augment data integrity.

Positive Ideal Solution: It is the solution that maximizes the benefit criteria and minimizes the cost criteria.

Vector: It is characterized as a line section having an explicit direction and a definite length.

Chapter 8
Prioritization of Farming Process by Considering Sustainability as Major Issue:
Sustainability in Farming Sector

Suchismita Satapathy
https://orcid.org/0000-0002-4805-1793
Kalinga Institute of Industrial Technology Bhubaneswar, India

Debesh Mishra
Kalinga Institute of Industrial Technology Bhubaneswar, India

ABSTRACT

Agriculture assumes an imperative role in the development of Indian economy, and it additionally contributes around 15% to the nation's GDP, offering work chances to around half of its population. Diverse devices and supplies implied for farming machines are utilized in farming processes which are either manually or mechanically operated. In spite of the fact that there have been advancements in new technologies, sustainability is the most important issue in farming. Modern farming process and advanced machineries have solved OHS (occupational health and safety) problems of farming. But modern equipment's smoke, dust, chemicals, and fertilizers both in manual-driven farming and modern farming are major environmental issues. So in this chapter, sustainability issues in farming are prioritized such that the policies, equipment, and process must be modified.

DOI: 10.4018/978-1-5225-8579-4.ch008

INTRODUCTION

Indian agriculture sector is expected to be the most important driver of Indian Economy within few years because of high investments for agricultural facilities, warehousing and cold storage. The utilization of genetically modified crops and Organic farming will improve the fertility of land and improve the crop production rate of Indian farmers. But still the small and medium agricultural sector are very poor and neglected, and are following the traditional method of crop production .Due to high cost of equipments unable to purchase and the conventional method of farming gives them many physical problems like lungs problem due to exposure to dust, and musculoskeletal disorders .Extreme weather condition, heavy work load during their working procedure gives them early old age, bone and muscle problems. So to attain better efficiency of performance and to improve productivity of the worldwide farmers in the agricultural sector it is essential to design the tools and equipments keeping in consideration the farmer's capabilities and limitations. The tools and equipments design should be able to provide more human comfort, of good quality, more output focused and reduce the musculoskeletal injury. Occupational safety is a big issue of discussion for Agricultural workers. The methods of working in field in extreme climate (heat, rain), contact with the chemicals (pesticides, fertilizers), the exposure to soil, dust, the contamination due to bacteria, exposure to animals, cattle, injury due to hand tools and musculoskeletal disorders are the most important injuries faced by all Agriworkers. Agricultural workers need sufficient precaution and safety measures at the time of field and machine work, such that no physical damage occurs to them .Most of the Agricultural injuries are resulted from the improper selection and use of hand tools. In agricultural sector the traditional hand tools play a major role in performing the farming activities. The conventional hand tools like spade/hoe, sickle, hammer, shovel, knife etc. have been used since the ancient though some modifications are found now a day. As most of the farmers in India are from poor economic background, they usually prefer the conventional methods in farming instead of using the developed power operated machineries. The hand tools are mostly used in all farming activities like land preparation, weeding, harvesting of crops etc. But tractors and other machineries are definitely solve injury and safety problems compared to conventional tool. Modern equipments are not sustainable due to high noise, vibration and pollution. Farming equipment modification, system design is essential to provide better life to farmers but without environmental protection, social and economic stability is having no meaning. When all over world is concerned about pollution farming policies must be framed to avoid pollution with improving productivity. As safety and sustainability like two sides of a coin, so both parts are important for farming sector. So in this paper an effort is taken to prioritize the farming processes by focusing sustainability issue, so that due precautions may be followed in the agricultural sectors for effective and more productive farming activities.

BACKGROUND

Agricultural workers need sufficient precaution and safety measures at the time of field and machine work, such that no physical damage occurs to them. The fatality rate in agriculture is far higher than any other economic sector. A large proportion of all fatal workplace accidents occur in agriculture, even though a small proportion of the workforce is employed in farming. The level of farm accidents is not decreasing. Similar accidents occur each year (Hope et al., 1999).

Most of the Agricultural injuries are resulted from the improper selection and use of hand tools. In agricultural sector the traditional hand tools play a major role in performing the farming activities. The conventional hand tools like spade/hoe, sickle, hammer, shovel, knife etc. have been used since the ancient though some modifications are found now a day. As most of the farmers in Odisha, India, are from poor economic background, they usually prefer the conventional methods in farming instead of using the developed power operated machineries. The hand tools are mostly used in all farming activities like land preparation, weeding, harvesting of crops etc. The lack of ergonomic considerations in the design of hand tools and equipment, lack of expertise, lack of the knowledge of safety measures and negligence of farmers, adverse environment may lead to agricultural accidents or injuries (Mishra et al., 2018). Murphy (1992) has defined farming as the occupation which depends on the farming awareness skill of individual and the capability to do complicated and repetitive jobs. Voaklander et al. (2006) have suggested having all the skills as defined by Murphy (1992), enhance the ability, performance of individuals and also the safety level in the working place. Further it is recommended that absence of any of the skill can result in farming injury. Karthikeyan et al. (2009) have reported that most of the conventional tools and equipment used in the agricultural sector were manufactured locally by the available materials like wood, iron or stone, respectively. Moreover, after these nominal farmers of Odisha, the large Agri-investors also faces many type of respirotary problems due to gas of large machinaries. So Govt should frame policies against accidents, frame traing programme and take sufficient safety measures such that injuries of farming sector can be avoided. Advancement of technology to modify tool design can also play an important role to avoid occupational injury for farmers. From the literature review, it was found that although various studies have been carried out about the farmers injuries across various regions in the world, but still no such study is being done in the state of Odisha particularly. Hence an attempt was made to study, collect information and analyze the farmers injuries in the South Odisha region in India.

The literature on agricultural injuries and accidents in India are limited to a few. Many authors and researchers have reported the rate of agricultural accidents as higher than the industrial sectors. Lack of infrastructural facility, medical facility, training and rare use & maintenance of machines and lack of defining the work according to gender or age, are the causes of agricultural injuries (Knapp, 1965, 1966). Mohan and Patel (1992) have reported 576 agricultural related injuries in 1 year in 9 villages of Haryana in India. 87% injuries were reported as minor, 11% as moderate and remaining as severe. It is found that most of the injuries were due to hand tools, i.e. 24% by spade, 23% by sickles, 6% by bullock carts, 6% by manually operated fodder cutters, 5% each by power operated fodder cutters, diesel engines and tractors. Zhou & Roseman (1994) found limbs as the most injured body part for 1000 farmers in Alabama by agricultural accidents. Mittal et al. (1996) have reported 36 agricultural injuries, i.e. 8.3% as fatal and 91.7% as non fatal, in 12 villages of Punjab in India during 1 year. The injury rate per thousand machines per year was found to be 23.7 for tractors, 15.5 for sprayers, 7.1 for electric motors, 5.7 for threshers and 2.2 for fodder cutters. DeMuri and Purschwitz, (2000) have reported that the agricultural injuries are due to lack of proper supervision, irrational expectations, economic difficulty and lack of proper safety device. Huiyun Xiang et al. (2000) *have* considered 1500 Chinese farmers from 14 villages and conducted face to face interview of 1358 farmers between July 1997 and September 1997, to evaluate the agricultural related injuries. Rautiainen and Reynolds (2002) have carried out agricultural survey in USA and reported a high fatality rate of 0.22 per 1000 workers per year and the injuries rate as 5 to 166 per 1000 workers per year. A survey was conducted in selected villages of four states viz. Madhya Pradesh, Tamil Nadu, Orissa and Punjab on accidents happened during the year 1995-99 (Gite and Kot, 2003). The limited data collected indicated that the fatalities due to agricultural accidents were 21.2 per 100,000 workers per year. The major source of accidents were farm machines namely tractors, threshers, chaff cutters, cane crushers, sprayers and electric motors, and other sources namely snake bites, drowning in wells/ponds and lightening. Tiwari et al. (2002) have found the agricultural accident rate as 1.25 per thousand workers per year in Madhya Pradesh district in India. It is reported that 77.6% of agricultural accidents were because of farming machines, 11.8% because of hand tools and 10.6% due to others. Helkamp & Lundstrom (2002) have reported the injuries of farmers to be higher than industrial workers. In India, agricultural workers constitute as one of the important sources of farm power. Besides, they also operate animal drawn equipment, tractors, power tillers, self propelled and power operated machines. Table 1 gives the population dynamics of Indian agricultural workers which shows that by 2020 the population of agricultural workers in the country will be about 242 million of which 50% will be the female workers. Thus, there is going to be a significant role of farm workers in country's agriculture and due attention needs to be given to their safety and occupational health issues so as to have higher productivity, less accidents, and minimum occupational health problems.

Adarsh Kumar et al. (2008) have found 576 agricultural related injuries with 332 i.e. 58% hand tool related, from 9 villages with 19,723 persons in 1st phase. Further, in 2nd phase with more 21 villages of 78,890 persons, it was reported of 54 i.e.19% of hand tool related out of 282 injuries. It was also recommended to have interventions development and training at block levels about the safety measures of equipments. Prasanna Kumar et al. (2009) have investigated the agricultural accident for six years i.e. between years 2000-2005 of 42 villages of 4 districts in Arunachal Pradesh in India. It was reported to have the accident rate as 6.39 per thousand workers per year with 40% farm implement related injury. S.K. Patel et al. (2010) have reported the agricultural accident rate as 0.8 per 1000 workers per year in Etawah district of Uttar Pradesh in India. Also it is reported of lack of study in agricultural injuries in developing countries due to non availability of compiled information. Nilsson et al. (2010) have analyzed the responses from 223 injured farmers collected by the Swedish Farm Registry as part of a survey sent to 7,000 farms by the Swedish University of Agricultural Sciences and Statistics Sweden in 2004. These data showed that there were no significant differences in injuries incurred between the age groups, but that senior farmers seemed to suffer longer from their injuries. This study highlighted

Table 1. Population dynamics of Indian agricultural workers (Number in millions)

Serial Number	Particulars	2001	2007	2012	2020
1	Country's population	1029	1130	1210	1330
2	Number of workers as percentage (%) of population	39	41	42	45
3	Total number of workers	402	463	508	605
4	Percentage (%) of agricultural workers to the total workers	58.2	52.0	47.5	40.0
5	Number of agricultural workers	234	241	241	242
6	Percentage (%) of female agricultural workers	39	42	45	50
	a) Number of male agricultural workers	143	140	132	121
	b) Number of female agricultural workers	91	101	109	121

Source (Banthia, 2004; GOI, 2002; GOI, 2006)

the importance of advising senior farmers to bear in mind that their bodies are no longer as young and strong as before. All age groups of farmers should, of course, be careful and consider the risks involved in their work, but since aging bodies need longer to heal, senior farmers probably need to be even more careful and review their work situation and work environment in order to avoid injuries during the final years of their working life. Then, It is recommended to educate senior farmers about the risks of injuries causing increasing damage due to their age. Cem Copuroglu et al. (2012) have studied the agricultural injuries of 41 patients for a period of 3 years and found hand as the most commonly injured part compared to the lower limb and foot. Robertoes Koekoeh K. Wibowo et al. (2016) have studied the farmers of East Java and Indonesia. Most of the farmer's injury was found in hand and the fatigue was reported as 92.8% in upper back, 93.6% in mid back and 91.8% in lower back. Safe, good and fit in hand were suggested as design criteria for hand tools. The recommended tool handles length was 12.4 cm and diameter was 3cm, respectively. Many researches are found on injury and equipment but very less research are found on Sustainability in Agri industry. Mark(2008) has explored that factory farming's method of crowding and confining animals in warehouse-like conditions before killing them and mass-producing both "meat" from cows, pigs and chickens as well as dairy and eggs poses "an unacceptable level of risk to public health and damage to the environment. Mark (2008) has explored that tens of millions of tons of animal waste and agricultural chemicals into the environment every year—driving land, water and air pollution in the process:factory farms typically concentrate tens or hundreds of thousands of animals in one area, and a large operation can produce as much excrement as a small city. Heather (2007) has used EPA statistics, that means California's 1.8 million dairy cows produce as much waste as 36–72 million people. So, when taking into consideration tens or hundreds of thousands of animals, it's not surprising that this amounts to about 130 times more excrement than is produced by the entire human population every year.From many years, farmers have used animal manure to fertilize their fields, but factory farms produce far more waste than the land can absorb, turning disposal of this toxic by-product into a big problem for both the agriculture industry and society.

Wiley et.al (2004) have explained that like human waste, animal excrement from factory farms is not processed as sewage, making it about 500 times more concentrated than treated human waste· while leaving pathogens and volatile chemicals intact. Farmers typically spray some liquidized manure onto the food being grown for animals using giant sprinkler jets, and store the rest in open-air cesspools that can be as large as several football fields[and hold millions of gallons of waste. However, neither of these dispersal techniques is environmentally safe or sustainable.

The United Nations: Food and Agriculture Organization. have reported that of all the agricultural chemicals applied in the U.S., about 37 percent were used to grow crops for animals raised for food.Agricultural chemicals (or agrichemicals) refer to the wide variety of chemical products used in agriculture, such as pesticides (including insecticides, herbicides and fungicides), as well as synthetic fertilizers, hormones and antibiotics. Farmers spray agricultural chemicals onto food grown for animals in order to kill bugs, rodents, weeds, and other organisms that would otherwise supplant or eat the grain grown for the animals. They also apply these substances directly to animals' skin, fur or feathers to combat insect infestation.

RESEARCH METHODOLOGY

A total of 145 farmers from five villages were selected randomly from Odisha state in India. Prior consent was obtained from the respective village heads or leaders. The author and co-author visited the farmers personally and collected the information and data collected using likert scale and by standardized nordic questionnaires as shown in the Appendix. Then, seven experts from different areas of expertise were consulted for the subsequent decision analysis.

Through extensive literature review and using expert opinion different processes involved in rice farming, and different farming pollution related which are disorders and discomforts were identified as Different processes in farming.

- Land preparation process (LPP)
- Seed soaking and scattering/Fertilizer applying process (SFP)
- Pesticides mixing and spraying process (PMSP)
- Weed pulling process (WPP)
- Rice harvesting process (RHP)
- Threshing process (TP)

The Questionnaire designed to find pollution occurs in above processes of rice farming that produces air, water and land pollution. Then 145 data is collected by using likert scale 1 to 5(1 = totally disagree, 2 = partially disagree, 3 =No opinion, 4 =Partially agree, 5 = totally agree) and analysed it is found that fertilizers and chemicals produces maximum pollution. Then the farming processes are ranked by promethee-2, to select most polluting process, such that the policies must be framed by stakeholders to avoid it and make sustainable farming process.

Table 2 shows farming process in Indian rice farming.

Table 2. Farming process

NO.	Farming process and symptoms	Number of farmers (%)
	Land preparation process	
A.	Irritation from smoke during burning rice stalks, if any.	05 (3.47)
B.	Exposure to diesel fumes from plowing and harrowing tractors.	08 (5.55)
C.	Exposure to loud noise and vibration from plowing and harrowing tractors.	08 (5.55)
D.	Dust spreads due to tractor	09 (6.20)
	Seed soaking and scattering/Fertilizer applying process	
E.	Exposure to wet and humid soil.	28 (19.31)
	Pesticides mixing and spraying process	
F.	Exposure to vibration of pesticide applying machines.	02 (1.38)
G.	chemical pollution from pesticide exposures:	44(30.34)
H.	Fertilizers create air pollution	48(33.10)
I.	Animal waste also creates air pollution.	46(31.72)
	Rice harvesting process	
J.	Exposure to vibration and noise of harvesting machines.	32(22.06)
	Threshing process	
K.	Exposure to vibration and noise of threshing machines.	20 (13.79)
L.	Exposure to grain dusts.	18 (12.41)

RESULT AND DISCUSSION

Then to prioritize multicriteria decision making like Promethee-II is implemented. Preference function based outranking method is a special type of Multi Criteria Decision Making tool that can provide a ranking ordering of decision offers.

Step 1: Decision matrix was normalized using the following equation:

$$R_{ij} = [X_{ij}\text{-min}(X_{ij})]/[\max(X_{ij})\text{-min}(X_{ij})] \ (i=1, 2..n; j=1,2..m)$$

where R_{ij} is the performance measure of ith processes with respect to j^{th} criterion.

All these evaluated results are shown in Table 3.

$$R_1 = [X_1\text{-min}(X_1)]/[\max(X_1)\text{-min}(X_1)]$$

$$R_2=[X_2\text{-min}(X_2)]/[\max(X_2)\text{-min}(X_2)]$$

Table 3. Normalized decision matrix

Farming activities that cause pollution	Normalized value
A	0.801205
B	0.39759
C	0.698795
D	0.198795
E	0.096385
F	0.60241
G	0.198795
H	0.801205
I	0.295181
J	0
K	1
L	0.39759

Step 2: Evaluative differences of ith farming processes were calculated with respect to others. This step involved the calculation of differences in criteria values between different processes pair-wise.

Step 3: Preference function Pj (i, i') was calculated using the following equation:

$$P_j(i, i') = 0 \text{ If } R_{ij} \leq R_{i'j}$$

$$P_j(i, i') = (R_{ij} - R_{i'j}) \text{ If } R_{ij} > R_{i'j}$$

These values are shown in Table . Preference Functions Values:

$$P(i, i') = 0 \text{ If } R_i \leq R_{i'}$$

$$P(i, i') = (R_i - R_{i'}) \text{ If } R_i > R_{i'}$$

Step 4: Aggregated preference function was calculated taking into account the weight criterion. Aggregated preference function:

$$\Pi(i, i') = [\sum_{j=1}^{m} Wj \times Pj(i, i')]/\sum_{j=1}^{m} Wj$$

where W_j was the relative importance (weight) of ith criterion. Π (i, i') = [W x $P\left(i,\ i'\right)$]/W (Here W = 3)

$(P_{11} - P_{12})*W = 0$

$(P_{11} - P_{13})*W = 0$

The results are shown in Table 5.

Step 5: Leaving and Entering outranking flows were determined as follows. Leaving (or positive) flow for i^{th} farming activity:

$$\varphi+ (i) = \frac{1}{n-1}\sum_{i'=1}^{n}\Pi\left(i,\ i'\right)(i \neq i')$$

Entering (or negative) flow for i^{th} farming activity.

$$\varphi- (i) = \frac{1}{n-1}\sum_{i'=1}^{n}\Pi\left(i',\ i\right)(i \neq i')$$

Here n was the number of farming Processes.

Step 6: Net outranking flow for each process was calculated using:

$\varphi (i) = \{\varphi^{+} (i)\} - \{\varphi^{-} (i)\}$

Step 7: Ranking of all the considered processes was determined depending on the values of ϕ (i). The higher value of ϕ (i), the more important was the barrier to be looked upon. Thus, the most important process to be looked upon was the one having the highest ϕ (i) value. Table 3 provides Normalized decision matrix and Table 4 shows preference function for all the pairs of alternatives.

Table 5 gives aggregated function and Table 6 provides ranking of farming activities From Table 6 it is found that K (Exposure to vibration and noise of threshing machines is t rank, A (Irritation from smoke during burning rice stalks),.second rank and H (Fertilizers create air pollution) shows Third rank. These processes are not only essential but also common in all types of farming processes. So, systems and equipments must be improved and avoid to make farming process pollution free and

Table 4. Preference functions for all the pairs of alternatives

Farming activities that cause pollution	Preference function	Farming activities that cause pollution	Preference function	Farming activities that cause pollution	Preference function
AB	0.404	EA	0	IA	0
AC	0.103	EB	0	IB	0
AD	0.603	EC	0	IC	0
AE	0.705	ED	0	ID	0.097
AF	0.199	EF	0	IE	0.199
AG	0.602	EG	0	IF	0
AH	0	EH	0	IG	0.096
AI	0.506	EI	0	IH	0
AJ	0.801	EJ	0.096	IJ	0.295
AK	0	EK	0	IK	0
AL	0.403	EL	0	IL	0
BA	0	FA	0	JA	0
BC	0	FB	0.205	JB	0
BD	0.199	FC	0	JC	0
BE	0.301	FD	0.404	JD	0
BF	0	FE	0.506	JE	0
BG	0.198	FG	0.403	JF	0
BH	0	FH	0	JG	0
BI	0.102	FI	0.307	JH	0
BJ	0.397	FJ	0.602	JI	0
BK	0	FK	0	JK	0
BL	0	FL	0.204	JL	0
CA	0	GA	0	KA	0.199
CB	0.301	GB	0	KB	0.603
CD	0.5	GC	0	KC	0.302
CE	0.602	GD	0.001	KD	0.802
CF	0.096	GE	0.103	KE	0.904
CG	0.499	GF	0	KF	0.398
CH	0	GH	0	KG	0.801
CI	0.403	GI	0	KH	0.199
CJ	0.69	GJ	0.199	KI	0.705

continued on following page

Table 4. Continued

Farming activities that cause pollution	Preference function	Farming activities that cause pollution	Preference function	Farming activities that cause pollution	Preference function
CK	0	**GK**	0	**KJ**	1
CL	0.3	**GL**	0	**KL**	0.602
DA	0	**HA**	0	**LA**	0
DB	0	**HB**	0.404	**LB**	0.001
DC	0	**HC**	0.103	**LC**	0
DE	0.102	**HD**	0.603	**LD**	0.2
DF	0	**HE**	0.705	**LE**	0.302
DG	0	**HF**	0.199	**LF**	0
DH	0	**HG**	0.602	**LG**	0.199
DI	0	**HI**	0.506	**LH**	0
DJ	0.198	**HJ**	0.801	**LI**	0.103
DK	0	**HK**	0	**LJ**	0.398
DL	0	**HL**	0.403	**LK**	0

Table 5. Aggregated preference functions for all the pairs of alternatives

	A	B	C	D	E	F	G	H	I	J	K	L
A	NA	0.61	0.43	0.49	0.53	0.30	0.21	0.07	0.30	0.35	0.07	0.28
B	0	0	0	0.071	0.10	0.03	0.07	0	0.03	0.14	0	0
C	0	0.17	0	0.17	0.21	0.05	0.17	0	0.14	0.24	0	0.10
D	0.025	0.21	NA	0	0.16	0.14	0	0	0.07	0.07	0	0
E	0.123	0.31	0.24	0.098	0.031	NA	0.24	0.098	0.17	0.13	0	0.04
F	0	0.30	0.16	0.215	0.019	0.25	0.31	0	0.13	0.21	0	0.07
G	0.025	0.49	0.42	0.28	0.10	0.25	NA	0.071	0.14	0.14	0.071	0.14
H	0.025	0.56	0.39	0.42	0.27	0.13	0.071	NA	0.25	0.28	0	0.21
I	0	0.34	0.28	0.24	0.30	0.18	0.35	0	NA	0.10	0	0.07
J	0.025	0.42	0.35	0.20	NA	0.48	0	0	0.07	NA	0	0.07
K	0.24	0.78	0.61	0.64	0.58	0.16	0.43	0.22	0.47	0.50	NA	0.38
L	0.07	0.40	0.32	0.25	0.24	0.16	0.12	0.04	0.16	0.19	0	NA

NA= NOT AVAILABLE

Table 6. Net outranking flow values for different Process alternatives

Farming activities that cause pollution	$\Phi^{+}(i)$	$\phi^{-}(i)$	$\Phi(i)$	Rank
A	3.687151	0.5453	3.141851	2
B	0.48648	4.630752	-4.2021	12
C	1.24082	3.389999	-2.10592	10
D	0.65935	3.123215	-2.50728	11
E	1.48424	3.242282	-1.76386	9
F	1.5735	2.032641	-0.52529	7
G	2.32964	1.514547	0.858417	4
H	3.12865	0.510411	2.592454	3
I	1.50268	1.976571	-0.4763	6
J	1.53715	2.40334	-0.84963	8
K	5.31155	0.142148	5.239007	1
L	2.01218	1.402569	0.598649	5

farming process will be sustainable. During farming The farmers were ascribed to hazards due to exposure to agrochemicals, psychosocial stresses, poor ergonomics and poor physical working conditions (White and Cessna, 1989). Tiwari et al. (2002) have revealed the farming mischance rate as 1.25 for every thousand farmers for every year in Madhya Pradesh (India). In addition it was accounted for that 77.6% of rural mischances were a direct result of cultivating machines, 11.8% due to hand apparatuses and 10.6% because of others. It was reported that the injuries of farmers were higher when compared to the industrial workers (Helkamp & Lundstrom, 2002). Kumar et al. (2008) have discovered 576 agricultural related injuries from 9 villages with 19,723 persons in their first phase of study. Where only hand tool related injuries were 332 (58%). Further, in their second phase of study considering more 21 villages with 78,890 persons, out of 282 injuries only hand tool related was reported as 54 (19%). Moreover both interventions development and training programs at block levels about the safety measures of equipments was recommended. Prasanna Kumar et al., (2009) have explored the farm related accidents considering a period of six years i.e. between years 2000-2005. A total of 42 villages of four districts in Arunachal Pradesh (India) were taken into consideration, where the accident rate was reported as 6.39 per thousand workers per year with 40% farm implement related. In view of the occupational health and safety issues during rice-cultivating process, it was observed that maximum number of farmers i.e. 09 (6.20%), reported to have dust spreads due to tractors during the land preparation operation (Table 2) In like manner for other rice-cultivating exercises the most extreme number of farmers who reported of occupational health and safety problems were as per the following: 28 (19.31%) farmers experienced exposure to wet and humid soils during seed soaking and scattering or fertilizer applying process, 48 (33.10%) farmers experienced air pollution due to fertilizers during pesticides mixing and spraying process, 32 (22.06%) farmers had exposure to vibration and noise of harvesting machines during harvesting process, and 20 (13.79%) farmers had exposure to vibrations and noises of threshing machines during crop threshing process.

CONCLUSION

Sustainability is a very critical issue in farming, if the farming is traditional manures of animal waste is used instead of chemicals and fertilizers to improve farming productivity. But use of wastes of animals also creates pollution. Two of the many possible practices of sustainable agriculture are crop rotation and soil amendment, both designed to ensure that crops being cultivated can obtain the necessary nutrients for healthy growth. Soil amendments would include using locally available compost from community recycling centers. These community recycling centers help produce the compost needed by the local organic farms. Sustainable agriculture is a type of agriculture that focuses on producing long-term crops and livestock while having minimal effects on the environment. This type of agriculture tries to find a good balance between the need for food production and the preservation of the ecological system within the environment.

REFERENCES

Agarwal, A., Shankar, R., & Tiwari, M. K. (2006). Modeling agility of supply chain. *Industrial Marketing Management*, *36*(4), 443–457. doi:10.1016/j.indmarman.2005.12.004

Banthia, J. K. (2004). *Census of India 2001 - Primary Census Abstracts*. Registrar General & Census Commissioner, Govt. of India.

Cooper, S. P., Burau, K. E., Frankowski, R., Shipp, E., Deljunco, D., Whitworth, R., ... Hanis, C. (2006). A cohort study of injuries in migrant farm worker families in South Texas. *Annals of Epidemiology*, *16*(4), 313–320. doi:10.1016/j.annepidem.2005.04.004 PMID:15994097

DeMuri, G. P., & Purschwitz, M. A. (2000). Fatal injuries in children: A review. *WMJ: Official Publication of the State Medical Society of Wisconsin*, *99*(9), 51–55. PMID:11220197

Gite, L. P., & Kot, L. S. (2003). Accidents in Indian Agriculture. *Technical Bulletin No. CIAE/2003/103*. Coordinating Cell, All India Coordinated Research Project on Human Engineering and Safety in Agriculture, Central Institute of Agricultural Engineering, Nabibagh, Bhopal.

GOI. (2002). *India Vision 2020. Planning commission*. New Delhi: Govt. of India.

GOI. (2006). *Population projections for India and States 2001-2006, Report of the technical group on population projections constituted by the National Commission on Population*. New Delhi: Govt. of India.

Helkamp, J., & Lundstrom, W. (2002). Tractor Related Deaths Among West Virginia Farmers. *Annals of Epidemiology*, *12*(7), 510. doi:10.1016/S1047-2797(02)00344-7

Hope, A., Kelleher, C., Holmes, L., & Hennessy, T. (1999). Health and safety practices among farmers and other workers: A needs assessment. *Occupational Medicine (Philadelphia, Pa.)*, *49*, 231–235. PMID:10474914

Huiyun, X., Zengzhen, W., Lorann, S., Thomas, J. K., Xuzhen, H., & Xianghua, F. (2000). Agricultural work-related injuries among farmers in Hubei, People's Republic of China. *American Journal of Public Health*, *90*(8), 1269–1276. doi:10.2105/AJPH.90.8.1269 PMID:10937008

Karthikeyan, C., Veeraragavathatham, D., Karpagam, D., & Ayisha, F. S. (2009). Traditional tools in agricultural practices. *Indian Journal of Traditional Knowledge*, *8*(2), 212–217.

Knapp, L. W. Jr. (1965). Agricultural Injuries Prevention. *Journal of Occupational Medicine.*, *7*(11), 553–745. doi:10.1097/00043764-196511000-00001 PMID:5831719

Knapp, L. W. (1966). Occupational and Rural Accidents. *Archives of Environmental Health*, *13*(4), 501–506. doi:10.1080/00039896.1966.10664604 PMID:5922020

Kumar, A., Singh, J. K., Mohan, D., & Varghese, M. (2008). Farm hand tools injuries: A case study from northern India. *Safety Science*, *46*(1), 54–65. doi:10.1016/j.ssci.2007.03.003

Kumar, A., Varghese, M., Pradhan, C. K., Goswami, A., Ghosh, S. N., & Nag, P. K. (1986). Evaluation of working with spade in agriculture. *The Indian Journal of Medical Research*, *84*, 424–429. PMID:3781600

Lombardi, M., & Fargnoli, M. (2018). Prioritization of hazards by means of a QFD-based procedure. *International Journal of Safety and Security Engineering.*, *8*(2), 342–353. doi:10.2495/SAFE-V8-N2-342-353

Mishra, D., & Satapathy, S. (2018). Drudgery Reduction of Farm Women of Odisha by Improved Serrated Sickle. *International Journal of Mechanical Engineering and Technology*, *9*(2), 53–61.

Mittal, V. K., Bhatia, B. S., & Ahuja, S. S. (1996). *A Study of the Magnitude, Causes, and Profile of Victims of Injuries with Selected Farm Machines in Punjab. Final Report of ICAR adhoc Research Project.* Ludhiana, India: Department of Farm Machinery and Power Engineering, Punjab Agricultural University.

Mohan, D. (2000). Equipment-related injuries in agriculture: An international perspective. *Injury Control and Safety Promotion*, *7*(3), 1–12.

Mohan, D., & Patel, R. (1992). Design of safer agricultural equipment: Application of ergonomics and epidemiology. *International Journal of Industrial Ergonomics*, *10*(4), 301–309. doi:10.1016/0169-8141(92)90097-J

Moore, H. (2007). You Can't Be a Meat-Eating Environmentalist. *American Chronicle*. Retrieved from http://www.americanchronicle.com/articles/view/24825

Murphy, D. J. (1992). *Safety and Health for Production Agriculture*. St. Joseph, MI: American Society of Agricultural Engineers.

Nag, P. K., & Nag, A. (2004). Drudgery, accidents and injuries in Indian agriculture. *Industrial Health*, *42*(4), 149–162. doi:10.2486/indhealth.42.149 PMID:15128164

Nilsson, K., Pinzke, S., & Lundqvist, P. (2010). Occupational Injuries to Senior Farmers in Sweden. *Journal of Agricultural Safety and Health*, *16*(1), 19–29. doi:10.13031/2013.29246 PMID:20222268

Patel, S. K., Varma, M. R., & Kumar, A. (2010). Agricultural injuries in Etawah district of Uttar Pradesh in India. *Safety Science*, *48*(2), 222–229. doi:10.1016/j.ssci.2009.08.003

Prasanna Kumar, G. V., & Dewangan, K. N. (2009). Article. *Safety Science*, *47*, 199–205. doi:10.1016/j.ssci.2008.03.007

Raj, T., & Attri, R. (2011). Identification and modeling of barriers in the implementation of TQM. *International Journal of Productivity and Quality Management.*, *28*(2), 153–179. doi:10.1504/IJPQM.2011.041844

Rautiainen, R. H., & Reynolds, S. J. (2002). Mortality and morbidity in agriculture in the United States. *Journal of Agricultural Safety and Health*, *8*(3), 259–276. doi:10.13031/2013.9054 PMID:12363178

Ravi, V., & Shankar, R. (2005). Analysis of interactions among the barriers of reverse logistics. *Technological Forecasting and Social Change*, *72*(8), 1011–1029. doi:10.1016/j.techfore.2004.07.002

Robertoes Koekoeh, K. W., & Peeyush, S. (2016). Article. *Agriculture and Agricultural Science Procedia.*, *9*, 323–332. doi:10.1016/j.aaspro.2016.02.142

Voaklander, K. D., Kelly, D. C., Rowe, B. H., Schopflocher, D. P., Svenson, L., Yiannakoulias, N., & Pickett, W. (2006). Pain, medication, and injury in older farmers. *American Journal of Industrial Medicine*, *49*(5), 374–382. doi:10.1002/ajim.20292 PMID:16526061

Wiley, K., Vucinich, N., Miller, J., & Vanzi, M. (2004, November). *Confined Animal Facilities in California*. Retrieved from http://sor.senate.ca.gov/sites/sor.senate.ca.gov/files/%7BD51D1D55-1B1F-4268-80CC-C636EE939A06%7D.pdf

Zhou, C., & Roseman, J. M. (1994). Agricultural injuries among a population based sample of farm operators in Alabama. *American Journal of Industrial Medicine*, *25*(3), 385–402. doi:10.1002/ajim.4700250307 PMID:8160657

KEY TERMS AND DEFINITIONS

Manual-Driven Farming: Hand tools used for farming.

Modern Farming: Machines like tractor and thresher used in farming.

OHS: Occupational health and safety; Occupational safety and health (OSH), also commonly referred to as occupational health and safety (OHS), occupational health, or workplace health and safety.

Sustainability, Sustainable Agriculture: Is a type of agriculture that focuses on producing long-term crops and livestock while having minimal effects on the environment.

APPENDIX

Table 7. Sustainability and health and safety problems during farming process

Farming process and symptoms
1. Irritation from smoke during burning rice stalks, if any.
2. Musculoskeletal problems from lifting and carrying tools or equipments.
3. Injuries from hand start of plowing and harrowing machines.
4. Exposure to diesel fumes from plowing and harrowing tractors.
5. Exposure to loud noise from plowing and harrowing tractors.
6. Injuries from plowing and harrowing blades.
7. Exposure to vibration of plowing and harrowing machines.
8. Injuries during operation of plowing and harrowing tractors.
9. Injuries from stepping on shell.
10. Musculoskeletal problems from carrying heavy pumping hoses or pipes.
11. Musculoskeletal problems from incorrect working postures.
1. Musculoskeletal problems from carrying heavy containers.
2. Puncture from bamboo sticks of seed containers.
3. Exposure to wet and humid soil.
1. Exposure to loud noise of pesticide applying machines.
2. Exposure to vibration of pesticide applying machines.
3. Injuries from falling of pesticide applying machines.
4. Musculoskeletal problems from carrying pesticides sprayers.
5. Symptoms from pesticide exposures:
a) Vomiting.
b) Skin irritations.
c) Nasal irritations.
d) Eye irritations.
e) Headache.
1. Musculoskeletal problems from incorrect weeding postures.
2. Musculoskeletal problems from weeding by hand.
3. Musculoskeletal problems from weeding by machine.
1. Injuries from manually operated harvester:
a) Using Sickles and knifes.
b) Using hand operated mechanical harvester.

continued on following page

Table 7. Continued

Farming process and symptoms
2. Exposure to vibration of harvesting machines.
3. Exposure to noise of harvesting machines.
4. Injuries while operating harvesting tractors.
5. Musculoskeletal problems from incorrect postures in harvesting.
6. Musculoskeletal problems from lifting heavy rice stacks.
1. Injuries from manually operated thresher:
a) Striking on wooden frame.
b) Using hand operated mechanical thresher.
c) Using Draft animal driven thresher.
2. Exposure to vibration of threshing machines.
3. Exposure to noise of threshing machines.
4. Musculoskeletal problems from incorrect postures in threshing.
5. Exposure to grain dusts.

Chapter 9
An Integrated MCDM and Ergonomic Approach for Agricultural Sectors of Odisha in India:
A Critical Analysis for Farming Sustainability

Debesh Mishra
Kalinga Institute of Industrial Technology Bhubaneswar, India

Suchismita Satapathy
https://orcid.org/0000-0003-2665-1365
Kalinga Institute of Industrial Technology Bhubaneswar, India

ABSTRACT

An attempt was made in this chapter to explore the agricultural hazards in the farming sectors of Odisha in India. There were three main contributions. At first the agricultural hazard factors were identified by the use of literature and standard Nordic questionnaires. In the second part, the Best Worst Method (BWM) was used to rank the different hazards based on the risk factors involved. Finally, in the third part, an ergonomic evaluation was made by using both REBA and OWAS ergonomic tools, considering different postures taken by farmers in selected farming activities, and accordingly, the corrective measures (if any) were recommended.

DOI: 10.4018/978-1-5225-8579-4.ch009

INTRODUCTION

Workers in various occupations are subjected to various sufferings and pains. As most of the workers spend a lot of time in their workplaces, so the workplace is supposed to be the second home of the workers. But unfortunately due to the ignorance of the workers by not adhering to the guidelines in the occupation, the workers are seriously affected by some of the health hazards. Occupational health and safety refers to factors and conditions affecting the health and safety of workers engaged in different workplaces. The issues related with this can be viewed as infections, mishaps or different risks emerging from workplace or circumstances that emerge in the endeavor to perform undertakings in any occupation. It is surely, a compensable illness that occurs during the span of work. Agriculture has been considered as one of the most hazardous occupation in the world. There is a progressive deterioration of health of agricultural workers. The fatal accident rate in agricultural sectors of several countries has been reported by different studies as to be twice than the average for all other industries. The developing nation like India is facing with traditional public health problems such as under-nourishment, poor environmental cleanliness, infectious diseases, and inadequate medical facilities. For improving the occupational health research still much is to be done like creation of advanced research facilities, human resources developments, environmental and occupational health cells (Habibullah et al., 2004).

BACKGROUND

The health and safety in the workplaces are dealt with all aspects of occupational health and is focused primarily on prevention of hazards (Khan et al., 2012). Serious social and economic consequences of public health are resulted by occupational health hazards (Tadesse et al., 2007). The occupational diseases occur due to physical, chemical, biological, social, and psychosocial factors that are found to be present during the course of employment in workplaces (Sambo et al., 2012). The incidence of 317 million non-fatal occupational accidents and 160 million non-fatal work-related diseases were reported of occurring annually (ILO, 2013). Thangaraj et al. (2017) have considered the workers engaged in automobile tasks to assess their awareness of personal protective equipments and recommended the need for proper education on the types, properties of the different protective devices, and the workplace hazards for protecting their health. The risk of exposure to chemical & biological risks, and incidence of work-related diseases were reported to be highly concentrated on four major types of occupational groups such as workers in elementary occupations, technicians, operators, and agricultural workers. Moreover, the health impact of production processes as well as services on workers' health was reported to be focused on an urgent basis (Montano, 2014). In the developing world the occupational exposure to health related risks was reported to be common in view of overload in work, lacking of task control and role ambiguities, including some other factors like inequitable management practices, poor organization, economic and human factors (Askenazy, 2006; Annalee et al., 2007). Nwankwo et al. (2018) have reported that the health workers are mostly affected by physical, chemical, biological and psychological agents. Some of the hazards reported were back-ache and accidents in work, violence and molestation related health hazards, lack of commitment to policy & policy enforcement, safety activities linked with health facility, participation of employees in safety programs, and post exposure compliance. Agricultural farmers are subjected to whole body vibration because of the physical risks and activities like ploughing, planting, weeding, and pesticide spraying (Kaergaard et al., 2000); carrying heavy loads of more than 30kg (Strazdins et al., 2004). Adverse health effects are caused due to the lack of pesticide safety practices among farmers and mostly in developing countries, it has been reported of the lack & absence of access to the personal protective equipments and proper safety measures while mixing & spraying of pesticides (Earle, 1930; Kibret et al., 2010; Manh et al., 2010; Sharp et al., 2007). Also, in the developing countries the importing, selling and utilization of pesticides and chemicals were reported to be largely unregulated (Curl et al., 2002). Splansky et al. (2007) have reported of a variety of biological hazards among agricultural workers such as cholera, malaria, anthrax and brucellosis. Edmund et al. (2015) have studied different hazards faced by workers in various occupations such as agriculture, construction, building, energy/ power, transport, education, and communications, in Enugu Metropolis. It was found that no policy guidelines were followed in any of these occupations. Further, it was recommended for the employers to improve the working conditions, organize seminars, educational plans and ergonomic interventions to ensure healthy workers. Rim et al. (2014) have made an extensive study of literatures on biological hazards to workers, and the measures to protect against these biological factors. Oluoch et al. (2017) have revealed that both biological and chemical risks were high, while ergonomics, physical and psychological risks were low among the employees working in the water service industry in Kisumu County. Accordingly, it was recommended to take adequate measures to reduce other risks.

The main objective of the present study is to evaluate the risk factors causing different hazards in the farming sectors of Odisha in India. However, finding the risk factors in the agricultural sectors are not enough for further analysis, unless these are ranked based on the level of risks involved. And application of a suitable multi-criteria decision making method plays a vital role in this regard. In any MCDM problem for the selection of the best alternatives, a number of alternatives are evaluated with respect to a number of criteria. Different existing MCDM methods are as following: the weighted sum model (WSM), weighted product model (WPM), and analytical hierarchy process (AHP) (Triantaphyllou, 2000). The present study utilized the Best Worst Method (BWM) which was developed by Rezaei, 2015. The major advantages of BWM, in comparison with other existing MCDM methods are: it requires less comparison of data and also it leads to more consistent comparisons. Moreover, the BWM has been used by several scientific papers. For instance: selection of suppliers (Gupta et al., 2017), power grid enterprise evaluations (You et al., 2017), performance evaluation of R&D firms (Salimi et al., 2018), evaluation of supply chains sustainability (Ahmadi et al., 2017), and evaluation of sustainable product package design (Rezaei et al., 2018).

The steps involved in BWM method are as discussed below (Rezaei, 2015):

Step 1: Determination of a set of decision criteria:

In this step, the decision maker identifies n criteria $\{C_1, C_{2,} \ldots\ldots, C_n\}$ based on literature review and expert opinion that are used to make a decision.

Step 2: Determining 'the best and the worst criteria", such as "most desirable or most important" and "least desirable or least important criteria", respectively.

Step 3: Determining the preference of the best criteria's over all other criteria by using a number between 1 and 9. The obtained best to others (BO) vector would be:

$$A_B = \{a_{B1}, a_{B2}, \ldots., a_{Bn}\},$$

where, a_{Bj} denotes the preference of the best criterion B over criterion j, and it is clear that $a_{BB} = 1$.

Step 4: Determining the preference of all the criteria's over the worst criteria by using a number between 1 and 9. The obtained others to worst (OW) vector would be:

$$A_W = \{a_{1W}, a_{2W}, \ldots., a_{nW}\}^T,$$

where, a_{jW} denotes the preference of the criterion j over the worst criterion W, and it is clear that $a_{WW} = 1$.

Step 5: Calculation of the optimal weights as $\{w_1^*, w_2^*, \ldots, w_n^*\}$.

The main objective in this step is to determine the optimal weights of the criteria such that the maximum absolute differences $\{|w_B - a_{BjWj}|, |w_j - a_{jw}w_w|\}$ for all j is minimized. This can be translated to the following minimax model:

Minimax $\{|w_B - a_{BjWj}|, |w_j - a_{jw}w_w|\}$

Subjected to,

$$\sum_j w_j = 1 \qquad w_j \geq 0, \text{ for all } j \tag{1}$$

Model (1) can be converted to following linear model as:

Min ξ^*

Subjected to,

$$\left|w_B - a_{Bj}w_j\right| \leq \xi^*, \text{ for all } j \left|w_j - a_{jW}w_W\right| \leq \xi^*, \text{ for all } j \sum_j w_j = 1 \qquad w_j \geq 0, \text{ for all } j \tag{2}$$

Solving the above model (2) optimized weights $\{w_1^*, w_2^*, \ldots, w_n^*\}$ and optimal value ξ^* can be obtained. Consistency (Ksi*) of comparisons also needs to be determined. According to Rezaei, (2016) a value closer to 0 is required for consistency.

Ergonomic Analysis in Agriculture

The agricultural sectors supply sustenance as well as help to utilize various laborers (Naeini et al., 2014). Very few applications of ergonomic aspects in the design of agricultural equipments have been reported (Vyavahare et al., 2012). Henceforth, abusing the accessible assets and advances at ideal dimension by presenting best ergonomic practices in the farming parts is exceedingly wanted to improve the profitability and agricultural yields. Roman-Liu, (2014) has reported of high workload as a major risk factor for the development of MSDs. For that reason reduction of workload is an appropriate way for risk reduction (Brandl et al., 2017a). Workload has been defined as a function of force exerted, working postures and sequences of time. Among these the main influencing factors for workload are reported as force exerted and working postures (Brandl et al., 2017a; Roman-Liu, 2014; van Nieuwenhuyse et al., 2006). A number of methods are available now a days for evaluating and analyzing the working postures and for assessing the workloads, such as the "Ovako Working Posture Analysing System" (OWAS) which was developed by the "Finish steel industry" (Karhu et al., 1977, 1981), and successfully applied in the ergonomic fields (Brandl et al., 2017b; Hellig et al., 2018); the "Rapid Upper Limb Assessment" (RULA) (McAtamney and Corlett, 1993); the "Rapid Entire Body Assessment" (REBA) (Hignett and McAtamney, 2000); and the "Postural Loading on the Upper Body Assessment" (LUBA) (Kee and Karwowski, 2001). Soheili-Fard et al. (2017) have used both REBA and OWAS methods for analyzing the postures of 48 workers in six tea factories. It was reported that 75% of various postures taken by the workers needed corrective measures. Moreover, 34% of different postures were found to be at high and very high risk levels based on the OWAS results. The agricultural tools and equipments which are designed with due considerations on various ergonomic aspects, enhances the production output as well as improves the working solaces and efficiency of agriculturists without trading off on their wellbeing and safety. Thus, these issues require a critical consideration for enhancing the connection among farmers and their workplaces. Therefore, an attempt was further made in this study to conduct an ergonomic risk assessment of different working postures of farmers during different selected farming activities, and accordingly to find for any corrective measures (if any) using both REBA and OWAS ergonomic tools.

RESEARCH METHODOLOGY

Materials

The following materials were used in the present study:

1. Standardized Nordic questionnaire
2. Laptop with Windows 7
3. Digital Camera
4. BWM-Solver.xlsx software (Rezaei, 2016)
5. ErgoFellow 2.0 Software

Method

The total number of farmers selected for the present analysis was 145 from different regions in Odisha. While selecting the farmers, it was ensured that they were free from any illness or diseases. At first using a yes and no type standardized nordic questionnaire as shown in Appendix 1, the information was collected about the demographic characteristics and the number of agricultural hazards or injuries of farmers because of different risk factors in the agricultural sectors as identified by the extensive study of literature and opinion of experts. The characteristics of the decision making experts were as shown in Table 1. Using a 5-point rating likert scale (1 = strongly disagree, 2 = disagree, 3 = neither disagree nor agree, 4 = agree, 5 = strongly agree) as shown in Appendix 2, different feedbacks were collected on the levels of consideration by different age groups of farmers, on the impact of agricultural hazards in the farming sectors of Odisha. Further by using Minitab 2017 version, the correlation between age as the demographic characteristic, and the agricultural hazards were obtained which was followed by the analysis of residual plot for age as the response variable. Subsequently, the BWM ranking was performed for different agricultural hazards based on the extent of risk factors involved, in consultation with the experts. Finally, by the use of "ErgoFellow 2.0 Software", the Rapid Entire Body Assessment (REBA) tool, and the Ovako Working Posture Analyzing System (OWAS) tool was used to assess the working postures of farmers in different selected farming activities, which were captured as still photographs using digital camera. And based on the obtained REBA and OWAS scores with the necessary recommended actions, a comparative analysis was done.

Table 1. The characteristics of the decision making experts

Area of Expertise	Number of Experts
Agriculture	01
Engineering	01
Industry	01
Medicine	01
Economic	01
Law	02

RESULTS AND DISCUSSION

The demographic characteristics of farmers indicated that out of 145 selected farmers, 102 farmers were males and 43 farmers were females (Table 2). Maximum number of farmers was found between the age group of 26-49 years which was 96 (66.20%). While, a total of 16 i.e. 11.03% farmers were found with less than or equal to 25 years of age.

Considering the agricultural hazards or injuries of farmers as illustrated in Table 3, it was observed that maximum number of farmers i.e. 93 (64.13%) farmers reported to have headache, shoulder, and elbow pain in the category of physical hazards, 85 (58.62%) farmers reported to suffer from fever in the category of biological hazards, 92 (63.44%) farmers reported of suffering headache, in the category of chemical hazards, and 76 (52.41%) farmers reported of physical tiredness in the category of psychological hazards, respectively.

The strength of linkage between two variables is depicted by their correlation coefficient which is in between -1 to 1. A positive correlation indicates a positive relationship, while a negative correlation indicates a negative relationship between the variables. The pearson correlation coefficients between age, physical hazards, biological hazards, chemical hazards, and psychological hazards were as shown in Table 4. The correlation coefficients for these parameters along with the corresponding p-values were obtained by using the pearson product moment correlation using Minitab 2017 version. It was observed that very high correlation coefficient i.e. 0.950 was in between the psychological hazards and chemical hazards. All the correlation coefficients were found significant at $p<0.01$. Also, it may be noted that the correlation of age with all the four types of hazards were found to be

Table 2. Demographic characteristics of farmers (n=145)

Parameters		Number of Farmers (%)
Gender	Male	102 (70.34)
	Female	43 (29.65)
Civil status	Married	98 (67.58)
	Un-married	47 (32.41)
Age in Years	≤ 25	16 (11.03)
	26-49	96 (66.20)
	≥ 50	33 (22.75)

Table 3. Agricultural hazards or injuries of farmers (n=145)

Agricultural Hazards or Injuries	Number of farmers (%)
Physical Hazards	
1. Sudden fall during work	05 (3.44)
2. Trauma	08 (5.51)
3. Crushing/attack	03 (2.06)
4. Animal kicks/scratches	03 (2.06)
5. Hearing loss	08 (5.51)
6. Eye injury	02 (1.38)
7. Headache, shoulder, and elbow pain	93 (64.13)
8. Sunstroke	07 (4.82)
9. Body fatigue	06 (4.13)
10. Cardiovascular strain	19 (13.10)
11. Swollen hand	56 (38.62)
12. Neck pain	38 (26.20)
13. Slip disc-back pain	26 (17.93)
14. Insecticide illness	19 (13.10)
15. Animal bites	06 (4.13)
Biological Hazards	
1. Fever	85 (58.62)
2. Conjunctivitis	25 (17.24)
3. Amoebiasis	22 (15.17)
4. Tuberculosis	08 (5.51)
5. Biting by rabid dog	08 (5.51)
6. Ringworm infection	33 (22.75)
7. Infestation with internal and external parasites	07 (4.82)
Chemical Hazards	
1. Eczema	05 (3.44)
2. Allergies	25 (17.24)
3. Headache	92 (63.44)
4. Eye Irritation	58 (40)
5. Nausea and Vomiting	48 (33.10)
6. Pesticides intoxication	33 (22.75)
7. Respiratory problems	27 (18.62)

continued on following page

Table 3. Continued

Agricultural Hazards or Injuries	Number of farmers (%)
Psychological Hazards	
1. Physical Tiredness	76 (52.41)
2. Psychological Fatigue	72 (49.65)
3. Boredom	22 (15.17)
4. Aggression	08 (5.51)
5. Hostility	08 (5.51)
6. Depression	33 (22.75)
7. Lack of Sleep	07 (4.82)
8. Forgetfulness	08 (5.51)
9. Irritability	08 (5.51)
10. Low Energy	33 (22.75)
11. Low Concentration	05 (3.44)

Table 4. Correlation between age, physical hazards, biological hazard, chemical hazards, and psychological hazards

Parameters	Age	Physical Hazards	Biological Hazards	Chemical Hazards	Psychological Hazards
Age	1.000				
Physical Hazards	0.692* (p = 0.000)	1.000			
Biological Hazards	0.775* (p = 0.000)	0.845* (p = 0.000)	1.000		
Chemical Hazards	0.628* (p = 0.000)	0.931* (p = 0.000)	0.885* (p = 0.000)	1.000	
Psychological Hazards	0.615* (p = 0.000)	0.948* (p = 0.000)	0.837* (p = 0.000)	0.950* (p = 0.000)	1.000

*Correlation is significant at $p < 0.01$ (2-tailed).

significant at $p < 0.01$. The residual plot for age was obtained as shown in Figure 1, where age was taken as the response variable and all the four corresponding agricultural hazards were considered as the predicting variables, in the regression analysis. It was observed that in the residual plot most of the points are clustered around the line in the top left graph which indicated that the error terms/residuals

Figure 1. Residual plot for age

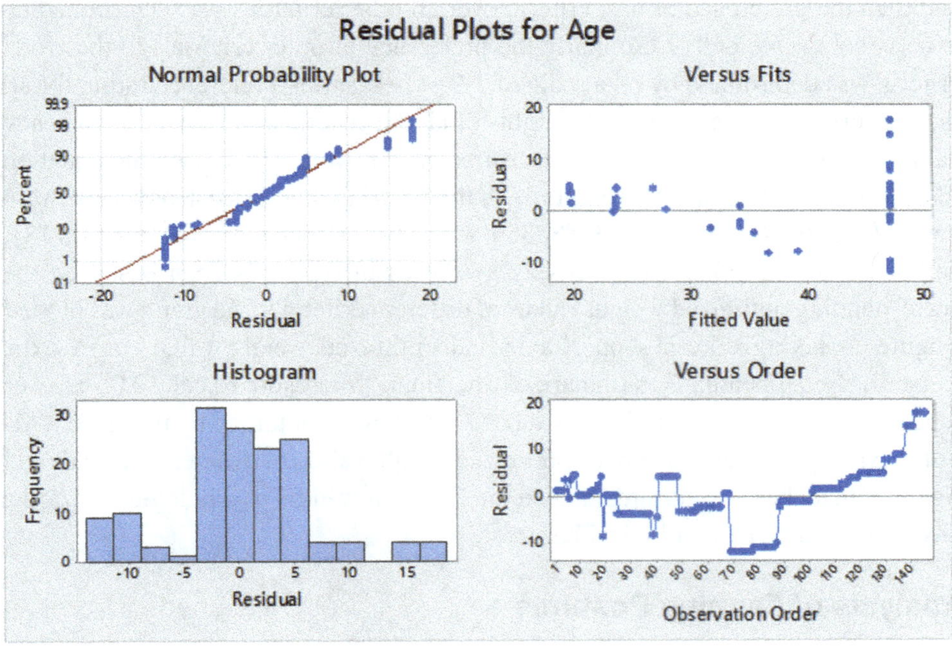

are approximately normal. The residuals plotted against the fitted values in the top right graph indicated that the assumption of residuals having mean zero is valid. The bottom left graph re-emphasized the normality assumption. The bottom right graph showed a closer clear cyclic pattern which indicated that the residuals were dependent on different age groups of farmers. The regression model summary indicated the R value as 67.64%, Adjusted R squared value as 66.72% as illustrated in Table 5, which indicated a good level of prediction.

BWM Ranking of Agricultural Hazards

From all the four types of agricultural hazards such as physical hazards, biological hazards, chemical hazards, and psychological hazards, the best and worst criteria

Table 5. Regression model summary

S	R-sq	R-sq(adj)	R-sq(pred)
6.77181	67.64%	66.72%	66.42%

were selected based on its impact on the farming sectors of Odisha by experts opinion and then the preference of best criteria over all other criteria's was determined by experts on a scale of 1-9. Similarly the preference of other criteria's to the worst criteria was determined by on a scale of 1-9. After getting preference rating for all the four criteria's, the optimized weights of all the criteria's as well as consistency value were obtained using "BWM-Solver.xlsx" software which used the equation (2). The obtained optimized weights of all the criteria's, and consistency value were $\{w_1^*=0.166666667, w_2^*=0.138888889, w_3^*=0.416666667, w_4^*=0.277777778\}$, and Ksi*= 0.833333333, respectively as shown in figure 2. Then based on the corresponding optimized weight values of different criteria's, the graph was plotted (Figure 3) taking criteria's on X-axis and optimized weight values on Y-axis, respectively. Appendix 3, summarized the final "Microsoft Excel 12.0" answer report, sensitivity report and limits report, which were obtained by using "BWM-Solver.xlsx". It was clear from the optimized weight values of different criteria's and figure 3, that chemical hazards ranked the first followed by psychological hazards, physical hazards, and biological hazards.

Analysis of Farming Postures

While manually preparing the land for farming purposes as shown in Figure 4, it was observed that the farmers neck was more than 20 degree, & was twisted, trunk

Figure 2. BWM ranking of agricultural hazards

Criteria Number = 4 Names of Criteria	Criterion 1 Physical Hazards	Criterion 2 Biological Hazards	Criterion 3 Chemical Hazards	Criterion 4 Psychological Hazards
Select the Best	Chemical Hazards			
Select the Worst	Psychological Hazards			

Best to Others	Physical Hazards	Biological Hazards	Chemical Hazards	Psychological Hazards
Chemical Hazards	5	6	2	3

Others to the Worst	Psychological Hazards
Physical Hazards	4
Biological Hazards	5
Chemical Hazards	2
Psychological Hazards	7

	Physical Hazards	Biological Hazards	Chemical Hazards	Psychological Hazards
Weights	0.166666667	0.138888889	0.416666667	0.277777778
Ksi*	0.833333333			

Figure 3. Weight distribution of agricultural hazards

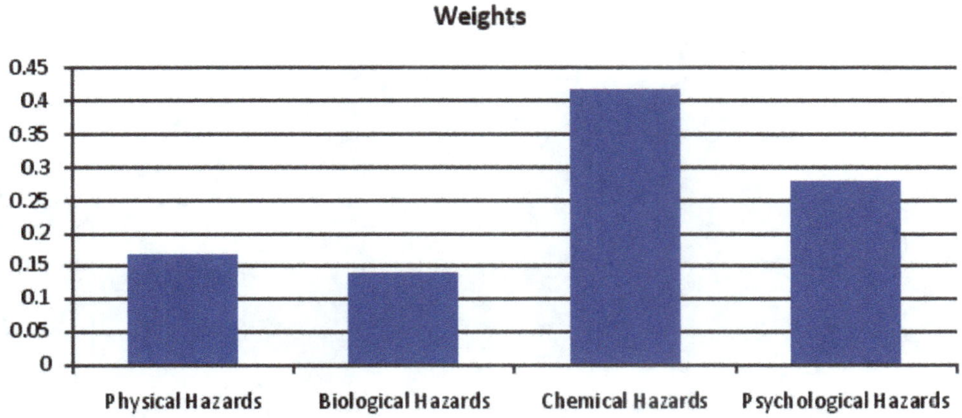

Figure 4. Manually preparing the land using country plough

was at 20 to 60 degree, and was twisted, legs were supported in the two legs, and knee was bent at more than 60 degree (Figure 5), load was more than 10kg (Figure 6), upper arm was at 45 to 90 degree, & was abducted, shoulder was raised, and arm was supported & the farmer was leaning, lower arm was at 0 to 60 degree, wrist was at more than 15 degree up and down, and it was bent from midline (Figure 7), coupling was fair (Figure 8), activity was repeated, and there was a rapid change in

Figure 5. REBA inputs for neck, trunk and legs for manually land preparation

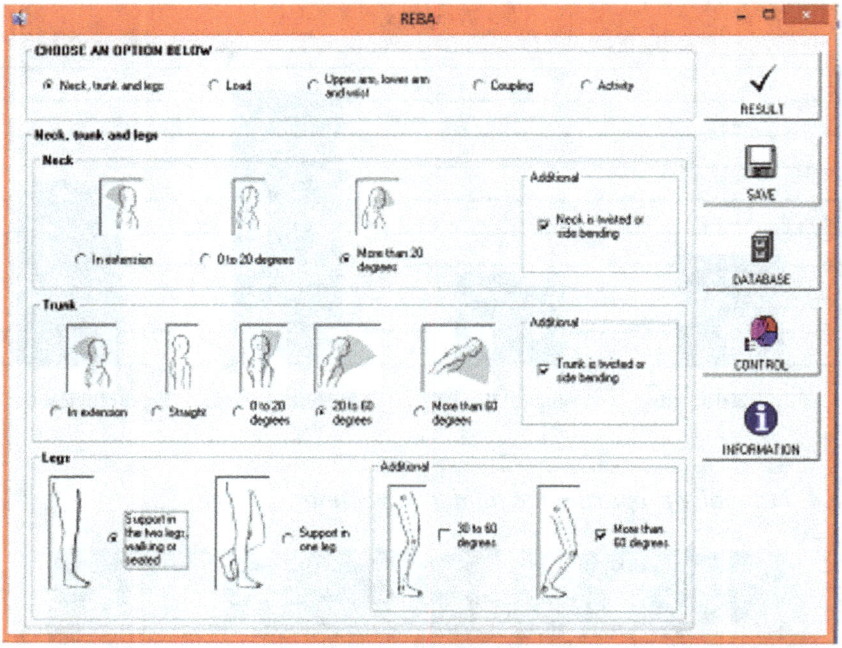

Figure 6. REBA inputs for load in manually land preparation

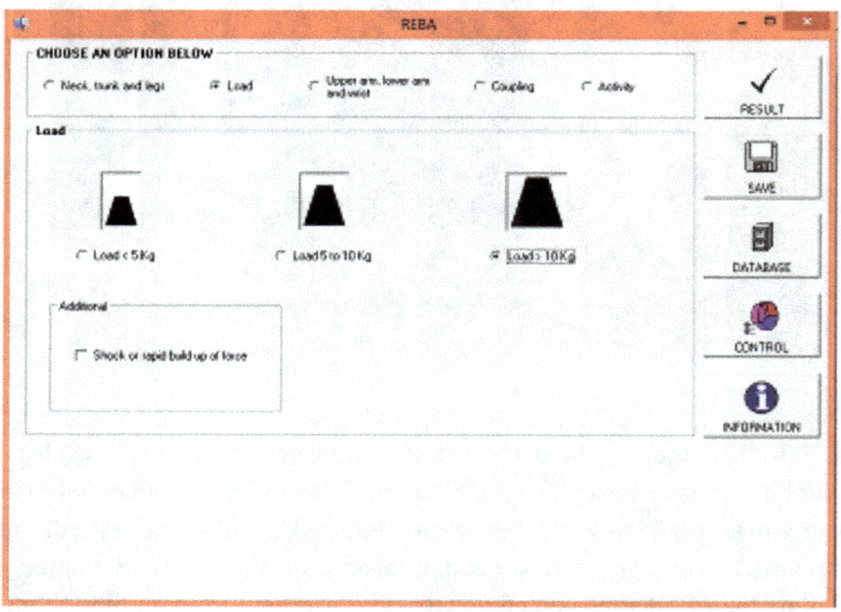

Figure 7. REBA inputs for upper and lower arms in manually land preparation

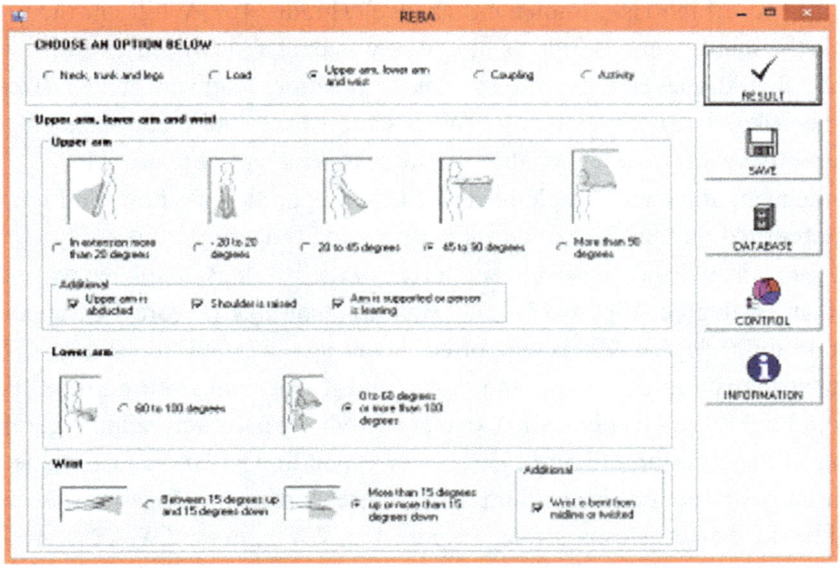

Figure 8. REBA inputs for type of coupling in manually land preparation

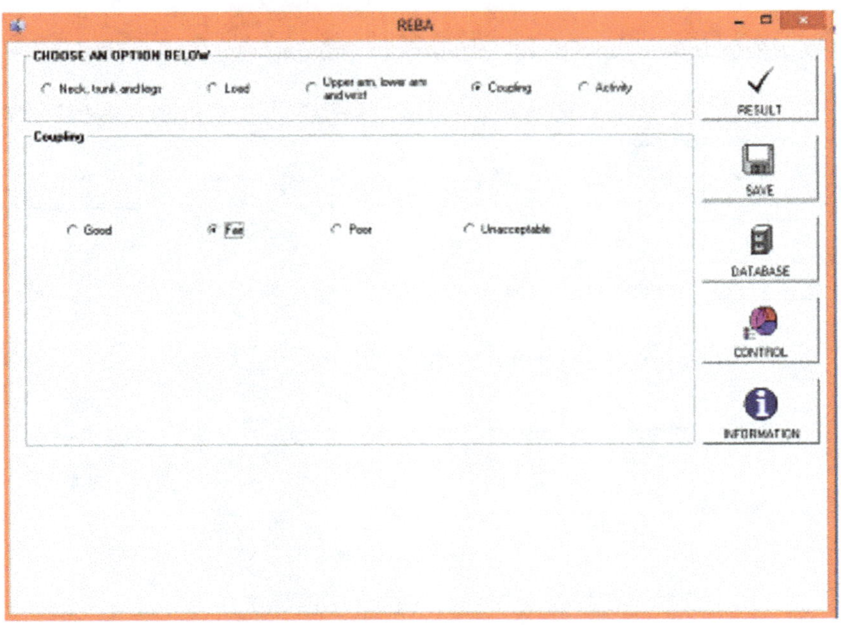

postures (Figure 9). The corresponding REBA score obtained was 14, with suggested action as "very high risk, implement change" (Figure 10). While, in the OWAS analysis the inputs were as follows: the back was bent and twisted, one arm was at or above shoulder level, legs were in walking position, load was greater 10kg and less than 20kg. The corresponding OWAS score was 3 with recommended action as "corrective actions should be done as soon as possible" (Figure 11).

While using the tractor implements for land preparation as shown in Figure 12, it was observed that the farmers neck was at 0 to 20 degree, trunk was at 0 to 20 degree and was twisted, legs were supported in the two legs, and knee was bent at more than 60 degree (Figure 13), load was less than 5kg (Figure 14), upper arm was at 45 to 90 degree & was supported, lower arm was at 0 to 60 degree, wrist was at more than 15 degree up and down, and it was bent from midline (Figure 15), coupling was good (Figure 16), one or more body parts was held longer than 1 minutes in static posture (Figure 17). The corresponding REBA score obtained was 6, with suggested action as "medium risk, requires further investigation and change should be done as soon as possible" (Figure 18). While, in the OWAS analysis the inputs were as follows: the back was straight, both arms were below the shoulder level, legs were in sitting position, load less than or equal to 10kg. The corresponding OWAS score was 1 with recommended action as "no actions required" (Figure 19).

Figure 9. REBA inputs for type of activities in manually land preparation

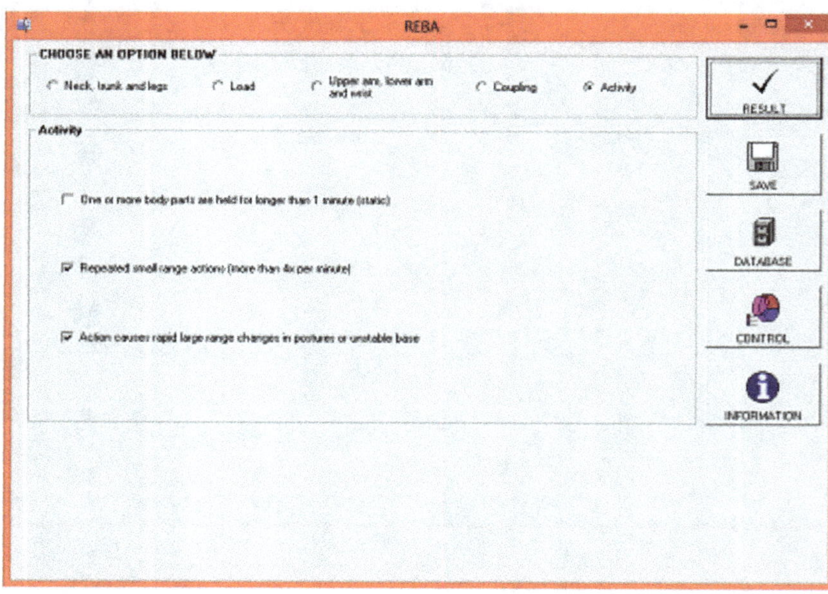

Figure 10. REBA output score with proposed action for manually land preparation

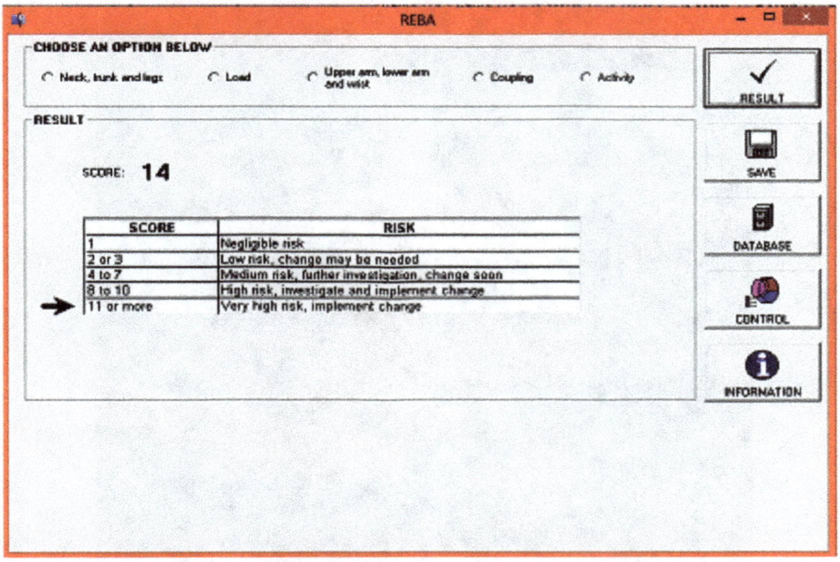

Figure 11. OWAS inputs and output for manually land preparation

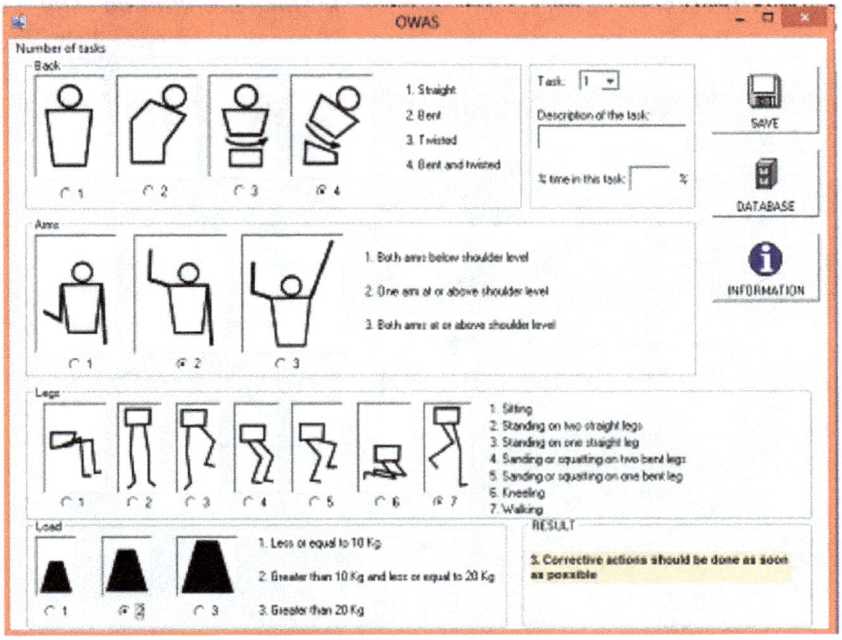

Figure 12. Use of tractor implements for land preparation

Figure 13. REBA inputs for neck, trunk and legs for land preparation using tractor implements

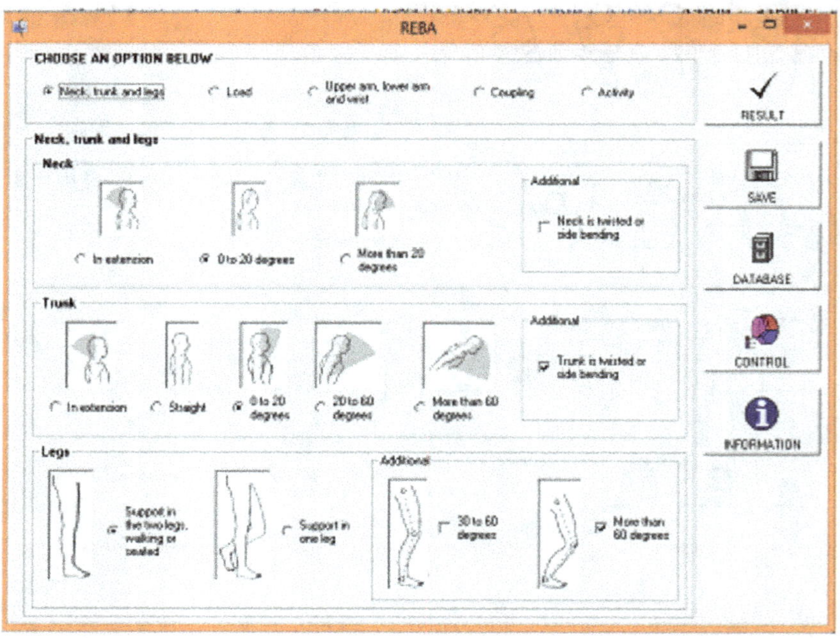

Figure 14. REBA inputs for loads in land preparation using tractor implements

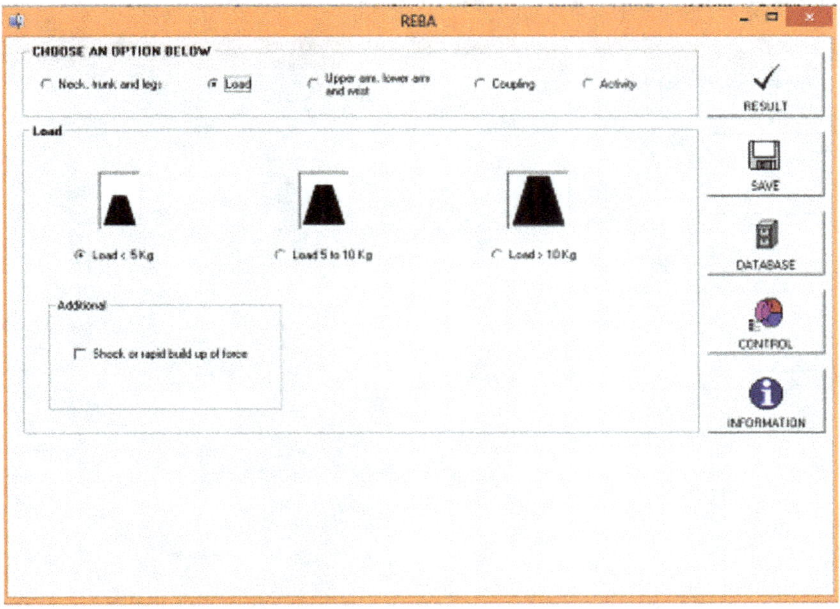

Figure 15. REBA inputs for upper and lower arms in land preparation using tractor implements

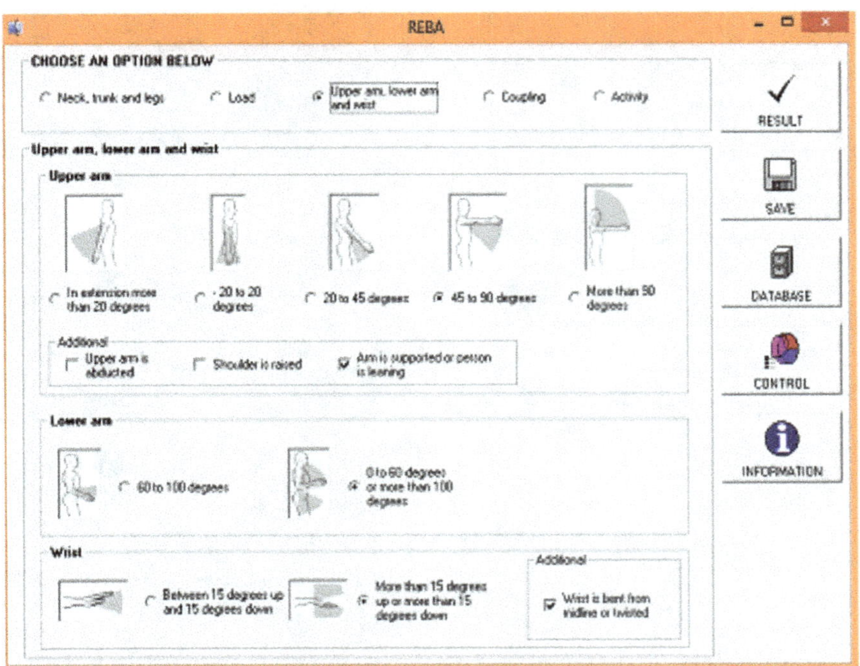

Figure 16. REBA inputs for type of coupling in land preparation using tractor implements

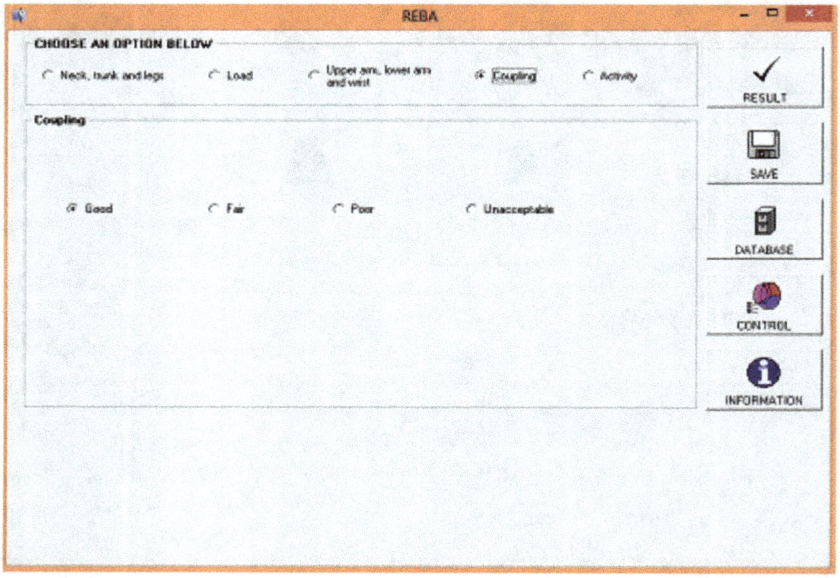

Figure 17. REBA inputs for type of activities in land preparation using tractor implements

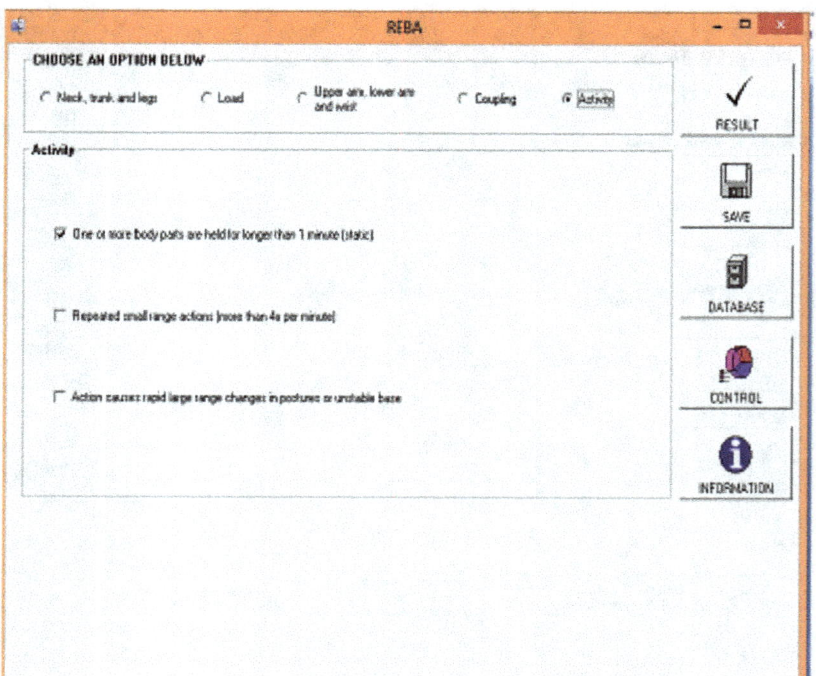

Figure 18. REBA output score and proposed action for land preparation using tractor implements

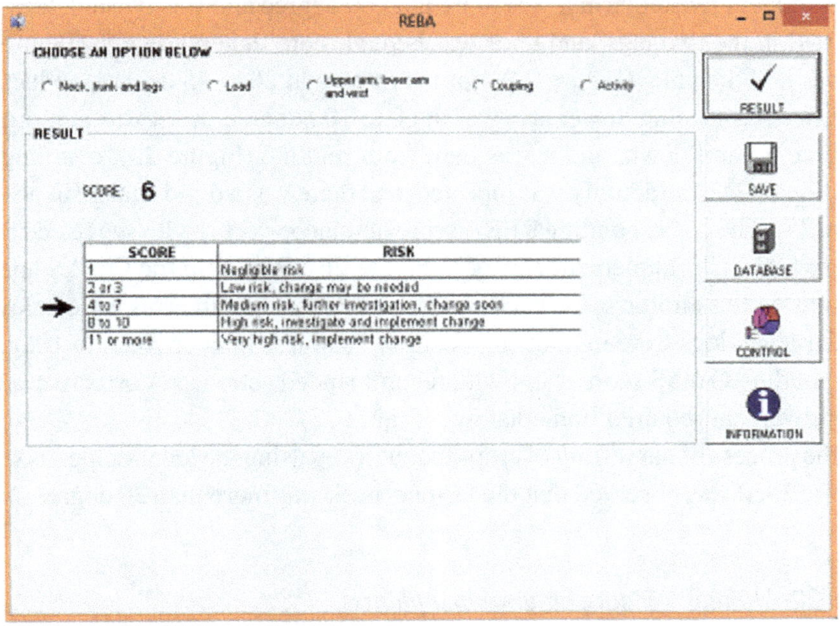

Figure 19. OWAS inputs and output for land preparation using tractor implements

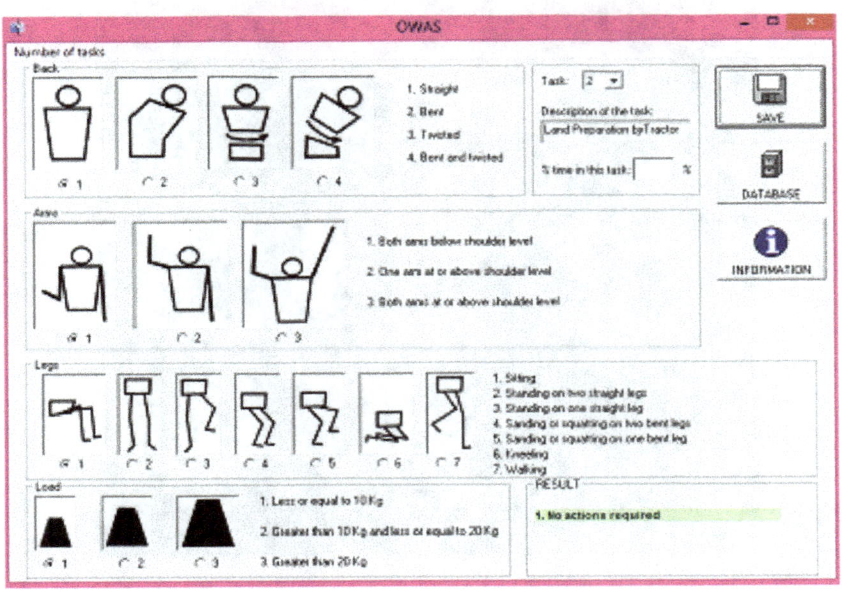

While weeding manually for removing the unwanted plants in the cultivated areas as shown in Figure 20, it was observed that the farmers neck was more than 20 degree and was twisted, trunk was more than 60 degree and was twisted, legs were supported in the two legs, and knee was bent at more than 60 degree (Figure 21), load was less than 5kg (Figure 22), upper arm was at 20 to 45 degree, abducted & the farmer was leaning, lower arm was at 60 to 100 degree, wrist was at more than 15 degree up and down, and it was bent from midline (Figure 23), coupling was good (Figure 24), the activity was repeated, and there was a rapid change in postures (Figure 25). The corresponding REBA score obtained was 11, with suggested action as "very high risk, implement change" (Figure 26). While, in the OWAS analysis the inputs were as follows: the back was bent & twisted, both arms were below the shoulder level, legs were in kneeling position, load less than or equal to 10kg. The corresponding OWAS score was 4 with recommended action as "corrective actions for improvement required immediately" (Figure 27).

In the process of harvesting of crops manually by using the local sickles as shown in Figure 28, it was observed that the farmers neck was more than 20 degree, & was

Figure 20. Manual weeding of unwanted plants

Figure 21. REBA inputs for neck, trunk and legs for manual weeding

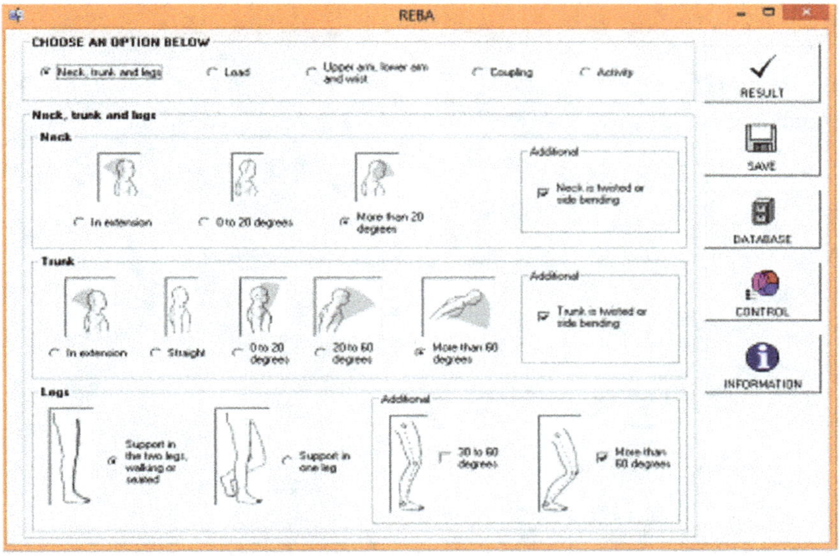

Figure 22. REBA inputs for loads in manual weeding

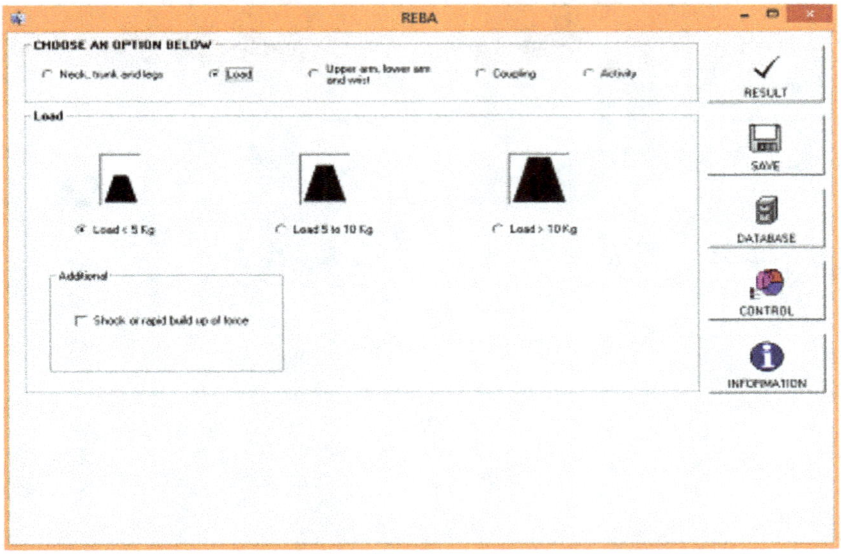

Figure 23. REBA inputs for upper and lower arms in manual weeding

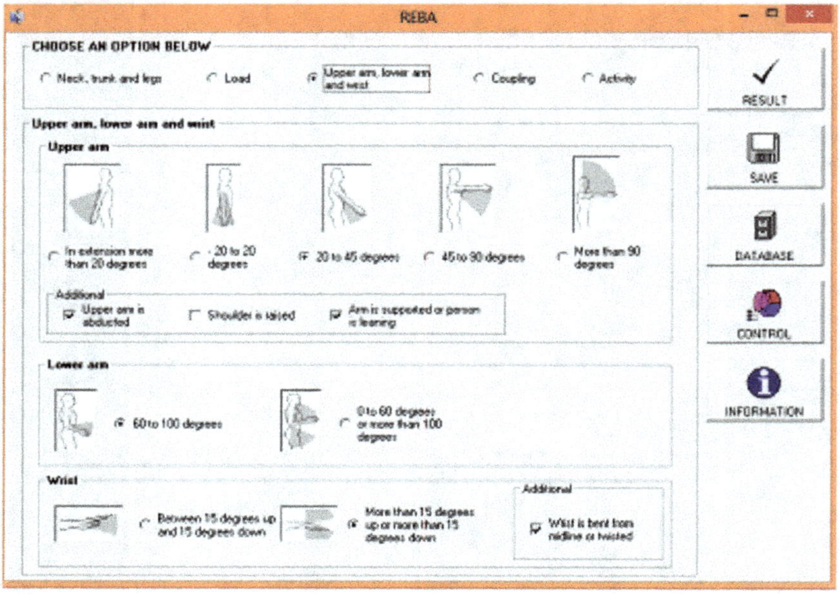

Figure 24. REBA inputs for type of coupling in manual weeding

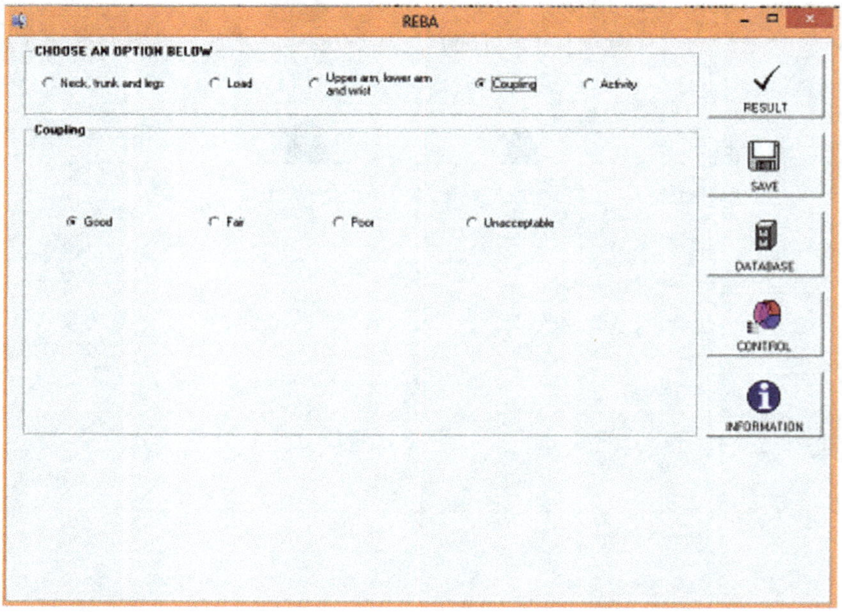

Figure 25. REBA inputs for type of activities in manual weeding

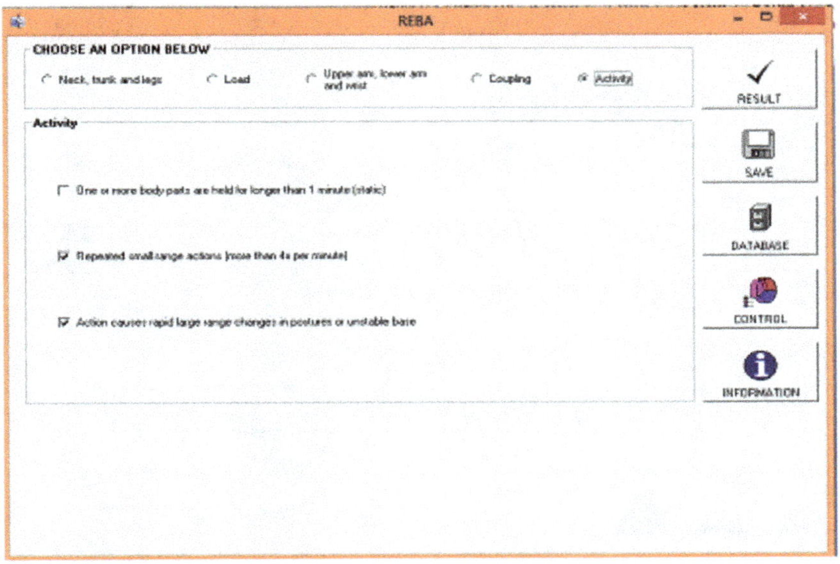

Figure 26. REBA output score and proposed action for manual weeding

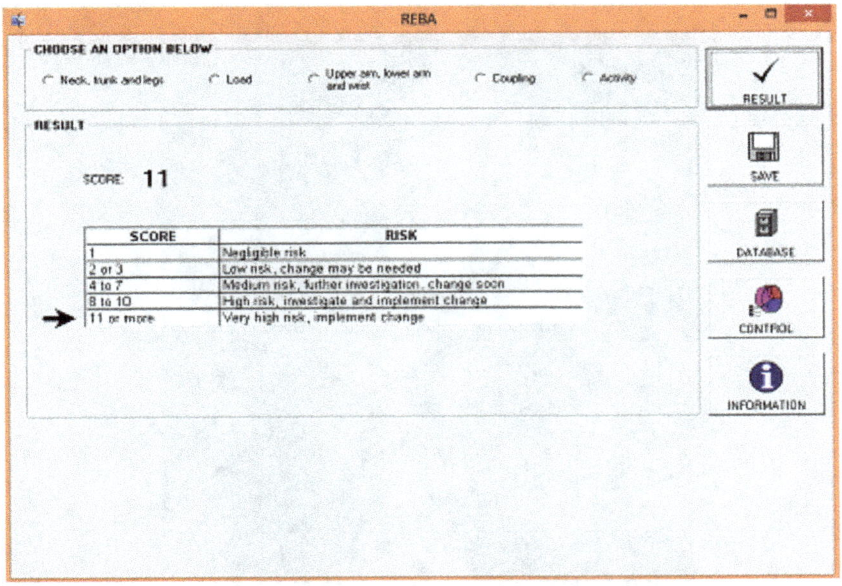

Figure 27. OWAS inputs and output for manual weeding

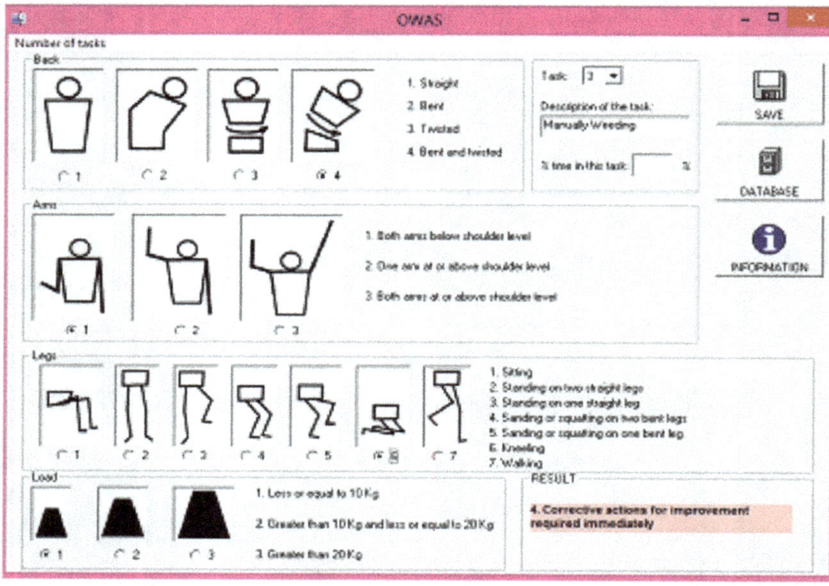

Figure 28. Harvesting manually by using sickle

twisted, trunk was at 20 to 60 degree, and was twisted, legs were supported in the two legs, and knee was bent at more than 60 degree (Figure 29), load was less than 5kg (Figure 30), upper arm was at 20 to 45 degree, it was abducted, shoulder was raised, and arm was supported & the farmer was leaning, lower arm was at 0 to 60

Figure 29. REBA inputs for neck, trunk and legs for manually harvesting

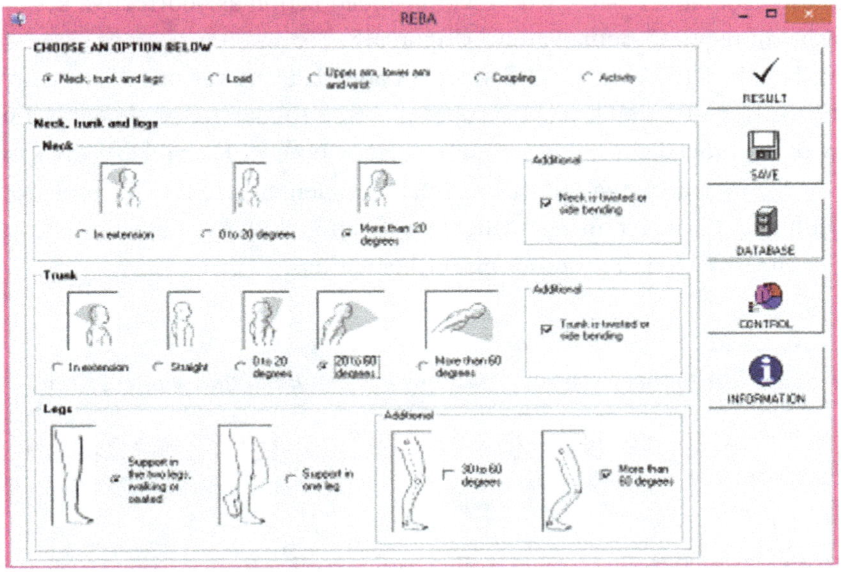

Figure 30. REBA inputs for loads in manually harvesting

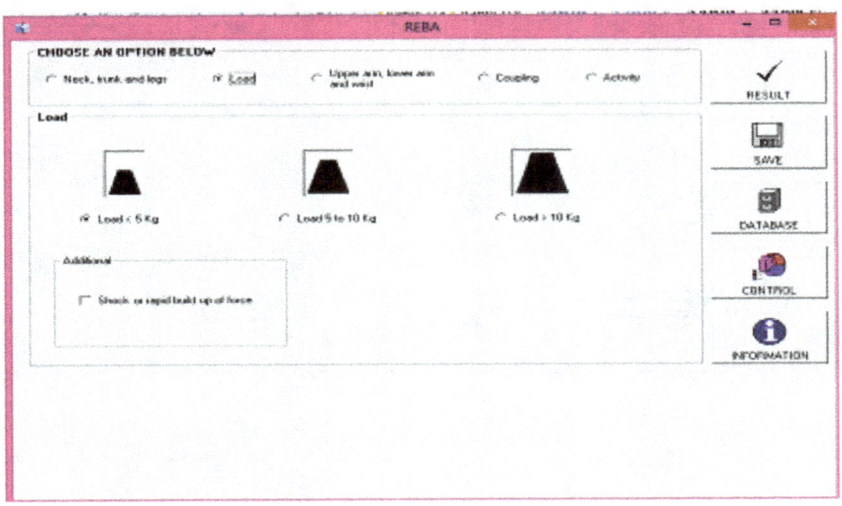

degree, wrist was at more than 15 degree up and down, and it was bent from midline (Figure 31), coupling was fair (Figure 32), activity was repeated, and there was a rapid change in postures (Figure 33). The corresponding REBA score obtained was 12, with suggested action as "very high risk, implement change" (Figure 34). While, in the OWAS analysis the inputs were as follows: the back was bent and twisted, both arms were below the shoulder level, the farmer was standing or squatting on two bent legs, and the load was less than or equal to 10kg. The corresponding OWAS score obtained was 4 with recommended action as "corrective actions for improvement required immediately" (Figure 35).

Based on the REBA and OWAS scores obtained during various selected farming activities, different action levels as proposed were summarized as shown in Table 6. In view of the proposed actions recommended by both REBA and OWAS scores, it was seen that for land preparation using tractor implements, the REBA tool proposes it as medium risk, and requiring further investigation & changes as soon as possible, where as the OWAS tool proposes no further actions.

Figure 31. REBA inputs for upper and lower arms in manually harvesting

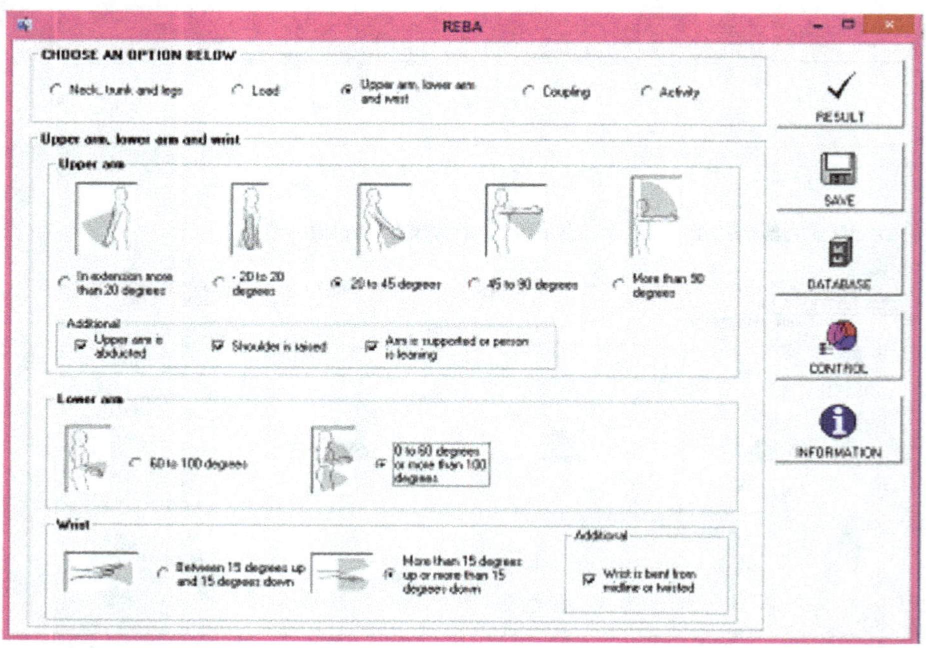

Figure 32. REBA inputs for type of coupling manually harvesting

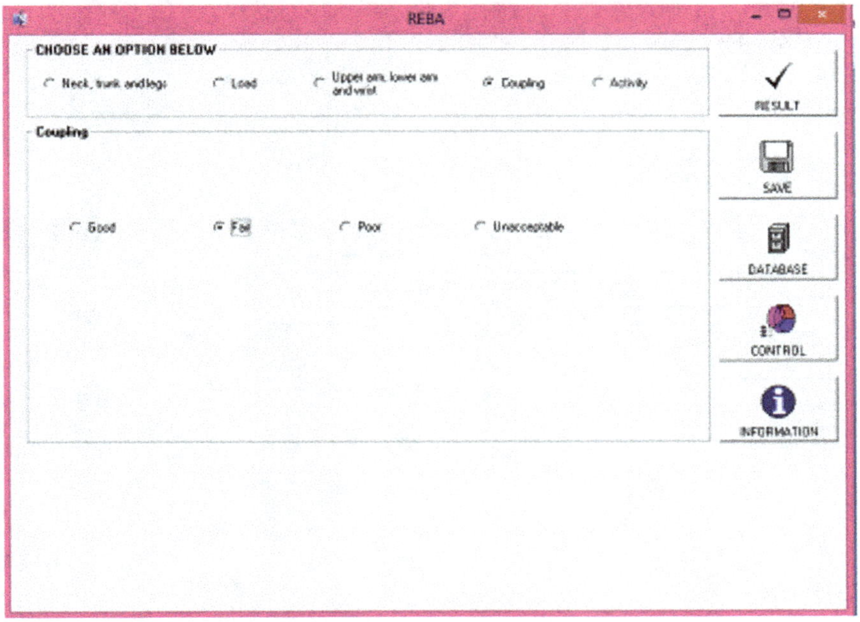

Figure 33. REBA inputs for type of activities in manually harvesting

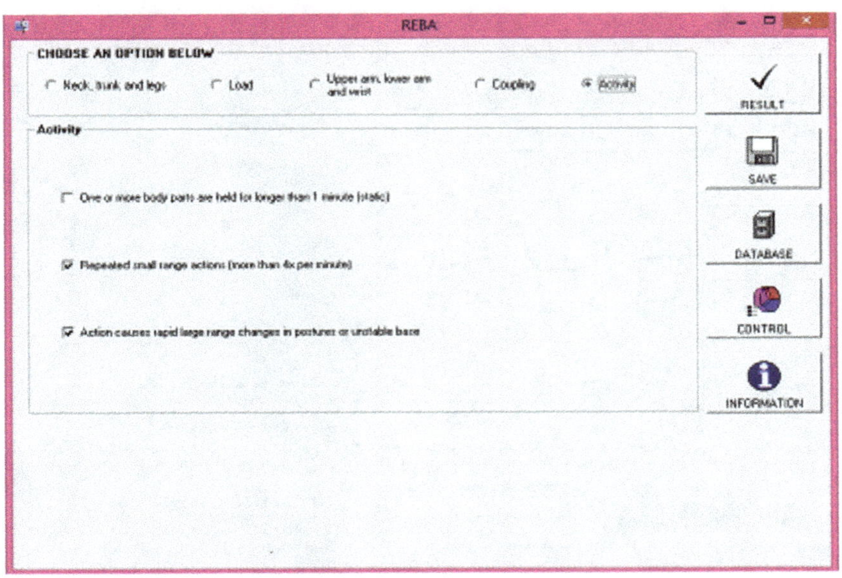

Figure 34. REBA output score and proposed action for manually harvesting

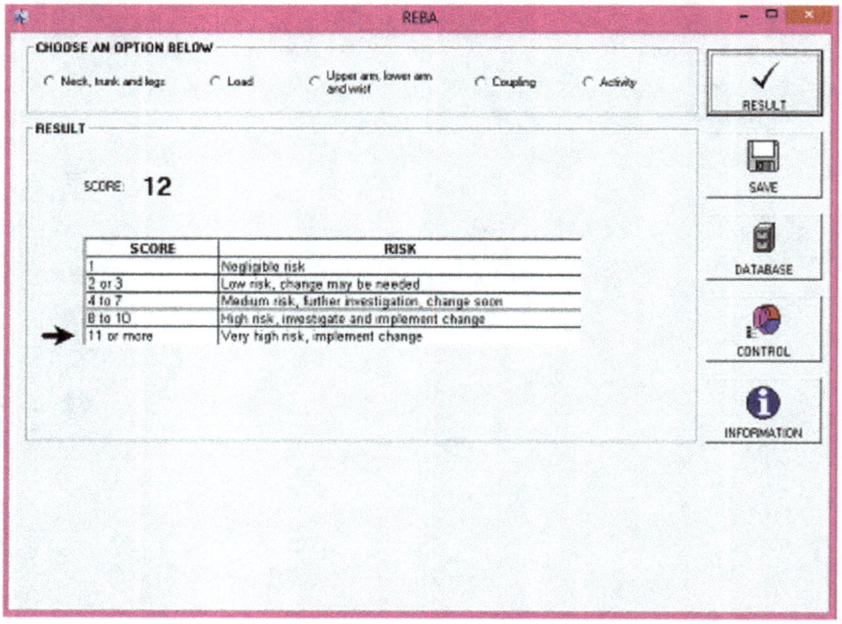

Figure 35. OWAS inputs and output for manually harvesting

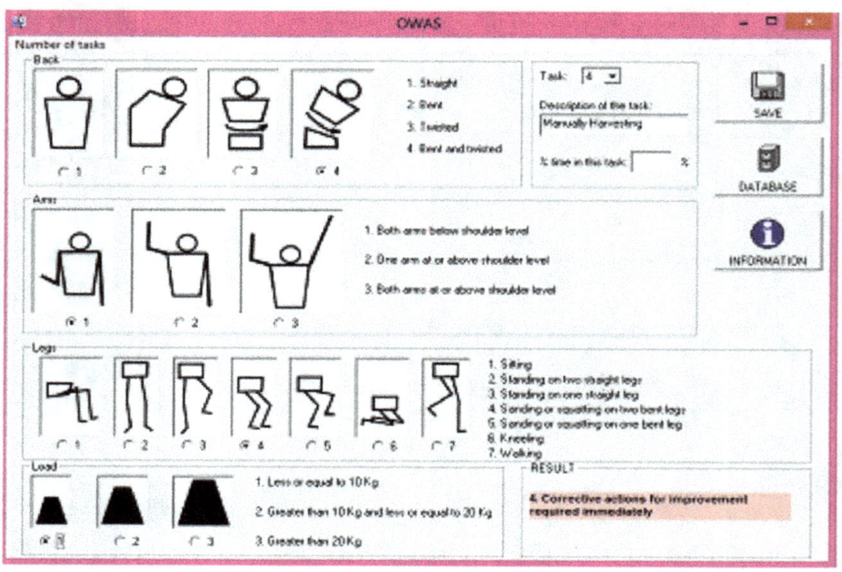

Table 6. Result summary of obtained REBA and OWAS scores

Activity	REBA Score	Proposed Actions	OWAS Score	Proposed Actions
Manually preparation of land	14	Very high risk, implement change	3	Corrective actions should be done as soon as possible
Land preparation using tractor implements	6	Medium risk, requires further investigation and change should be done as soon as possible	1	No actions required
Manually weeding for removing unwanted plants	11	Very high risk, implement change	4	Corrective actions for improvement required immediately
Manually harvesting	12	Very high risk, implement change	4	Corrective actions for improvement required immediately

CONCLUSION

Considering the agricultural hazards or injuries of farmers, it was observed that maximum number of farmers i.e. 93 (64.13%) farmers reported to have headache, shoulder, and elbow pain in the category of physical hazards, 85 (58.62%) farmers reported to suffer from fever in the category of biological hazards, 92 (63.44%) farmers reported of suffering headache, in the category of chemical hazards, and 76 (52.41%) farmers reported of physical tiredness in the category of psychological hazards, respectively.

A very high correlation i.e. 0.950 was obtained in between the psychological hazards and chemical hazards. Also, the correlation of age with all the four types of hazards was found to be significant at $p < 0.01$. From the BWM method, chemical hazards ranked the first followed by psychological hazards, physical hazards, and biological hazards. Further, based on the proposed actions recommended by both REBA and OWAS scores, it was seen that for land preparation using tractor implements, the REBA tool proposes it as medium risk, and requiring further investigation & changes as soon as possible, where as the OWAS tool proposes no further actions.

REFERENCES

Ahmadi, H. B., Kusi-Sarpong, S., & Rezaei, J. (2017). Assessing the social sustainability of supply chains using best worst method. *Resources, Conservation and Recycling*, *126*, 99–106. doi:10.1016/j.resconrec.2017.07.020

Annalee, Y., Karen, L., Ray, C., Mickey, K., & Marc Corbiere, E. (2007). Determinants of Healthcare Workers' Compliance with Infection Control Procedures. *Healthcare Quarterly*, *10*(1), 44–52. doi:10.12927/hcq.2007.18648 PMID:17326369

Askenazy, P. (2006). *Some determinants of reporting workplace accidents in France: The role of labour contract*. Paris: Centre Pour La Recherche Economique Et ses Applications.

Brandl, C., Mertens, A., & Schlick, C.M. (2017a). Ergonomic analysis of working postures using OWAS in semi-trailer assembly, applying an individual sampling strategy. *Int. J. Occup. Saf. Ergon, JOSE,* 1-8. . doi:10.1080/10803548.2016.1191224

Brandl, C., Mertens, A., & Schlick, C. M. (2017b). Effect of sampling interval on the reliability of ergonomic analysis using the Ovako working posture analysing system (OWAS). *International Journal of Industrial Ergonomics*, *57*, 68–73. doi:10.1016/j.ergon.2016.11.013

Curl, C. L., Fenske, R. A., Kissel, J. C., Shirai, J. H., Moate, T. F., Griffith, W., ... Thompson, B. (2002). Evaluation of take-home organophosphorus pesticide exposure among agricultural workers and their children. *Environmental Health Perspectives*, *110*(12), A787–A792. doi:10.1289/ehp.021100787 PMID:12460819

Earle, W. C. (1930). Malaria in Puerto Rico and its relation to the cultivation of Sugar Cane. *Southern Medical Journal*, *23*(5), 499–452. doi:10.1097/00007611-193005000-00021

Edmund, E. E. (2015). Analysis of Occupational Hazards and Safety of Workers in Selected Working Environments within Enugu Metropolis. *Journal of Environmental & Analytical Toxicology*, *5*(6), 1000337. doi:10.4172/2161-0525.1000337

Gupta, H., & Barua, M. K. (2017). Supplier selection among SMEs on the basis of their green innovation ability using BWM and fuzzy TOPSIS. *Journal of Cleaner Production*, *152*, 242–258. doi:10.1016/j.jclepro.2017.03.125

Habibullah, N. S., & Rajnarayan, R. T. (2004). Occupational Health Research in India. *Industrial Health*, *42*(2), 141–148. doi:10.2486/indhealth.42.141 PMID:15128163

Hellig, T., Mertens, A., & Brandl, C. (2018). The interaction effect of working postures on muscle activity and subjective discomfort during static working postures and its correlation with OWAS. *International Journal of Industrial Ergonomics*, *68*, 25–33. doi:10.1016/j.ergon.2018.06.006

Hignett, S., & McAtamney, L. (2000). Rapid entire body assessment (REBA). *Applied Ergonomics*, *31*(2), 201–205. doi:10.1016/S0003-6870(99)00039-3 PMID:10711982

International Labour Organisation. (2013). *ILO calls for urgent global action to fight occupational diseases*. Available from: http://www.ilo.org/global/about-the-ilo/media-centre/pressreleases/WCMS_211627/lang--en/index.htm

Kaergaard, A., & Andersen, J. H. (2000). Musculoskeletal disorders of the neck and shoulders in female sewing machine operators: Prevalence, incidence, and prognosis. *Occupational and Environmental Medicine*, *57*(8), 528–534. doi:10.1136/oem.57.8.528 PMID:10896959

Karhu, O., Härkönen, R., Sorvali, P., & Vepsäläinen, P. (1981). Observing workin postures in industry: Examples of OWAS application. *Applied Ergonomics*, *12*(1), 13–17. doi:10.1016/0003-6870(81)90088-0 PMID:15676393

Karhu, O., Kansi, P., & Kuorinka, I. (1977). Correcting working postures in industry: A practical method for analysis. *Applied Ergonomics*, *8*(4), 199–201. doi:10.1016/0003-6870(77)90164-8 PMID:15677243

Kee, D., & Karwowski, W. (2001). LUBA: An assessment technique for postural loading on the upper body based on joint motion discomfort and maximum holding time. *Applied Ergonomics*, *32*(4), 357–366. doi:10.1016/S0003-6870(01)00006-0 PMID:11461037

Khan, M. R., & Sudha, S. (2012). Evaluation of genotoxicity in automobile mechanics occupationally exposed to polycyclic aromatic hydrocarbons using micronuclei and other nuclear abnormalities. *Iranian Journal of Cancer Prevention*, *5*(2), 87–92. PMID:25628826

Kibret, S., Alemu, Y., Boelee, E., Tekie, H., Alemu, D., & Petros, B. (2010). The impact of a small-scale irrigation scheme on malaria transmission in Ziway area, Central Ethiopia. *Tropical Medicine & International Health*, *15*, 41–50. PMID:19917039

Manh, C. D., Beebe, N. W., Van, V. N., Quang, T. L., Lein, C. T., Nguyen, D. V., ... Cooper, R. D. (2010). Vectors and malaria transmission in deforested, rural communities in north-central Vietnam. *Malaria Journal*, *9*(1), 259. doi:10.1186/1475-2875-9-259 PMID:20846447

McAtamney, L., & Corlett, N. E. (1993). RULA: A survey method for the investigation of work related upper limb disorders. *Applied Ergonomics*, *24*(2), 91–99. doi:10.1016/0003-6870(93)90080-S PMID:15676903

Montano, D. (2014). Chemical and biological work-related risks across occupations in Europe: A review. *Journal of Occupational Medicine and Toxicology (London, England)*, *9*(28). Retrieved from http://www.occup-med.com/content/9/1/28 PMID:25071862

Naeini, H. S., Karuppiah, K., Tamrin, S. B., & Dalal, K. (2014). Ergonomics in agriculture: An approach in prevention of work-related musculoskeletal disorders (WMSDs). *J Agriculture and Environmental Sciences*, *3*, 33–51.

Nwankwo, C. M., Karanja, S., & Vasanthakaalam, H. (2018). The occurrence of occupational health hazards in districts health facilities in Kigali, Rwanda. *Int J Community Med Public Health*, *5*(1), 21–29. doi:10.18203/2394-6040.ijcmph20175509

Oluoch, I., Njogu, P., & Ndeda, J.O.H. (2017). Effects of Occupational Safety and Health Hazards' Exposure on Work Environment in the Water Service Industry within Kisumu County - Kenya. *IOSR Journal of Environmental Science, Toxicology and Food Technology, 11*(5-1), 46-51. doi:10.9790/2402-1105014651

Rezaei, J. (2015). Best-worst multi-criteria decision-making method. *Omega*, *53*, 49–57. doi:10.1016/j.omega.2014.11.009

Rezaei, J., Papakonstantinou, A., Tavasszy, L., Pesch, U., & Kana, A. (2018). Sustainable product-package design in a food supply chain: A multi-criteria life cycle approach. *Packaging Technology & Science*, 1–17. doi:10.1002/pts.2418

Rim, K.-T., & Lim, C.-H. (2014). Biologically Hazardous Agents at Work and Efforts to Protect Workers' Health: A Review of Recent Reports. *Safety and Health at Work*, *5*(2), 43–52. doi:10.1016/j.shaw.2014.03.006 PMID:25180133

Roman-Liu, D. (2014). Comparison of concepts in easy-to-use methods for MSD risk assessment. *Applied Ergonomics*, *45*(3), 420–427. doi:10.1016/j.apergo.2013.05.010 PMID:23849897

Salimi, N., & Rezaei, J. (2018). Evaluating firms' R&D performance using best worst method. *Evaluation and Program Planning*, *66*, 147–155. doi:10.1016/j.evalprogplan.2017.10.002 PMID:29091790

Sambo, M. N., Idris, S. H., & Shamang, A. (2012). Determinants of occupational health hazards among roadside automobile mechanics in Zaria, Northern Western Nigeria. *BOMJ*, *9*(1), 5–9.

Sharp, B. L., Kleinschmidt, I., Streat, E., Maharaj, R., Barnes, K. I., Durrheim, D. N., ... Grange, L. A. (2007). Seven years of regional malaria control collaboration--Mozambique, South Africa, and Swaziland. *The American Journal of Tropical Medicine and Hygiene*, *76*(1), 42–47. doi:10.4269/ajtmh.2007.76.42 PMID:17255227

Soheili-Fard, F., Rahbar, A., & Marzban, A. (2017). Ergonomic investigation of workers in tea factories using REBA and OWAS methods -case study: (Langroud region, Guilan, Iran). *Agricultural Engineering International: CIGR Journal*, *19*(3), 112–119.

Splansky, G. L., Corey, D., Yang, Q., Atwood, L. D., Cupples, L. A., Benjamin, E. J., ... Levy, D. (2007). The Third Generation Cohort of the National Heart, Lung, and Blood Institute's Framingham Heart Study: Design, recruitment, and initial examination. *American Journal of Epidemiology*, *165*(11), 1328–1335. doi:10.1093/aje/kwm021 PMID:17372189

Strazdins, L., & Bammer, G. (2004). Women, work and musculoskeletal health. *Social Science & Medicine*, *58*(6), 997–1005. doi:10.1016/S0277-9536(03)00260-0 PMID:14723897

Tadesse, T., & Kumie, A. (2007). Prevalence of factors affecting work-related injury among workers engaged in small and medium-scale industries in Gondar Wereda, North Gondar zone, Amhara Regional state, Ethiopia. *The Ethiopian Journal of Health Development*, *21*(1), 25–34. doi:10.4314/ejhd.v21i1.10028

Thangaraj, S., & Shireen, N. (2017). Occupational health hazards among automobile mechanics working in an urban area of Bangalore - a cross sectional study. *International Journal of Medical Science and Public Health*, *6*(1), 18. doi:10.5455/ijmsph.2016.31052016540

Triantaphyllou, E. (2000). Multi-criteria decision making methods. In *Multi-Criteria Decision Making Methods: A comparative Study*. Boston, MA: Springer. doi:10.1007/978-1-4757-3157-6_2 van Nieuwenhuyse, A., Somville, P. R., Crombez, G., Burdorf, A., Verbeke, G., Johannik, K., ... Moens, G. F. (2006). The role of physical workload and pain related fear in the development of low back pain in young workers: Evidence from the BelCoBack Study; results after one year of follow up. *Occupational and Environmental Medicine*, *63*(1), 45–52. doi:10.1136/oem.2004.015693 PMID:16361405

Vyavahare, R. T., & Kallurkar, S. P. (2012). Anthropometric and strength data of Indian agricultural workers for equipment design: A review. *Agricultural Engineering International: CIGR Journal*, *14*(4), 102–114.

You, P., Guo, S., Zhao, H., & Zhao, H. (2017). Operation performance evaluation of power grid Enterprise using a hybrid BWM-TOPSIS method. *Sustainability*, *9*(12), 23–29. doi:10.3390u9122329

KEY TERMS AND DEFINITIONS

Biological Hazard: A biological hazard refers to the biological substances that pose a threat to the health of human beings and other living organisms.

Chemical Hazard: It refers to the occupational hazard which is caused due to the exposure to chemicals in the workplaces.

Hazard: A hazard is something that causes harm.

Injury: An injury damages to a person's feelings.

Physical Hazard: A physical hazard is an agent, factor, or circumstance that causes harm with or without contact.

Psychological Hazard: It refers to those hazards that affect the mental wellbeing of the workers and affecting the ability of workers healthy and safety manner of working.

APPENDIX 1

Table 7. Questionnaire to farmers on agricultural hazards

Name (optional): Age: Gender: Male () / Female () Civil status: Married () / Un-married ()		
Agricultural Hazards/ Injury	Please mark Tick (√) where applicable	
Physical Hazards		
1. Sudden fall during work	Yes ()	No ()
2. Trauma	Yes ()	No ()
3. Crushing/attack	Yes ()	No ()
4. Animal kicks/scratches	Yes ()	No ()
5. Hearing loss	Yes ()	No ()
6. Eye injury	Yes ()	No ()
7. Headache, shoulder, elbow pain	Yes ()	No ()
8. Sunstroke	Yes ()	No ()
9. Body fatigue	Yes ()	No ()
10. Cardiovascular strain	Yes ()	No ()
11. Swollen hand	Yes ()	No ()
12. Neck pain	Yes ()	No ()
13. Slip disc-back pain	Yes ()	No ()
14. Insecticide illness	Yes ()	No ()
15. Animal bites	Yes ()	No ()
Biological Hazards		
1. Fever	Yes ()	No ()
2. Conjunctivitis	Yes ()	No ()
3. Amoebiasis	Yes ()	No ()
4. Tuberculosis	Yes ()	No ()
5. Biting by rabid dog	Yes ()	No ()
6. Ringworm infection	Yes ()	No ()
7. Infestation with internal and external parasites	Yes ()	No ()
Chemical Hazards		
1. Eczema	Yes ()	No ()
2. Allergies	Yes ()	No ()
3. Headache	Yes ()	No ()

continued on following page

Table 7. Continued

4. Eye Irritation	Yes ()	No ()
5. Nausea and Vomiting	Yes ()	No ()
6. Pesticides intoxication	Yes ()	No ()
7. Respiratory problems	Yes ()	No ()
Psychological Hazards		
1. Physical Tiredness	Yes ()	No ()
2. Psychological Fatigue	Yes ()	No ()
3. Boredom	Yes ()	No ()
4. Aggression	Yes ()	No ()
5. Hostility	Yes ()	No ()
6. Depression	Yes ()	No ()
7. Lack of Sleep	Yes ()	No ()
8. Forgetfulness	Yes ()	No ()
9. Irritability	Yes ()	No ()
10. Low Energy	Yes ()	No ()
11. Low Concentration	Yes ()	No ()

APPENDIX 2

Table 8. Sample questionnaire to identify the impact of agricultural hazards in the farming sectors of Odisha (India). The figures in the box indicate rating values in Likert scale.

Strongly disagree	Disagree	Neither disagree nor agree	Agree	Strongly Agree				
1	**2**	**3**	**4**	**5**				
Serial Number	Type of agricultural hazards		Ratings					
1	The impact of physical hazards is more in the agricultural sectors of Odisha.		1	2	3	4	5	
2	The impact of biological hazards is more in the agricultural sectors of Odisha.		1	2	3	4	5	
3	The impact of chemical hazards is more in the agricultural sectors of Odisha.		1	2	3	4	5	
4	The impact of psychological hazards is more in the agricultural sectors of Odisha.		1	2	3	4	5	

APPENDIX 3

Figure 36. Answer report

Microsoft Excel 12.0 Answer Report
Worksheet: [BWM Solver.xlsx]C=4
Report Created: 9/2/2011 12:05:17 AM

Target Cell (Min)

Cell	Name	Original Value	Final Value
$C$21	Ksi* Physical Hazards	0	0.833333333

Adjustable Cells

Cell	Name	Original Value	Final Value
$C$19	Physical Hazards	0	0.166666667
$D$19	Biological Hazards	0	0.138888889
$E$19	Chemical Hazards	0	0.416666667
$F$19	Psychological Hazards	0	0.277777778
$C$21	Ksi* Physical Hazards	0	0.833333333

Constraints

Cell	Name	Cell Value	Formula	Status	Slack
$C$25	Sum of weights Physical Hazards	1	$C$25=1	Not Binding	0
$C$27	Constraint 1 Physical Hazards	-0.833333333	$C$27<=$C$21	Not Binding	1.666666667
$D$27	Constraint 1 Biological Hazards	-0.833333333	$D$27<=$C$21	Not Binding	1.666666667
$E$27	Constraint 1 Chemical Hazards	-0.833333333	$E$27<=$C$21	Not Binding	1.666666667
$F$27	Constraint 1 Psychological Hazards	-0.833333333	$F$27<=$C$21	Not Binding	1.666666667
$C$28	Physical Hazards	0.833333333	$C$28<=$C$21	Binding	0
$D$28	Biological Hazards	0.833333333	$D$28<=$C$21	Binding	0
$E$28	Chemical Hazards	0.833333333	$E$28<=$C$21	Binding	0
$F$28	Psychological Hazards	0.833333333	$F$28<=$C$21	Binding	0
$C$30	Constraint 2 Physical Hazards	0.166666667	$C$30<=$C$21	Not Binding	0.666666667
$D$30	Constraint 2 Biological Hazards	0.138888889	$D$30<=$C$21	Not Binding	0.694444444
$E$30	Constraint 2 Chemical Hazards	0.416666667	$E$30<=$C$21	Not Binding	0.416666667
$F$30	Constraint 2 Psychological Hazards	0.277777778	$F$30<=$C$21	Not Binding	0.555555556
$C$31	Physical Hazards	-0.166666667	$C$31<=$C$21	Not Binding	1
$D$31	Biological Hazards	-0.138888889	$D$31<=$C$21	Not Binding	0.972222222
$E$31	Chemical Hazards	-0.416666667	$E$31<=$C$21	Not Binding	1.25
$F$31	Psychological Hazards	-0.277777778	$F$31<=$C$21	Not Binding	1.111111111
$C$19	Physical Hazards	0.166666667	$C$19>=0	Not Binding	0.166666667
$D$19	Biological Hazards	0.138888889	$D$19>=0	Not Binding	0.138888889
$E$19	Chemical Hazards	0.416666667	$E$19>=0	Not Binding	0.416666667
$F$19	Psychological Hazards	0.277777778	$F$19>=0	Not Binding	0.277777778

Figure 37. Sensitivity report

Microsoft Excel 12.0 Sensitivity Report
Worksheet: [BWM-Solver.xlsx]C=4
Report Created: 9/2/2011 12:05:27 AM

Adjustable Cells

Cell	Name	Final Value	Reduced Gradient
$C$19	Physical Hazards	0.166666667	0
$D$19	Biological Hazards	0.138888889	0
$E$19	Chemical Hazards	0.416666667	0
$F$19	Psychological Hazards	0.277777778	0
$C$21	Ksi* Physical Hazards	0.833333333	0

Constraints

Cell	Name	Final Value	Lagrange Multiplier
$C$25	Sum of weights Physical Hazards	1	0.833333333
$C$27	Constraint 1 Physical Hazards	-0.833333333	0
$D$27	Constraint 1 Biological Hazards	-0.833333333	0
$E$27	Constraint 1 Chemical Hazards	-0.833333333	0
$F$27	Constraint 1 Psychological Hazards	-0.833333333	0
$C$28	Physical Hazards	0.833333333	-0.166666667
$D$28	Biological Hazards	0.833333333	-0.138888889
$E$28	Chemical Hazards	0.833333333	-0.416666667
$F$28	Psychological Hazards	0.833333333	-0.277777778
$C$30	Constraint 2 Physical Hazards	0.166666667	0
$D$30	Constraint 2 Biological Hazards	0.138888889	0
$E$30	Constraint 2 Chemical Hazards	0.416666667	0
$F$30	Constraint 2 Psychological Hazards	0.277777778	0
$C$31	Physical Hazards	-0.166666667	0
$D$31	Biological Hazards	-0.138888889	0
$E$31	Chemical Hazards	-0.416666667	0
$F$31	Psychological Hazards	-0.277777778	0

Figure 38. Limits report

Microsoft Excel 12.0 Limits Report
Worksheet: [BWM-Solver.xlsx]Limits Report 1
Report Created: 9/2/2011 12:05:33 AM

Cell	Target Name	Value
$C$21	Ksi* Physical Hazards	0.833333333

Cell	Adjustable Name	Value	Lower Limit	Target Result	Upper Limit	Target Result
$C$19	Physical Hazards	0.166666667	0.166666667	0.833333333	0.166666667	0.833333333
$D$19	Biological Hazards	0.138888889	0.138888889	0.833333333	0.138888889	0.833333333
$E$19	Chemical Hazards	0.416666667	0.416666667	0.833333333	0.416666667	0.833333333
$F$19	Psychological Hazards	0.277777778	0.277777778	0.833333333	0.277777778	0.833333333
$C$21	Ksi* Physical Hazards	0.833333333	0.833333333	0.833333333	#N/A	#N/A

Chapter 10
Reducing Clinical Laboratory Footprints on the Environment With Intuitionistic Fuzzy Distance Measure

Vijay Kumar
Manav Rachna International Institute of Research and Studies, India

Jyoti Chawla
Manav Rachna International Institute of Research and Studies, India

Rajeev Kumar
Manav Rachna International Institute of Research and Studies, India

ABSTRACT

Medical diagnosis with the help of computational techniques is a very useful tool for doctors for the purpose of better diagnosis. In this chapter, an attempt has been made to reduce the number of tests required for diagnosis using generalized fuzzy sets for initial decision making as per the characteristics of ingested water in normal routine. The quality of water in terms of type and concentration of contaminants varies from region to region. The analogy between concentration of different contaminants in drinking water and risk analysis based on intuitionistic fuzzy sets (IFSs) have been investigated. Hypothetical data was processed in view of the finite set of heavy metals, diseases, and places to investigate the effect of selected heavy metal on the human health. The findings will not only help in diagnosis but also offer cost-cutting and ecofriendly strategy by avoiding unnecessary clinical laboratory examinations.

DOI: 10.4018/978-1-5225-8579-4.ch010

INTRODUCTION

Clinical laboratory investigations are one of the primary requirements in health care sector. Water is one of the fundamental requirements for supporting healthy life. Due to the industrialization, water is polluted by various contaminants. Water pollution due to cadmium, copper, lead, chromium, silver, thorium and arsenic is one of important concern throughout the world (Tao, 2016). The presence of such contaminants in water poses a potential health hazard to human beings from long term exposure by increasing the risk of certain diseases (Alves, 2017). Several studies have reported the occurrence of these contaminants in environmental water samples and their concentration in water was found to be more than the limits deemed safe by the World Health Organization (WHO) and US Environmental Protection Agency (EPA) (EPA, 2018; WHO, 2008). On 7th Jan, 2016 Flint, Michigan, was declared to be in state of emergency due to its drinking water contamination with lead (CNN, 2016). In Kolkata (India), 150 out of 250 blood samples of kids were tested positive for high levels of lead. Daily use items such as lead pencils, toys, cheap colors and crayons, cheap plastic mugs and wall paints are to be blamed for this lead poisoning (The Times of India, 2013). As per the survey dated January, 2015 conducted by the Punjab government on the contamination of vegetables reveals that they are contaminated with toxic metals like cadmium, chromium, lead, nickel etc (Dawn, 2015). In hospitals the number of patient with different kind of complaints are increasing day by day and contaminated drinking water is one of the main cause of diseases. Health problems due to the presence of contaminants in water include gastrointestinal infection, kidney damage, neurological disorders and even cancer (Kaur, 2016; Kumar, 2014; Kumar, 2015). Infants, young children, pregnant women, and the old people are particularly vulnerable to illness from some contaminants.

It is important to mention that only one contaminant type in drinking water may not be responsible for symptoms found in the patient as well as the diagnosis is always not about decision of yes or no because of involvement of numerous factors. Health care advisor/doctor will therefore be having many options after comprehensive analysis and may take the help of laboratory examinations in order to reach at a conclusion regarding the state of patient. In a recent study by Freedman 2015 it has been anticipated that $6.8 billion of medical investigations in the US entails avoidable testing that do not play any role in patient's improvement. However, many strategies are available to handle the laboratory waste but the steps considered necessary to put into practice for environmental management are either complex or very expensive. The decrease in the number of laboratory test will also give a significant contribution for conservation energy as well as water resources. Medical diagnosis with the help of computational techniques is very useful tool for doctors for the purpose of better diagnosis. The idea of multiple attribute decision making

(MADM) has been extensively used in real life situations. Moreover, it can also provide more precise and reliable information that may help the diagnostic process. When it is taken with fuzzy sets, it gives more strength to the concept. Among these, intuitionistic fuzzy set is considered as a promising technique for the purpose of better decision making.

BACKGROUND

The application of fuzzy sets in multiple criteria decision making (MCDM) problems is increasing in healthcare sector as the decision/output can help the health care advisors/physicians to execute regular tasks that need an extensive training and more time (Muthukumar 2016; Khattibi 2009; Atannassov 1986; Szmidt 2000; Zadeh 1965; Kumar 2016; Kumar 2013; Kumar 2012). Fuzzy set theory and fuzzy logic find many applications in medical field. Various models like fuzzy matrices have been proposed for medical diagnosis. Researchers have proposed different diagnostic models for fuzzy matrices to elucidate the relationship between symptoms and diseases. Burillo and Bustince, 1986 defined the distance measure between IFSs, and they first gave the axiomatic definition of IF entropy to characterize it. Szmidt and Kacprzyk (2001) proposed a new entropy method for IFS. Wang & Zhang (2009) and Wang (2008) proposed algorithms flavored with intuitionistic fuzzy information theory for multi-criteria decision making problems to execute right decision making with incomplete and imprecise information. Wan (2010) proposed various methods for the problems of Multi-criteria decision making. Yager (2004) modeled various MCDM techniques to streamline decision making. Wei (2012) elaborated the generalized concept of MCDM as explained by Yager (2004) and Zhang (2014) proposed some intuitionistic fuzzy operators to deal with MCDM problems.

Atanassov (1986) introduced the basic concepts of intuitionistic fuzzy sets (IFSs), and many researchers began to conduct studies for entropy measures on IFS from many kinds of viewpoints. Atanassov (1986) proposed IFS, which has the potential to deal with vagueness in some special situations and is defined as:

For a fixed set X, an IFS of A is defined as:

$$A=\left\{< x, \mu_A(x), \nu_A(x) > | \, x \in X\right\} \tag{1}$$

where, $\mu_A(x)$ *and* $\nu_A(x)$ define the degree of membership and non-membership of the element $x \in X$ to the set A. For every $x \in X$, $0 \leq \mu_A(x) + \nu_A(x) \leq 1$ and the amount $\pi_A(x) = 1 - \mu_A(x) - \nu_A(x)$ is termed as intuitionistic index or hesitation index and is the degree of indeterminacy concerning the membership of x in A. Also, $0 \leq \mu_A(x) + \nu_A(x) + \pi_A(x) = 1$, if $\pi_A(x) = 0$, IFS reduces automatically to fuzzy set. Szmidt (2000) introduced the popular distances between two fuzzy sets by taking into account the membership values $\mu(x)$ and non-membership values $\nu(x)$ of each element x in $X = \{x_1, x_2, \ldots\ldots, x_n\}$. He further extended the concept of distances to the intuitionistic fuzzy sets. He inserted a hesitancy degree or intuitionistic index π, in the intuitionistic fuzzy set by taking into account the membership values $\mu(x)$ and non-membership values $\nu(x)$ and intuitionistic index π of each element x.

MAIN FOCUS OF THE CHAPTER

The objective is to investigate the analogy between concentration of different contaminants in drinking water and risk analysis based on dynamic intuitionistic fuzzy weighted averaging (DIFWA) operator under intuitionistic fuzzy environment. The results may also be analyzed for investigating the health needs as well as effects in the different regions and important in many ways (Figure 1). The flow chart of methodology is given in Figure 2.

Figure 1. Utility of MCDM techniques in health care sector

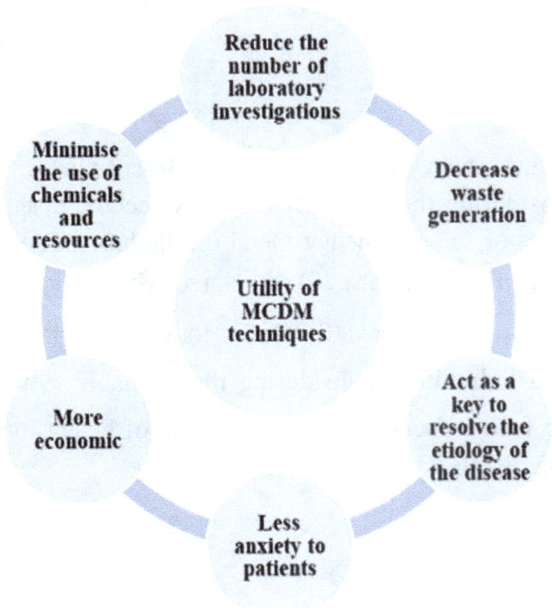

Figure 2. Flow chart of the methodology

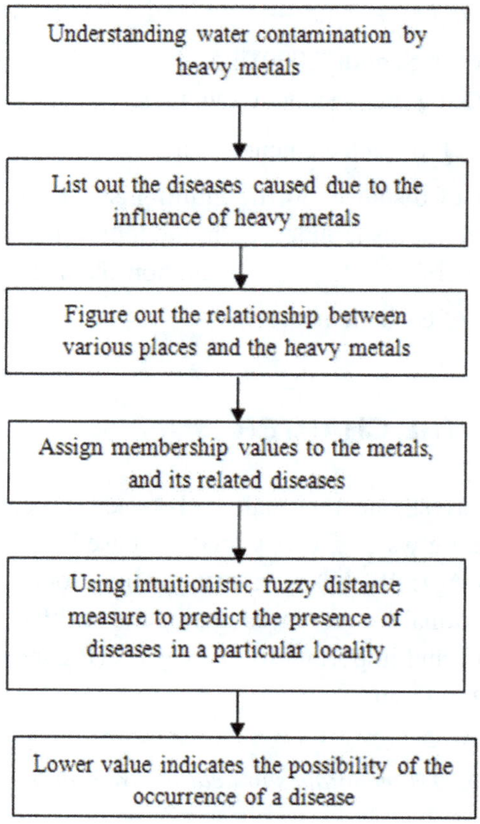

DISCUSSION

We have selected seven heavy metals and five different places. Hypothetical data was processed considering the finite set of heavy metals, diseases and places to investigate the effect of selected heavy metal on the human health. Decision was made for each place p_i $(i = 1, 2......, 5)$ associated with the set of heavy metals in different concentrations s_j $(j = 1, 2......, 7)$ to characterize the diseases d_k, $(k = 1, 2......, 7)$ in individuals considering the available symptoms. Individual fuzzy membership with respect to places, presence of heavy metals in water and their corresponding health effects are generated.

Distances Based on Fuzzy Sets and Intuitionistic Fuzzy Set

Let $\mu_A(x)$ and $\mu_B(x)$ be the membership function, $\nu_A(x)$ and $\nu_B(x)$ be the non-membership function, π_A and π_B be the intuitionistic index of fuzzy set A, B in $X = \{x_1, x_2, \ldots\ldots, x_n\}$. The distances between the two intuitionistic fuzzy sets A, B in $X = \{x_1, x_2, \ldots\ldots, x_n\}$ are:

Intuitionistic Normalized Euclidean Distance

$$INE(A, B) = \left[\frac{1}{n}\sum_{i=1}^{n}\left[\left(\mu_A(x_i) - \mu_B(x_i)\right)^2 + \left(\nu_A(x_i) - \nu_B(x_i)\right)^2 + \left(\pi_A(x_i) - \pi_B(x_i)\right)^2\right]\right]^{\frac{1}{2}} \tag{2}$$

With the condition that,

$$\mu_A(a_{ij}) + \nu_A(a_{ij}) + \pi_A(a_{ij}) = 1,$$

$$\mu_B(b_{ij}) + \nu_B(b_{ij}) + \pi_B(b_{ij}) = 1$$

A Hypothetical Study

To see the application of the method, let us frame a hypothetical case study: Let

$$S = \{ s_1, s_2, \ldots\ldots\ldots\ldots\ldots\ldots, s_m \}; \quad D = \{ d_1, d_2, \ldots\ldots\ldots\ldots\ldots\ldots, d_n \};$$

$$P = \{ p_1, p_2, \ldots\ldots\ldots\ldots\ldots\ldots, p_q \};$$

be the finite set of heavy metals, diseases and places respectively.

The fuzzy relations (FR), Q and R are defined as:

$$Q = \left\{< (p, s), \mu_Q(p, s), \nu_Q(p, s) > \mid (p, s) \in P \times S \right\};$$

$$R = \left\{ < (s,d)\,,\, \mu_R(s,d)\,,\, \nu_R(s,d) > \mid (s,d) \in S \times D \right\}$$

where, $\mu_Q(p,s)$ indicate the degree to which the symptom s appears in patient p and $\nu_Q(p,s)$ indicate the degree to which the symptom s does not appears in patient p. Similarly, $\mu_R(s,d)$ indicate the degree to which the symptom s confirm the disease d and $\nu_R(s,d)$ indicate the degree to which the symptom s does not confirm the disease d. We can estimate the level of heavy metals present in the water using the information obtained from the Table 1 of given case study. This information plays a vital role in the decision making process.

Let

$$Q = \{ Cd, Cu, Pb, Cr, Ag, Th, As \}$$

$$P = \{ P_1, P_2, P_3, P_4, P_5 \}$$

be the set of places where we drawn the sample of the water, be the set of inorganic impurities responsible present in the water and let

$$D = \left\{ Kidney\ disease, \ldots\ldots\ldots\ldots, Y \right\}$$

be the set of diseases. Suppose the IFR $Q(P \to S)$ is given by (hypothetically):

Table 1. Membership and non-membership values of places from where sample is drawn and the corresponding presence of heavy metals

Q	Cd		Cu		Pb		Cr		Ag		Th		As	
	μ_R	ν_R	μ_R	ν_R	μ_R	ν_R	μ_R	ν_R	μ_R	ν_R	μ_R	ν_R	μ_R	ν_R
P1	0.8	0.2	0.1	0.5	0.7	0.1	0.6	0.2	0.4	0.3	0.6	0.3	0.7	0.2
P2	0.0	0.7	0.6	0.4	0.8	0.1	0.4	0.4	0.7	0.1	0.5	0.2	0.4	0.5
P3	0.7	0.1	0.4	0.3	0.6	0.2	0.3	0.4	0.2	0.6	0.8	0.1	0.5	0.5
P4	0.5	0.3	0.3	0.6	0.2	0.5	0.0	0.6	0.3	0.4	0.5	0.5	0.2	0.6
P5	0.6	0.2	0.2	0.4	0.1	0.6	0.1	0.6	0.6	0.2	0.4	0.6	0.1	0.4

Suppose the IFR $R(S \to D)$ is given by (hypothetically): From the Table 1 and 2, our task is to do making decision making for each place p_i, $(i = 1, 2......, 5)$ associated with the set of heavy metals s_j, $(j = 1, 2......, 7)$, characterize for each disease d_k, $(k = 1, 2......, 7)$. The lowest distance obtained points out correct decision making. Using equation 1, we get the distances for all the jth heavy metal of the ith place from the kth diseases is given as: *Calculation of risk score and related issues*

The risk score was calculated and given in table 3 on a scale of 0 to 1. The lowest membership value indicates maximum risk of particular health problem at a place. The data may be used to rank the places on the basis of water quality. On the basis of risk score, system provides the decision regarding the type of health effect. Probable health effects on individuals belonging to different places with reference to the decision measure are evaluated (Figure 3). As per the decision output, the

Table 2. Membership and non-membership values for heavy metals and related diseases

R	Kidney disease		Liver		High BP		Neurological		Cancer		Allergy		Intestinal disease	
	μ_R	ν_R	μ_R	ν_R	μ_R	ν_R	μ_R	ν_R	μ_R	ν_R	μ_R	ν_R	μ_R	ν_R
Cd	0.4	0.4	0.3	0.6	0.2	0.7	0.1	0.7	0.4	0.4	0.5	0.4	0.2	0.6
Cu	0.3	0.6	0.1	0.8	0.4	0.5	0.6	0.2	0.2	0.3	0.1	0.7	0.6	0.2
Pb	0.2	0.7	0.6	0.3	0.2	0.6	0.5	0.3	0.7	0.2	0.6	0.2	0.4	0.5
Cr	0.1	0.8	0.2	0.7	0.7	0.2	0.4	0.4	0.3	0.6	0.4	0.5	0.7	0.1
Ag	0.4	0.6	0.6	0.3	0.3	0.5	0.5	0.4	0.5	0.4	0.2	0.7	0.8	0.1
Th	0.5	0.2	0.4	0.5	0.1	0.7	0.3	0.6	0.4	0.3	0.7	0.1	0.1	0.7
As	0.2	0.6	0.5	0.5	0.6	0.2	0.2	0.7	0.1	0.6	0.2	0.6	0.7	0.1

Table 3. Membership values of places and related diseases

Places	Kidney disease	Liver	High BP	Neurological	Cancer	Allergy	Intestinal disease
P1	0.3257	0.1971	0.26	0.2942	0.1828	0.1571	0.2714
P2	0.26	0.1428	0.2542	0.0999	0.1514	0.2256	0.1914
P3	0.1914	0.2199	0.3257	0.2285	0.1428	0.0914	0.3828
P4	0.0799	0.1285	0.2314	0.16	0.1542	0.1399	0.3442
P5	0.1428	0.2028	0.2628	0.2157	0.1514	0.2428	0.2942

Figure 3. Region wise comparison of health effects

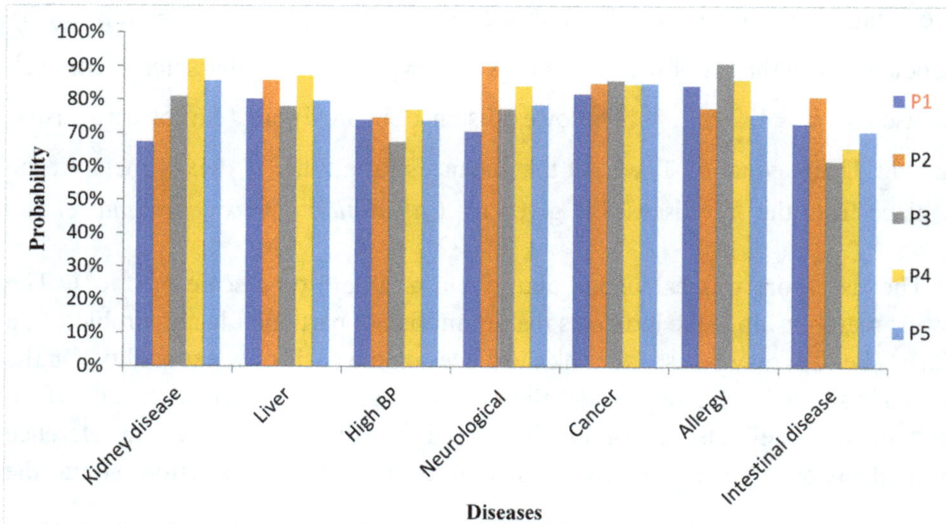

individual may be further advised for specific laboratory tests for confirmation. From the results it may be concluded that people drinking water from place 1& 3 have more probability of allergic disease followed by cancer and similarly disease wise ranking may be done for other regions as depicted in Figure 3. The work can be further extended to incorporate more contaminants and also be used for other environmental components for holistic risk assessment. It is important to mention that these types of tools can be used only if concentration of contaminants in environmental or drinking water samples is known. Therefore, regular monitoring of water quality is recommended in order to get the value of input parameters which is a fundamental requirement for MCDM tool. Therefore more monitoring studies are required to use MCDM as exploratory tools in order to define the trends for specific regions in terms of risk analysis.

CONCLUSION

The proposed method is proficient to compute the risk posed due to ingestion of heavy metal contaminated water. Fuzzy evaluation matrices for hazard parameters (Cd, Cu, Pb, Cr, Ag, Th and As) are established for different hypothetical places. Correspondingly, fuzzy evaluation matrices for health effects are also developed. Risk was calculated at these places using fuzzy compositional rules. Finally, these places are compared based on risk score for related diseases. On the basis of risk score, system provides the decision support regarding the type of health effect that will help the health care professional. The method may be further utilized to rank the places on the basis of decision measure using real time data of different water quality parameters in different locations. These findings will offer cost-cutting and environment friendly strategy by avoiding unnecessary laboratory examinations. It will also provide sustainable tool that can contribute to reduce the environmental impact caused by laboratory evaluations with regard to chemical/biomedical waste. Future work may also focus on prediction of water quality as per the variation of concentration of other contaminants.

REFERENCES

Alves, C. M., Ferreira, C. M. H., Soares, E. V., & Soares, M. V. M. (2017). A multi-metal risk assessment strategy for natural freshwater ecosystems based on the additive inhibitory free metal ion concentration index. *Environmental Pollution, 223*, 517–523. doi:10.1016/j.envpol.2017.01.053 PMID:28159400

Atanassov, K. T. (1986). Intuitionistic fuzzy sets. *Fuzzy Sets and Systems, 20*(1), 87–96. doi:10.1016/S0165-0114(86)80034-3

Burillo, P., & Bustince, H. (1986). Construction theorems for intuitionistic fuzzy sets. *Fuzzy Sets and Systems, 84*(3), 271–281. doi:10.1016/0165-0114(95)00313-4

Chowdhury, S., Mazumder, M. A. J., Al-Attas, O., & Husain, T. (2016). Heavy metals in drinking water: Occurrences, implications, and future needs in developing countries. *The Science of the Total Environment, 569–570*, 476–488. doi:10.1016/j.scitotenv.2016.06.166 PMID:27355520

EPA. (2008). The provision and quality of drinking water in Ireland: A report for the year. Environmental Protection Agency, 102.

Freedman, D. B. (2015). Towards better test utilization–strategies to improve physician ordering and their impact on patient outcomes. *The Electronic Journal of the International Federation of Clinical Chemistry and Laboratory Medicine, 26*(1), 15–30. Retrieved from https://edition.cnn.com/2016/03/04/us/flint-water-crisis-fast-facts/index.html https://timesofindia.indiatimes.com/city/kolkata/Lead-poisoning-affects-20-Kolkata-kids/articleshow/21062303.cms https://www.dawn.com/news/1157836

Jaishankar, M., Tseten, T., Anbalagan, N., Mathew, B. B., & Beeregowda, K. N. (2014). Toxicity, mechanism and health effects of some heavy metals. *Interdisciplinary Toxicology, 7*(2), 60–72. doi:10.2478/intox-2014-0009 PMID:26109881

Kaur, I., Mandiyal, D., Singh, B. P., Kumar, R., & Chawla, J. (2016). Amino-functionalized mesoporous MCM-41: An efficient adsorbent for the removal of chromium (III) ions from aqueous solution. *Journal of Water Supply: Research & Technology - Aqua, 65*(6), 480–493. doi:10.2166/aqua.2016.118

Khatibi, V., & Gholam Ali, M. (2009). Intuitionistic fuzzy set vs. fuzzy set application in medical pattern recognition. *Artificial Intelligence in Medicine, 47*(1), 43–52. doi:10.1016/j.artmed.2009.03.002 PMID:19375899

Kumar, R., & Chawla, J. (2014). Removal of cadmium metal from water/wastewater by nano-metal oxide: A review. *Water Quality, Exposure, and Health, 5*(4), 215–226. doi:10.100712403-013-0100-8

Kumar, R., Chawla, J., & Kaur, I. (2015). Removal of cadmium metal from water by carbon based nanosorbents: A review. *Journal of Water and Health, 13*(1), 18–33. doi:10.2166/wh.2014.024 PMID:25719462

Kumar, V., Arora, H., & Pal, K. (2016). Intuitionistic trapezoidal fuzzy prioritized weighted geometric operator: An algorithm for the selection of suitable treatment for lung cancer. *Proceedings of Fifth International Conference on Soft Computing for Problem Solving, 436*, 605-615.

Kumar, V., Arora, H. D., & Pal, K. (2016). Intuitionistic Trapezoidal Fuzzy Prioritized Weighted Average Operators: An Algorithm for the Suitable Treatment for Lung Cancer. *Journal of Applied Probability and Statistics, 11*(1), 91–104.

Kumar, V., Bharti, I., & Sharma, Y. K. (2012). Fuzzy diagnosis procedure of the types of glaucoma. *International Journal of Applied Information Systems, 1*(6), 42–45. doi:10.5120/ijais12-450191

Kumar, V., & Vashist, D. (2013). Fuzzy procedure for the selection of car among various brands. *International Journal of Engineering Research and Technology, 6*(3), 337–342.

Muthukumar, G. P., & Krishnan, S. S. (2016). A similarity measure of intuitionistic fuzzy soft sets and its application in medical diagnosis. *Applied Soft Computing, 41*, 148–156. doi:10.1016/j.asoc.2015.12.002

Szmidt, E., & Kacprzyk, J. (2000). Distances between intuitionistic fuzzy sets and Systems. *Information and Control, 114*, 505–518.

Szmidt, E., & Kacprzyk, J. (2001). Entropy for intuitionistic fuzzy sets. *Fuzzy Sets and Systems, 118*(3), 467–477. doi:10.1016/S0165-0114(98)00402-3

Tao, W., Chen, G., Zeng, G., Yan, M., & Wang, L. (2016). Influence of silver nanoparticles on heavy metals of pore water in contaminated river sediments. *Chemosphere*, *162*, 117–124. doi:10.1016/j.chemosphere.2016.07.043 PMID:27494311

Wan, S. P., & Dong, J. Y. (2010). Method of trapezoidal intuitionistic fuzzy number for multi-attribute group decision. *Control Decision*, *25*(5), 773–776.

Wang, J. Q., & Zhang, Z. H. (2008). Programming method of multi-criteria decision-making based on intuitionistic fuzzy number with incomplete certain information. *Control Decision*, *23*, 1145–1148.

Wang, J. Q., & Zhang, Z. H. (2009). Multi-criteria decision-making method with incomplete certain information based on intuitionistic fuzzy number. *Control Decision*, *24*, 226–230.

Wei, G. W. (2012). Hesitant fuzzy prioritized operators and their application to multiple attribute decision making. *Knowledge-Based Systems*, *31*, 176–182. doi:10.1016/j.knosys.2012.03.011

WHO. (2008). Guidelines for drinking water quality: Recommendations. World Health Organization.

Yager, R. R. (2004). Modeling prioritized multi-criteria decision making. *IEEE Transactions on Systems, Man, and Cybernetics. Part B, Cybernetics*, *34*(6), 2396–2404. doi:10.1109/TSMCB.2004.837348 PMID:15619938

Zadeh, L. A. (1965). Fuzzy sets. *Information and Control*, *8*(3), 338–353. doi:10.1016/S0019-9958(65)90241-X

Zhang, Z. (2014). Intuitionistic trapezoidal fuzzy prioritized operators and their application to multiple attribute group decision making. *Brit. J. Math. Comput. Sci.*, *4*(14), 1951–1998. doi:10.9734/BJMCS/2014/10133

Chapter 11
A Hybrid MCDM Method for Optimization of VAWT Performance Parameters

Agnimitra Biswas
National Institute of Technology Silchar, India

Jagadish
https://orcid.org/0000-0001-5699-5674
National Institute of Technology Raipur, India

Rajat Gupta
National Institute of Technology Mizoram, India

ABSTRACT

Vertical axis wind turbines (VAWT) have an inherent limitation of power performance for its low efficiency. However, the design of VAWT can be tailor-made to work in its built environment due to its varied advantages compared to other designs. Thus, the performance optimization problem entails a multitude framework of designs and operating conditions that must be satisfied for harnessing effective wind power from the turbine design. Furthermore, optimization of the performance of VAWT is considered to be an MCDM optimization problem. In this context, the chapter proposes a hybrid MCDM consisting of entropy and VIKOR-based methods for performance parameter optimizations of VAWT. To show the strength and applicability, a real-life case of optimizing performance parameters (Savonius type VAWT [SVAWT]) is used. The results show that SVWAT provides optimal results at an overlap ratio (16%), Vfree (34.763 m/s), N (3796 rpm), V (26.499 m/s), TSR (0.732).

DOI: 10.4018/978-1-5225-8579-4.ch011

INTRODUCTION

Vertical Axis Wind Turbine (VAWT)

Wind energy is not a new thing, it was also in vogue in the past. Wind turbines are the heart of the wind energy conversion system, as wind turbines converts wind energy into mechanical power. Wind turbines are of two types- horizontal axis wind turbine (HAWT) and vertical axis wind turbine (VAWT). VAWT's rotational axis is vertical, and in vertical orientation rotational plane of its blades is horizontal. VAWT is gaining much popularity and getting much attention especially due to possibility of installing in the built environment. Built environment is considered as the defined locations in urban areas, for e.g. building roof top, building front or back sides, and even location between two consecutive buildings. VAWTs are of various types; out of which Savonius rotor configuration has fascinated the research community and wind turbine project implementers due to varied reasons for e.g. ability to start on its own from any direction, simple configuration of two or three semi-circular cup shaped blades mounted on the vertical shaft in 'S' fashion (alphabet 'S' bifurcated at the middle giving two semi-circular sections of the blades), less noise, low starting torque etc. However, the major problem confronting this configuration is its poor efficiency compared to other types of VAWT or HAWT. To determine the best geometry, S.J. Savonius tested more than 30 different models of the Savonius turbine in wind tunnel and reported very encouraging results. Further, Savonius conducted tests in natural wind also. He reported that the S-rotor would run at a higher speed in natural wind than in the wind tunnel for the same wind velocity, thus ideally suited for the built environment. The best of the rotor models had an efficiency of 31% while the maximum reported efficiency of the prototype was 37% (Savonius, 1931). Inspired by the work of Savonius, Bach (Bach, 1931) made some investigations of the Savonius turbine reporting a maximum efficiency 24%. Between Sixties and early 21st century, many researchers (Macpherson, 1972; Newman, 1974; Khan, 1975; Modi, et al. 1984; Sharma, 2005; Biswas, et al. 2007; Sivasegaram, 1978 and Khan, 1988) had worked on Savonius rotors, and evaluated their performance coefficients. There are various designs and operating attributes for assessing performance of the Savonius wind turbine, for that matter any vertical axis wind turbine. Specific to the Savonius turbine, the design attributes are blade overlap ratio (Savonius, 1931;Bach, 1931; Newman,1974; Khan, 1975; Modi, et al. 1984; Sharma, 2005; Biswas,2007; Sivasegaram,1978; Khan,1988; Mojola, 1985; Nobuyuki, 1992; Gupta et al. 2008; Kamoji et al. (2008) ; Roth, 1982 and Mahamarakkalage, 1980), number of blades

(Modi, 1989; Alexander, 1978; Saylers, 1985 and Blackwell, 1978), aspect ratio (i.e. ratio of turbine height to its diameter (Newman,1974; Khan, 1975; Modi, et al. 1984; Sharma, 2005; Biswas,2007; Sivasegaram,1978; Khan,1988; Mojola, 1985; Nobuyuki, 1992; Gupta et al. 2008; Kamoji et al. (2008) ; Roth, 1982 ; Mahamarakkalage, 1980 and Saha, 2008), curtain design near the turbine (Altan & Atılgan, 2008a; Atılgan & Altan, 2008b) etc. The various operating conditions are free-stream wind speed, wind speed relative to the blades, blade tip-speed ratio, and rotational speed, which are variable due to the operating features of the built environment (Savonius, 1931; Biswas, 2007; Sivasegaram, 1978; Khan, 1988; Mojola, 1985; Nobuyuki, 1992; Gupta et al. 2008; Kamoji et al. (2008); Roth, 1982 and Mahamarakkalage, 1980). All these works have led to the determination of the turbine efficiency in the range 0.15-0.42. The performance parameters that depend on these attributes are power coefficient (C_p) and torque coefficient (C_t). Scientists have had attempted to optimize the Savonius turbine performance through parametric investigations so as to understand the contribution of these design parameters for the given operating conditions. Various researchers evaluated the performance of this turbine and obtained power coefficient in the range of 0.15 to 0.50 at various simulated conditions; however none of the above works attempted to optimize the performance of the Savonius turbine.

Optimization Issues in VAWT

A built environment entails different wind patterns that need to be utilized for effective and usable power generation by the Savonius type VAWT. However the main issue facing this design is its low efficiency. Hence scientists and researchers have been trying to improve the same for a considerable period of time over the last few decades. However, there is a still a valid reason why its efficiency needs to be enhanced for the built environment. This is because if Savonius turbine design can be improved by improving its efficiency, then this turbine can produce electrical energy in a standalone fashion. This will suffice the energy demands of small nuclear families living in urban areas. To this end, the process of performance improvement is not so straight forward as it strongly depends on variety of design conditions, which requires an optimization approach to solve the problem.

Further, in the built environment, the operating conditions are fast changing due to fluctuations resulting from the interaction of wind with the surrounding buildings, trees, or any physical objects. Due to these interactions, wind speeds, turbine rotational speeds, tip speed ratios etc. also change for any design condition thus requiring optimization of turbine performance under different operating conditions.

Thus, under the different attributes, the performance optimization problem entails a multitude framework of design and operating conditions that must be fulfilled for harnessing effective wind power by the turbine design.

Furthermore, optimization of the performance of Savonius wind turbine under different design and operating attributes is considered to be a multi-criteria decision making method (MCDM) because it involves number of criteria i.e. design and operating attributes, and number of alternatives as its performance parameters. Improper selection of design and operating attributes leads to the error in the performance of the turbine. Hence, optimization of performance parameters of the Savonius wind turbine is essential. Although MCDM has been applied in the field of wind energy, but the same has mostly been limited to blade material selection for wind turbine (Suresh et al. 2006; Lahu, et al.2014), selection of wind turbine support structure (Athanasios et al.2016), site specific wind turbine selection (Chamanehpour & Ahmadizadeh, 2017; Heracles & Stefan, 2014; Shafiqur & Salman, 2017; He-Yau, et al. 2011), selection of wind turbine for wind firm (Amy et al. 2012; Salman, 2016; Amy et al. 2009). There is hardly any instance of MCDM being applied to the performance optimization of Savonius VAWT, to the best of the authors' knowledge.

PROPOSED MCDM METHOD

MCDM methods are not new; these were also used in the past years. Scientists have applied MCDM in the optimization problem of variety of engineering areas; for those, some of the literature (Chakraborty et al. 2018; Chatterjee & Chakraborty, 2016; Ranjan et al. 2016; Ranjan et al. 2015; Rai et al. 2013; Chatterjee & Chakraborty,2013; Athawale et al. 2012; Chatterjee, et al. 2011; Chatterjee et al. 2010; Chatterjee, et al. 2009) may be referred. An integrated MCDM method consisting of entropy (Caliskan et al. 2013 ; Emma et al. 2013) coupled with VIKOR (Caliskan et al. 2013; Mohamed & Santawy, 2012) is illustrated for modeling and optimization of vertical axis wind turbine (VAWT) performance parameters for the built environment. The detailed steps of entropy and VIKOR method are as follows:

Step 1: Formulation of decision matrix

In this step, development of decision matrix is done based on the number of criteria and number of alternatives. The decision matrix includes performance values of each of the criteria corresponding to the alternatives. The decision matrix is evaluated using the following expression (Mohamed & Santawy, 2012).

$$DM_{ij} = \begin{bmatrix} & C1 & C2 & \ldots & Cn \\ A1 & X_{11} & X_{12} & \ldots & X_{1n} \\ A2 & X_{21} & X_{22} & \ldots. & X_{2n} \\ \ldots & \ldots & \ldots & \ldots. & \ldots. \\ Am & X_{m1} & X_{m2} & \ldots.. & X_{mn} \end{bmatrix} \text{For}(i = 1, 2, \ldots.m,\ j = 1, 2, \ldots.n) \quad (1)$$

where DM_{ij} represents the response value of ith criteria on jth alternative; X_{11}, X_{12}, X_{13}.... X_{mn} denote the performance values of each of the criteria with respect to the corresponding alternative; m represents the number of criteria and n represents the number of alternatives.

Step 2: Determination of precise priority weights using entropy

In this step, precise priority weights for each of the criteria are determined using Equations (2-4). Here, first, degree of divergence (d_j) for each of the criteria is calculated using Equations (2-3), which measures the distance of each criteria with respect to the comparable sequence (Caliskan et al. 2013). After that, precise priority weights are determined using Equation (4).

$$\Psi_j = -k \sum_{i=1}^{m} P_{ij} \ln\left(P_{ij}\right) \quad j = 1, 2, \ldots, n. \quad (2)$$

$$d_j = \left|1 - \Psi_j\right|. \quad (3)$$

$$\zeta_j = \frac{d_j}{\sum_{j=1}^{n} d_j}. \quad (4)$$

where k is Boltzmann's constant, $[k = 1 / \ln m, 0 \leq d_j \leq 1]$, ζ_j are the precise priority weights for each of the criteria, Ψ_j.

Step 3: Determination of best (B+), worst (B-) values and values of E_i and F_i

After the criteria weights, best (B^+), worst (B^-) and values of E_i and F_i for each of the criteria are determined using the following expressions (Caliskan et al. 2013; Mohamed & Santawy, 2012).

$$B^+ = Max\left(X_{ij}\right) \tag{5}$$

$$B^- = Min\left(X_{ij}\right) \tag{6}$$

$$E_i = \sum_{j=1}^{n}\left\{\zeta_j \times \left(\frac{(X_{ij})_{Max} - (X_{ij})}{(X_{ij})_{Max} - (X_{ij})_{Min}}\right)\right\} \tag{7}$$

$$F_i = Max^n \sum_{j=1}^{n}\left\{\zeta_j \left(\frac{(X_{ij})_{Max} - (X_{ij})}{(X_{ij})_{Max} - (X_{ij})_{Min}}\right)\right\} \tag{8}$$

where, B^+ and B^- represents the best and worst criteria from the decision matrix, ζ_j are the criteria weights.

Step 4: Determination of performance index (*PI*) values

Next, determination of overall performance index (*PI*) values for each of the alternatives is done in order to assess the optimal criteria. This step converts the multi-criteria into a single criteria subjected to certain constraints with consideration of both best/beneficial and worst/non-beneficial criteria. The performance index values for each of the alternatives are calculated using the following expressions (Caliskan et al. 2013; Mohamed & Santawy, 2012).

$$PI_i = \left\{v\left(\frac{E_i - Min(E_i)}{Max(E_i) - Min(E_i)}\right) + (1-v)\left(\frac{F_i - Min(F_i)}{Max(F_i) - Min(F_i)}\right)\right\} \tag{9}$$

where, PI_i represents the performance index values for ith alternative, Max(E_i) is the maximum value of E_i, and Min(E_i) is the minimum value of E_i, and Max(F_i) is the maximum value of F_i, and Min(F_i) is the minimum value of F_i; v is used as the weight strategy of the majority of criteria and its value is taken as 0.5, while its value is generally in the range 0 to 1.

At last, compromised ranking of the alternatives is carried out based on the PI_i values. The best alternatives are determined as the ones with the minimum value of PI_i in the ranking. The corresponding criteria values are considered as optimal criteria corresponding to the optimal alternatives for the system (Caliskan et al. 2013; Mohamed & Santawy, 2012).

CASE STUDY: MODELING AND OPTIMIZATION OF SAVONIUS TYPE VAWT (SVAWT) PERFORMANCE PARAMETERS

Background

As already discussed, performance optimization of Savonius type VAWT is an important task when it is intended to be installed in the built environment with the variability in operating and design conditions. Although estimation of small-scale performance of this turbine has been done by large number of scientists since the middle of the last century; however MCDM based optimization has not been solved by any author so far to the best of the present authors' knowledge.

The Savonius turbine design considered for this research is a three-bladed design as shown in Figure 1.

The blades are mounted on a vertical shaft using nut and bolt. The turbine unit is supported on a ball bearing fixed at the bottom of the shaft. The bearing is housed within a mild steel cylindrical cross-section base. The individual blade diameter is 8 cm and height is 20 cm. Each blade is of 3 mm thickness. The shaft is made up of mild-steel with 1.5 cm diameter. The blades are made from lightweight aluminium alloy. Aluminum has been selected due to its low weight, high strength, easy workability, and excellent corrosion resistant property. It has a density of 2700 kg/m^3, which is about one third that of steel. The aluminium sheet is first cut into the dimensions of the required diameter and height of the rotor as per the design (Figure 1). Then these are bent in the form of a semi-circular arc of the required diameter-using sheet bending machine. Washers & nuts having knurled surface have done variation of the overlap between semi-circular parts. By changing the number of nuts and washers between the blades, overlap has been increased or decreased. As the overlap has increased, the overall rotor diameter has decreased and vice versa.

Figure 1. Schematic diagram of a three-bladed Savonius turbine

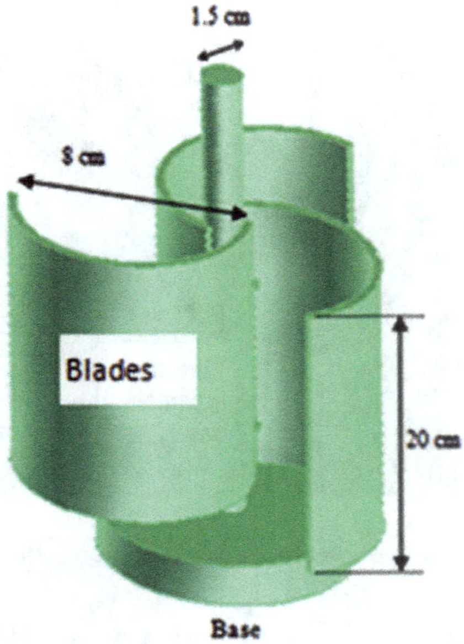

Overlap is the gap between the inner edges of the blades across the shaft. The overlap variations have been done in the range of 0 to 35% of the turbine diameter. These design parameters of the SVAWT, i.e. overall size (height and diameter) and overlap ratio are considered by referring to standard literature available in this field, for e.g. (Macpherson, 1972; Newman, 1974; Khan, 1975; Modi, et al. 1984; Sharma, 2005; Biswas, et al. 2007; Sivasegaram, 1978 and Khan, 1988).

Experimental Procedure

The experiments are carried out in an open circuit wind tunnel. It is a subsonic type of Wind- tunnel capable of producing a maximum velocity of 35 m/s. The tunnel is run by a centrifugal blower having an axial flow fan with variable pitch blades placed in a cylindrical casing. The capacity of the fan is 6700 cfm. A three phase 15 kW motor having rated rpm 2890 drives the fan. It consists of a test section of height and width 30 cm and length 300 cm respectively. At the exit side one gate opening is adjusted to achieve a wide range of velocities up to a maximum of 35 m/s. The wind tunnel consists of a honeycomb chamber at the entrance to break the large-scale disturbances and eddies followed by a settling chamber to create uniform flow in the test section. The picture of the wind tunnel is shown in Figure 2. As can be seen from Figure 2 (b), the Savonius turbine model is placed inside the test section to test its aerodynamic performance. The flow uniformity of the test section

Figure 2. (a) Convergent cone of the wind tunnel;(b) Test section of the wind tunnel of the wind tunnel with divergent cone at its end

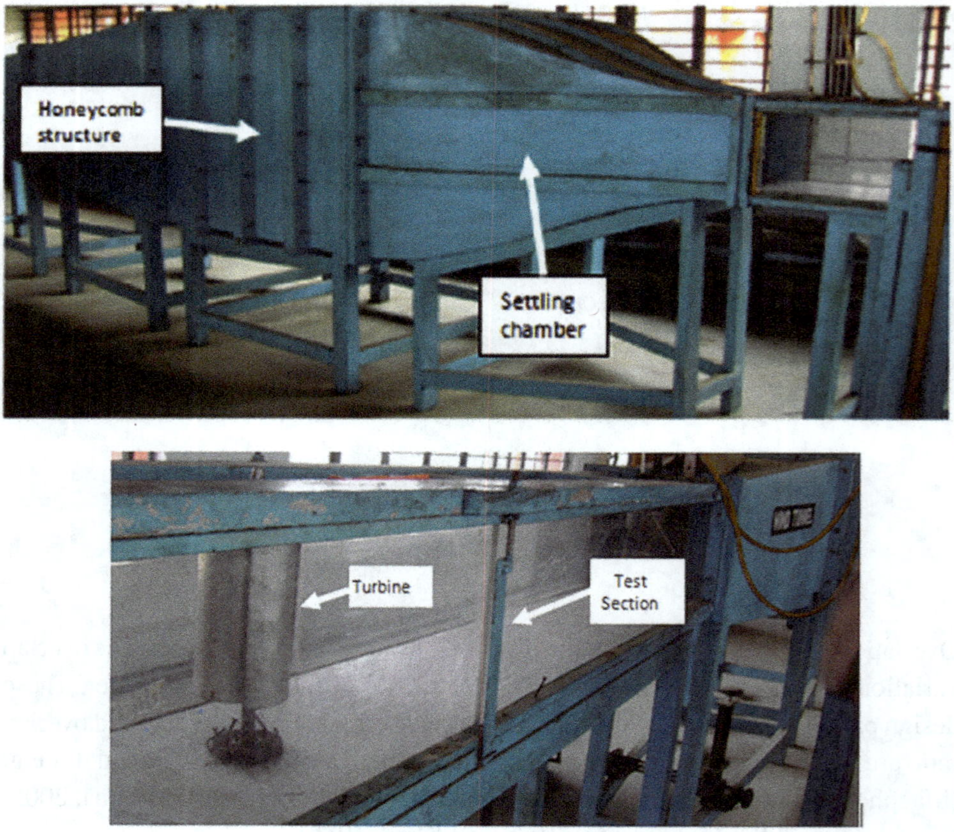

is within ± 1%. The turbulence intensity in the test section is less than ±1%. The wind velocity is measured using an anemometer with an accuracy of ±2%. The speed of the Savonius turbine is measured using a digital tachometer (range 10–99,999 rpm) with an accuracy of ±0.05%.

Modeling of SVAWT Using Proposed MCDM Method

In this section, modeling of vertical axis wind turbine in the built environment using proposed MCDM method (Caliskan et al. 2013; Mohamed & Santawy, 2012; Emma et al. 2013) is discussed. The proposed method used is the entropy method with VIKOR. Here, entropy method is used for priority weights calculation of different criteria while VIKOR is used for obtaining optimal operating conditions of the vertical axis wind turbine in order to get higher C_p and C_t. In this work, five input parameters with 30 numbers of trails are considered as alternatives and two output parameters i.e. C_p and C_t are considered as number of criteria.

In the modeling, first, formulation of decision matrix is done using Equation (1). In this work, there are 30 numbers of trails are performed by varying the five input parameters such as overlap ratio in %, free stream velocity (V_{free}) in m/s, speed (N) in rpm, approach velocity (V) in m/s and λ (tip-speed ratio, TSR) and corresponding output values like power coefficient (C_p) and torque coefficient (C_t) are determined and results are tabulated in Table. 1. Second, priority weights for each of the vertical axis wind turbine parameters or criteria such as power coefficient (C_p) and torque coefficient (C_t) are determined using entropy method. Here first, degree of divergence (d_j) for each of the vertical axis wind turbine parameters or criteria are calculated using Equations (2-3). The degree of divergence measures the distance of each turbine parameter with respect to the comparable sequence values. After that, precise priority weights for each of the vertical axis wind turbine parameters or criteria are determined using Equation (4) and the results are shown in Table 1. The precise priority weights (ζ) for the SVAWT response parameters such as C_p and C_t are found out to be 0.3890 and 0.6110 respectively (Caliskan et al. 2013). The result indicates that response parameter C_t is most influencing parameter / criteria compared to the C_p.

In the fourth step, determination of best (B^+), worst (B^-) values as well as values of E_i and F_i for each of the SVAWT parameters / criteria are done using Equations (5-7). This measures the separation distance of each of the SVAWT parameters from the reference values (Caliskan et al. 2013; Mohamed & Santawy, 2012). The SVAWT parameters or criteria with higher E_i and F_i values indicate the least influence while least E_i and F_i values indicate the most influence on the SVAWT performance. The results of E_i and F_i values for each of the SVAWT parameters are shown in Table 2. Thereafter, overall performance index (PI) values for SVAWT are determined using Equation (8). This step converts multi-criteria into a single criteria subjected to certain constraints with consideration of both best/beneficial and worst/non-beneficial criteria of the SVAWT parameters into account. The overall performance index values for the SVAWT are tabulated in Table 2.

Table 1. Priority weights of the VAWT response parameters

Priority Weights (ζ)	**Response parameters**	
	Power Coefficient (C_p)	Torque Coefficient (C_t)
	0.3890	0.6110

Table 2. E_i, F_i, and PI_i values for SVAWT parameters

Trail No.	E_i	F_i	PI
1	0.384	0.245	0.314
2	0.501	0.326	0.466
3	0.410	0.251	0.337
4	0.427	0.246	0.343
5	0.423	0.262	0.355
6	0.128	0.128	0.037
7	0.250	0.147	0.134
8	0.156	0.089	0.018
9	0.134	0.098	0.012
10	0.179	0.127	0.068
11	0.186	0.186	0.129
12	0.321	0.265	0.293
13	0.184	0.120	0.066
14	0.187	0.116	0.064
15	0.331	0.184	0.221
16	0.772	0.611	0.914
17	0.540	0.376	0.540
18	0.334	0.334	0.367
19	0.442	0.344	0.446
20	0.411	0.331	0.414
21	0.574	0.329	0.516
22	0.256	0.244	0.231
23	0.547	0.348	0.517
24	0.313	0.298	0.319
25	0.408	0.278	0.360
26	0.722	0.424	0.702
27	0.386	0.357	0.422
28	0.480	0.352	0.478
29	0.906	0.518	0.911
30	0.863	0.474	0.842

RESULTS AND DISCUSSIONS

The results generated from the experimental investigations of various parameters are shown in Table 3. In this table, V_{free} is free-stream wind velocity ahead of the turbine model in the test section whereas V is wind velocity relative to the turbine blades. The second velocity is lower than the first as the wind speed is harnessed by the turbine to produce power. The range of various parameters (overlap ratio, V_{free}, V, TSR, N) of Table 3 is selected from the available literature (Biswas et al. 2007; Gupta et al. 2008).

Optimization Results of SVAWT Parameters

The optimization of SVAWT parameters is discussed in this section. The overall performance index values are used for selection of optimal operating conditions for SVAWT. Ranking of each performance value for the 30 trails has been carried out and optimization results are tabulated in Table 4.

The optimization results as depicted in Table 4 shows that, the trail no. 9 yields least PI value among the 30 trails of experiments. The PI value obtained for the optimal setting or optimal trail no. 9 is 0.012 and shows least value among the other experiments. The optimal settings obtained are overlap ratio (16%), V_{free} (34.763 m/s), N (3796 rpm), V (26.499 m/s), λ (0.732 TSR). The optimal setting obtained via. Entropy-VIKOR method [24, 26] provides the most optimal response / output parameters for the SVAWT. The optimal conditions provide higher C_p and higher C_t as well as improve the performance and efficiency of the SVAWT. This work also suggests that the proposed hybrid MCDM method can be utilized as systematic framework model for modeling and optimization of other wind turbines.

Confirmatory Tests

Additionally, confirmatory tests are performed for the optimal conditions to confirm the results. The optimal conditions for the confirmatory tests are taken from the optimization using Entropy-VIKOR. The confirmatory test results are tabulated in Table 5. The results show that confirmatory experiments are obtained to be comparable and acceptable with the experimental results.

Table 3. Experimental readings of SVAWT

Trail No.	Input Parameters					Output Parameters	
	Overlap ratio (OR) (%)	Free stream velocity (V_{free}) (m/s)	Speed (N) (rpm)	Approach Velocity (V) (m/s)	Tip-speed ratio (λ) (TSR)	Power Coefficient (C_p)	Torque Coefficient (C_t)
1	0	32.580	2200	27.704	0.495	0.357	1.224
2	0	34.997	2360	28.575	0.494	0.329	1.176
3	0	34.879	2430	28.859	0.510	0.342	1.220
4	0	34.763	2459	27.108	0.518	0.325	1.223
5	0	35.043	2397	28.859	0.501	0.340	1.214
6	16	32.58	3308	23.901	0.680	0.366	1.369
7	16	34.997	3446	25.558	0.661	0.351	1.308
8	16	34.879	3741	26.499	0.718	0.414	1.316
9	16	34.763	3796	26.499	0.732	0.438	1.311
10	16	34.290	3752	26.499	0.733	0.425	1.294
11	20	32.580	3415	25.875	0.669	0.466	1.259
12	20	34.997	3761	28.859	0.686	0.422	1.212
13	20	34.879	3605	26.189	0.660	0.372	1.331
14	20	34.763	3600	26.189	0.661	0.375	1.327
15	20	35.043	3368	26.499	0.614	0.351	1.260
16	25	32.580	1890	32.328	0.343	0.340	1.007
17	25	34.997	1900	25.558	0.324	0.172	1.272
18	25	34.879	1945	27.407	0.332	0.205	1.369
19	25	34.763	1981	26.499	0.340	0.197	1.311
20	25	35.043	2130	26.499	0.362	0.207	1.322
21	30	32.580	2285	23.907	0.389	0.209	1.224
22	30	34.997	2607	28.575	0.413	0.275	1.362
23	30	34.879	2640	29.969	0.420	0.310	1.163
24	30	34.763	2700	25.558	0.431	0.233	1.360
25	30	35.043	2636	27.108	0.417	0.249	1.292
26	35	35.040	2306	29.140	0.337	0.233	1.118
27	35	34.997	2344	25.875	0.343	0.187	1.352
28	35	34.879	2176	26.957	0.320	0.191	1.293
29	35	34.763	1251	32.705	0.184	0.163	1.062
30	35	35.043	1311	32.202	0.192	0.162	1.088

Table 4. Performance index values (PI) values for SVAWT parameters

Trail No.	Input Parameters					Performance Index (PI) Values	Rank
	Overlap Ratio (OR) (%)	Free Stream Velocity (V_{free}) (m/s)	Speed (N) (rpm)	Approach Velocity (V) (m/s)	Tip-Speed Ratio, λ (TSR)		
1	0	32.58	2200	27.704	0.495	0.314	13
2	0	34.997	2360	28.575	0.494	0.466	23
3	0	34.879	2430	28.859	0.51	0.337	15
4	0	34.763	2459	27.108	0.518	0.343	16
5	0	35.043	2397	28.859	0.501	0.355	17
6	16	32.58	3308	23.901	0.68	0.037	3
7	16	34.997	3446	25.558	0.661	0.134	8
8	16	34.879	3741	26.499	0.718	0.018	2
9	**16**	**34.763**	**3796**	**26.499**	**0.732**	**0.012**	**1**
10	16	34.29	3752	26.499	0.733	0.068	6
11	20	32.58	3415	25.875	0.669	0.129	7
12	20	34.997	3761	28.859	0.686	0.293	11
13	20	34.879	3605	26.189	0.66	0.066	5
14	20	34.763	3600	26.189	0.661	0.064	4
15	20	35.043	3368	26.499	0.614	0.221	9
16	25	32.58	1890	32.328	0.343	0.914	29
17	25	34.997	1900	25.558	0.324	0.540	27
18	25	34.879	1945	27.407	0.332	0.367	19
19	25	34.763	1981	26.499	0.34	0.446	22
20	25	35.043	2130	26.499	0.362	0.414	20
21	30	32.58	2285	23.907	0.389	0.516	25
22	30	34.997	2607	28.575	0.413	0.231	10
23	30	34.879	2640	29.969	0.42	0.517	26
24	30	34.763	2700	25.558	0.431	0.319	14
25	30	35.043	2636	27.108	0.417	0.360	18
26	35	35.04	2306	29.14	0.337	0.702	28
27	35	34.997	2344	25.875	0.343	0.422	21
28	35	34.879	2176	26.957	0.32	0.478	24
29	35	34.763	1251	32.705	0.184	0.911	30
30	35	35.043	1311	32.202	0.192	0.842	29

Table 5. Confirmatory tests

Trail No.	Optimum Input Parameters					Output Parameters			
						Experimental Result		Confirmatory Result	
	Overlap ratio (OR) (%)	Free stream velocity (V_{free}) (m/s)	Speed (N) (rpm)	Approach Velocity (V) (m/s)	Tip-speed ratio, λ (TSR)	C_p	C_t	C_p	C_t
9	16	34.763	3796	26.499	0.732	0.438	1.311	0.387	1.236
8	16	34.879	3741	26.499	0.718	0.414	1.316	0.421	1.287

CONCLUSION AND FUTURE DIRECTION

Optimization of SVAWT performance in built environment is illustrated. A hybrid MCDM method i.e. entropy with VIKOR method is proposed and applied for optimization of SVAWT parameters. To show the strength and applicability of the proposed method, a real life case study is taken. The optimized result shows that trail no. 9 yields least performance index values compared to 30 trails and are the optimal combinations for the SVAWT.

The optimal conditions of the SVAWT parameters obtained are Overlap ratio (16%), V_{free} (34.763 m/s), N (3796 rpm), V (26.499 m/s), λ (0.732 TSR). The optimal setting obtained via Entropy-VIKOR method provides the most optimal response / output parameters for the SVAWT. The optimal conditions provide higher C_p and higher C_t as well as improve the performance and efficiency of the SVAWT. At last, confirmatory tests are conducted to verify the results with experimental results and found that confirmatory results are comparable and acceptable. Hence, it is concluded that the proposed hybrid MCDM method can be utilized as systematic framework model for modeling and optimization of other wind turbines.

Although the present chapter provided the application of Entropy-VIKOR based MCDM method for modeling and optimization of SVAWT performance parameters under built environment, however, development of other MCDM methods such as TOPSIS, PROMETHEE, OCRA, JAYA etc., and their applications in wind turbine optimization have not been investigated yet.

REFERENCES

Alexander, A. J., & Holownia, B. P. (1978). Wind tunnel test on a Savonius rotor. *Journal of Industrial Aerdynamics*, *3*(4), 343–351. doi:10.1016/0167-6105(78)90037-5

Altan, B. D., & Atılgan, M. (2008). An experimental and numerical study on the improvement of the performance of Savonius wind rotor. *Energy Conversion and Management*, *49*(12), 3425–3432. doi:10.1016/j.enconman.2008.08.021

Altan, B. D., Atılgan, M., & Özdamar, A. (2008). An experimental study on improvement of a Savonius rotor performance with curtaining. *Experimental Thermal and Fluid Science*, *32*(8), 1673–1678. doi:10.1016/j.expthermflusci.2008.06.006

Amy, H. I. L., Hsing, H. C., & He-Yau, K. (2009). Multi-criteria decision making on strategic selection of wind farms. *Renewable Energy*, *34*(1), 120–126. doi:10.1016/j.renene.2008.04.013

Amy, H. I. L., Meng-Chan, H., He-Yau, K., & Pearn, W. L. (2012). A wind turbine evaluation model under a multi-criteria decision making environment. *Energy Conversion and Management*, *64*, 289–300. doi:10.1016/j.enconman.2012.03.029

Athanasios, K., Varvara, M., Estivaliz, L. M., & Konstantinos, S. (2016). A Comparative Study of Multiple-Criteria Decision-Making Methods under Stochastic Inputs. *Energies*, *9*(566), 1–21.

Athawale, V. M., Chatterjee, P., & Chakraborty, S. (2012). Selection of industrial robots using compromise ranking method. *International Journal of Industrial and Systems Engineering*, *11*(1/2), 3–15. doi:10.1504/IJISE.2012.046651

Bach, G. (1931). *Investigation Concerning Savonius rotors and related Machines*. Quebec, Canada: Brace Research Institute.

Biswas, A., Gupta, R., & Sharma, K. K. (2007). Experimental Investigation of Overlap and Blockage Effects on Three-Bucket Savonius Rotors. *Wind Engineering*, *31*(5), 363–368. doi:10.1260/030952407783418702

Blackwell B.F. (1978) *Wind tunnel performance data for two- & three- bucket Savonius rotor*. Sandia Laboratories, Sand 76-0131 under act AT/29-11-789.

Caliskan, H., Kursuncu, B., Kurbanoglu, C., & Guven, S. Y. (2013). Material selection for the tool holder working under hard milling conditions using different multi criteria decision making methods. *Materials & Design*, *45*, 473–479. doi:10.1016/j.matdes.2012.09.042

Chakraborty, S., Chatterjee, P., & Prasad, K. (2018). An Integrated DEMATEL-VIKOR Method-Based Approach for Cotton Fibre Selection and Evaluation. Journal of the Institution of Engineers. doi:10.100740034-018-0113-8

Chamanehpour, E., & Ahmadizadeh, A. (2017). Site selection of wind power plant using multi-criteria decision-making methods in GIS: A case study. *Computational Ecology and Software*, *7*(2), 49–64.

Chatterjee, P., Athawale, V. M., & Chakraborty, S. (2009). Selection of materials using compromise ranking and outranking methods. *Materials & Design*, *30*(10), 4043–4053. doi:10.1016/j.matdes.2009.05.016

Chatterjee, P., Athawale, V. M., & Chakraborty, S. (2010). Selection of industrial robots using compromise ranking and outranking methods. *Robotics and Computer-integrated Manufacturing*, *26*(5), 483–489. doi:10.1016/j.rcim.2010.03.007

Chatterjee, P., & Chakraborty, S. (2013). Selection of materials using multi-criteria decision-making methods with minimum data. *Decision Science Letters*, *2*(3), 135–148.

Chatterjee, P., & Chakraborty, S. (2016). A comparative analysis of VIKOR method and its variants. *Decision Science Letters*, *5*(4), 469–486. doi:10.5267/j.dsl.2016.5.004

Chatterjee, P., Mukherjee, P., & Chakraborty, S. (2011). Supplier selection using compromise ranking and outranking methods. *Journal of Industrial Engineering International*, *7*(14), 61–73.

Emma, M., Kieran, S., & Vida, M. (2013). An assessment of sustainable housing affordability using a multiple criteria decision making method. *Omega, 41*, 270-279.

Gupta, R., Biswas, A., & Sharma, K. K. (2008). Comparative study of a three-bucket Savonius rotor with a combined three-bucket Savonius – three-bladed Darrieus rotor. *Renewable Energy*, *33*(9), 1974–1981. doi:10.1016/j.renene.2007.12.008

He-Yau, K., Meng-Chan, H., Pearn, W. L., Amy, H. I. L., & Mei-Sung, K. (2011). An Integrated Multi-Criteria Decision Making Model for Evaluating Wind Farm Performance. *Energies*, *4*(11), 2002–2026. doi:10.3390/en4112002

Heracles, P., & Stefan I. (2014). Micro-siting/positioning of wind turbines: introducing a multi-criteria decision analysis framework. *Proceedings of Engineering and Technological Innovation*, 23-27.

Kamoji, M. A., Kedare, S. B., & Prabhu, S. V. (2008). Experimental investigations on single stage modified Savonius rotor. *Applied Energy*, *86*(7-8), 1064–1073. doi:10.1016/j.apenergy.2008.09.019

Khan, M. H. (1975). *Improvement of Savonius Rotor-windmill* (M.S. thesis). University of the Phillipines, Lasbonas.

Khan, M. H. (1988). Model & prototype performance characteristics of S-rotor wind mills. *Journal of Wind Engineering*, *12*, 59–75.

Lahu, P. M., Gandigude, A. U., & Kale, S. A. (2014). Selection of Material for Wind Turbine Blade by Analytic Hierarchy Process (AHP) Method. *Applied Mechanics and Materials*, *612*, 145–150. doi:10.4028/www.scientific.net/AMM.612.145

Macpherson, R. B. (1972). *Design, Development and testing of Low Head High Efficiency Kinetic Energy Machine* (MSc thesis). University of Massachusetts, Amherst, MA.

Mahamarakkalage, F. (1980). *On the Performance and Wake Aerodynamics of the Savonius Wind Turbine* (PhD thesis). University of Peradenyia, Sri Lanka.

Modi, V. J., & Fernando, M. S. U. K. (1989). On the performance of the Savonius wind turbine. *ASME Journal of Solar Engineering*, *11*(1), 71–76. doi:10.1115/1.3268289

Modi, V. J., Roth, M. N. J., & Fernando, S. U. K. (1984). optimal configuration studies and prototype design of a wind energy operated irrigation system. *Journal of Wind Engineering and Industrial Aerodynamics*, *16*(1), 85–96. doi:10.1016/0167-6105(84)90050-3

Mohamed, F., & Santawy, E. (2012). A VIKOR Method for Solving Personnel Training Selection Problem. *International Journal of Computational Science*, *1*(2), 9–12.

Mojola, O. (1985). Aerodynamic design of the Savonius windmill rotor. *Journal of Wind Engineering and Industrial Aerodynamics*, *21*(2), 223–231. doi:10.1016/0167-6105(85)90005-4

Newman, B. G. (1974). Measurement on a Savonius rotor with variable gap. *Proceeding of Sherbrook University Symposium on wind energy*, 116-127.

Nobuyuki, F. (1992). On the torque mechanism of Savonius rotors. *Journal of Wind Engineering and Industrial Aerodynamics*, *40*(3), 277–292. doi:10.1016/0167-6105(92)90380-S

Rai, D., Jha, G. K., Chatterjee, P., & Chakraborty, S. (2013). Material selection in manufacturing environment using compromise ranking and regret theory-based compromise ranking methods: A comparative study. *Universal Journal of Materials Science*, *1*(2), 69–77.

Ranjan, R., Chatterjee, P., & Chakraborty, S. (2015). Evaluating performance of engineering departments in an Indian University using DEMATEL and compromise ranking methods. *Opsearch*, *52*(2), 307–328. doi:10.100712597-014-0186-1

Ranjan, R., Chatterjee, P., & Chakraborty, S. (2016). Performance evaluation of Indian railway zones using DEMATEL and VIKOR methods. *Benchmarking: An International Journal, 23*(1), 78-95.

Roth, N. J. (1982). *Prototype Design and Performance of Savonius Rotor Based on Irrigation System* (Master's thesis). University of British Columbia.

Saha, U. K., Thotla, S., & Maity, D. (2008). Optimum design configuration of Savonius rotor through wind tunnel experiments. *Journal of Wind Engineering and Industrial Aerodynamics*, *96*(8-9), 1359–1375. doi:10.1016/j.jweia.2008.03.005

Salman, A. K. (2016). Multi-criteria Decision-making for Selection of Wind Turbine with Rated Power of 1000 KW using Dubois and Prade Fuzzy Operator. *International Journal of Computing and Network Technology*, *4*(1), 27–32. doi:10.12785/ijcnt/040104

Savonius, S. J. (1931). The S-rotor and its application. *Mechanical Engineering (New York, N.Y.)*, *53*, 1–6.

Saylers, A. T. (1985). Blade configuration optimization & performance characteristics of a simple Savonius rotor. *Proceedings - Institution of Mechanical Engineers*, *199*(3), 185–191. doi:10.1243/PIME_PROC_1985_199_023_02

Shafiqur, R., & Salman, A. K. (2017). Multi-Criteria Wind Turbine Selection using Weighted Sum Approach. *International Journal of Advanced Computer Science and Applications*, *8*(6), 128–132.

Sharma, K. K., Gupta, R., Singh, S. K., & Singh, S. R. (2005). Experimental investigation of the characteristics of a Savonius wind turbine. *Wind Engineering*, *29*(1), 77–82. doi:10.1260/0309524054353674

Sivasegaram, S. (1978). Secondary parameters affecting the performance of resistance type vertical axis wind rotors. *Wind Engineering*, *2*, 49–58.

Suresh, K. B., Subbaraju, N. V., Reddy, S. M., & Nageswara, R. D. (2006). The material selection for typical wind turbine blades using a MADM approach & analysis of blades. *Conference on MCDM 2006*, 1-11.

Chapter 12
Product Prediction and Recommendation in Sustainable E-Commerce Using Association Rule Mining and K-Means Clustering:
A Novel Approach

Subro SantiRanjan Thakur
MCKV Institute of Engineering, India

Soma Bandyopadhyay
https://orcid.org/0000-0002-5987-3119
MCKV Institute of Engineering, India

Jyotsna Kumar Mandal
University of Kalyani, India

ABSTRACT

The tremendous growth of customers and products in recent years poses some key challenges for recommender systems. These are producing high quality recommendations and performing many recommendations per second for millions of customers and products. New recommender system technologies are needed that can quickly produce high quality recommendations, even for very large-scale problems. The authors address the performance issues by scaling up the neighborhood formation process through the use of clustering techniques. By using association rule learning, it has been observed that customers who purchase the items t-shirt and jeans have an increasing trend to buy shoes, etc. These systems, especially the k-means clustering-based ones, are achieving widespread success in e-commerce nowadays, and the results are encouraging (i.e., the category silver is preferable as purchasing amount is concerned). Enterprises can use the model to predict the stock and customer for their business sustainability.

DOI: 10.4018/978-1-5225-8579-4.ch012

INTRODUCTION

With the increasing popularity of Internet technology, there is a new trend of buying various products from E-commerce website which offers different products for sale. As the customers can acquire a lot of information on-line, they may get confused with information overload. Information and communication technology (ICT) infrastructures such as Personal computers, mobile phones, servers, routers etc. are needed for E-Commerce business. Extensive use of these technologies may lead to energy consumption and electronic waste. Virtual business network and digitization of E-commerce business has created an impact on transportation, land usage and productivity of resources which in turn may affect environment. E-Commerce business has also changed the lifestyle and consumption patterns of people that indirectly affect the environment (Niu, 2016). Miniaturization of this devices and its low power consumption rate, make the electronic gadgets very popular in market.

Clustering has been extensively used as an unsupervised learning method (Ester, Kriegel, Sander, & Xu, 1996, Pitsilis, Zhang & Wang, 2011) which creates a great impact on the researcher in the field of computer science. Clustering techniques has been used as a subject of research in the area of Recommender systems, still there is a need to explore this field. According to Pitsilis, Zhang & Wang (2011) to make future choice of neighbor it is needed to partition choices into smaller sets permanently. The research of Sarwar, Karypis, Konstan & Riedl (2001) is considered as the pioneer in the field of Recommendation system which uses clustering techniques. Sumathi et al. (2010) introduced the concept of recommendation systems, and is based on navigational patterns of the user's. They used model based clustering, where recommendations was provided to needs of the user based on their requirements (Thiyagarajan, Thangavel & Rathipriya, 2014). Irvin (1994) has done the life time value analysis to identify lucrative customers and to build up strategies for new customers. Choi, Yoo, Kim & Suh (2012) discover the buying pattern of customers of different categories, as it helps for the future prediction using K-means clustering algorithm. K-means clustering algorithm uses Euclidian distance, where the distance is computed by finding the square of the distance between each data points, and then summing the squares to get the desired result (Oyelade, Oladipupo & Obagbuwa, 2010).

Recommendation system plays a significant role to address these problems where various data mining techniques are incorporated, and by using the knowledge gathered about products actions can be taken which may benefit the users. Customers' preference rating for different products has also been successfully mined for automatic product recommendation. For the creation of recommendation rules, the association rule mining was applied based on the frequent purchase patterns obtained from different group of customers (Agrawal, Imielinski & Swami, 1993; Liu & Shih, 2005; Bandyopadhyay, Thakur, & Mandal, 2018). Using collaborative filtering, the frequent purchase patterns represents the purchasing behaviour of customers having similar taste for products (Aggarwal, Wolf, & Yu, 1999; Thakur, Kundu & Sing, 2011; Paul, Sarkar, Chelliah, Kalyan, & Nadkarni, 2017). In this work, Association Rule Learning (ARL) (Agrawal & Srikant, 1994) and traditional k- means clustering algorithm have been implemented and categorization has been done. This work uses Euclidean distance which is a measure of similarity has been chosen for categorization of customer in terms of their buying pattern.

So to make E-commerce business sustainable not only the supply chain management based on customers expenditure is sufficient, there is a need to determine design, use, and regulation so that it does not affect environment adversely. E-commerce and business basically covers the economic transactions (profit oriented or not) are the application of internet technologies, which is focusing on products and service. Hence there is a need of sustainable design, its usage and regulation of the digital economy. Whether an E-commerce business will sustain or not that depends on the financial growth of the enterprises. It has been observed that though the e-commerce business has improved their performance in short term through investing and financing, there are some financial risk even bankruptcy.

In our work, the customers' classification has been done in terms of their spending habits during festival season, which recommends products to the new customers in near future. By using a clustering technique, the customers have been group into different categories, which is based on their buying habits, as it is difficult to obtain a complete opinion of customers with traditional grouping. Association rule and K-means clustering algorithm are explained next, followed by the proposed model and implementation details. Experimental results, conclusion and future course of work are presented in details after the proposed model.

ASSOCIATION RULE

Item-to-item correlations in data mining helps the recommendation system to identify the items frequently found in association with other items, for which a user has shown interest or purchased the items earlier. Association can be applied based on co-purchase of items/data and the preference given by users, or any other measures. Item-to-item similarity can be used to identify matching items, such as in case of clothing items the item that will be purchased by a customer when he purchased a T-shirt. Ideal recommender systems match an entire set of data or items which is available in a customer's shopping cart.

Meaning and Uses

Association rule has been used and it analyzes the items of preference among different products, and suggests products to customers which are based on other products they have purchased earlier. The process of finding relationship about products is chosen along with other products using association rules. There is a definite relationship between the association rules with the number of products or items available in a database. However, effective search space within item sets can be reduced using confidence and support and by combining with algorithms available that make association rules. By effective usage of association rule it will definitely improve the efficiency and if shelf layout of a retail store is organized in this manner the sale performance of the store may be increased. As domain knowledge management is concerned and to recommend items to a customer, association rule may be used as a powerful tool which attempts to predict the products that may be most valuable to customers.

Mining Association Rule

Association rules are required which satisfies user-specified minimum support and confidence at the same time (Fang, 2013; Mandave, Mane, Patil, 2013).

Association rules generation is split into two separate steps:

1. By applying minimum support in a user database, the frequent item set is generated.
2. Using these frequent item sets and minimum confidence the rules are formed.

To find all frequent item sets in a database, it is important to search combinations of all items from a database. The set of possible itemsets is the power set over I and has size 2^n-1 and empty set is not a valid item set. Average size of the large items is denoted by I and number of items is denoted by n. The size of the powerset grows exponentially by items n in I, efficient search is possible using the downward-closure property of support and for a frequent item set all its subsets are also frequent. Similarly, in case of infrequent item set it is possible that all its supersets are also infrequent. Apriori algorithm uses this feature by finding all frequent item sets. Eclat and FP-Growth are the other algorithms which are used for generating association rule.

K-MEANS CLUSTERING ALGORITHM

K-means algorithm is used to classify objects based on their attributes/features which use unsupervised learning to group objects in K number of groups where K is a positive integer. This grouping is done by minimizing the sum of squares of distances, between data and the corresponding cluster centroid. This method uses K clusters which is fixed at the beginning, and is used to classify a given data set (Celebi, Kingravi & Vela, 2013).

Given a data set of n data points x_1, x_2, …, x_n such that each data point is in R^d, the problem of finding the minimum variance of the dataset into K clusters is that of finding k points $\{m_j\}$ (j=1, 2, …, k) in R^d such that

$$1/n\sum_{i=1}^{n}[min_j d^2\left(x_{i,} m_j\right)] \quad (1)$$

is minimized, where $d(x_i, m_j)$ denotes the Euclidean distance between x_i and m_j. The points $\{m_j\}$ (j=1, 2, …,k) are known as cluster centroids. The problem in equation (1) is to find K cluster centroids, such that the average squared Euclidean distance (mean squared error, MSE) between a data point and its nearest cluster centroid is minimized.

The basic step of K-means clustering (Kim & Ahn, 2008) is discussed below:

Step 1: Determine the number of cluster K at the beginning.

Step 2: Determine the centroid coordinate or centre of these clusters. Any random data point can be taken as the initial centroid or first K data point can also be used as the initial centroids.

Step 3: Determine the distance of each data point to the centroids.

Step 4: Group the object based on minimum distance (find the closest centroid). Each record is assigned to the nearest cluster using a measure of distance (Euclidean distance).

Step 5: Keeping the same number of clusters, the new centroid of each cluster is calculated, by taking the arithmetic mean of all the data points which belong to that cluster.

Step 6: Repeat step (2) to step (5) until the clusters stop changing or stop conditions are satisfied.

PROPOSED MODEL AND IMPLEMENTATION DETAILS

The block diagram of the proposed model is shown in Figure 1. The process of dividing customers into different categories, where individual customers in the same group are similar which is known as market segmentation or customer segmentation. The customers can be divided using different characters, attributes/features e.g. like spending habits, interests, age, etc. In this work the customers have been categorized in terms of spending money at the time of festival season.

The customers who have spent high amount above Rs. 12000 or more can be considered as Diamond Customers. The customers who have spent greater than Rs. 9000 to 12000 can be considered as Gold customers. Similarly the customers who have spent less than Rs. 3000 may be considered as value less customers or Copper customer. Table 1 shows the categorization of customer according to money spent by them. The retailers can try to make close relationship to the customers based on purchased amount. After finding the likeminded customers in terms of spending amount the association rule mining (ARM) has been applied to dataset of a particular cluster. It has been observed that the customers who are considered as Gold customers have purchased similar type of item while the value less customers have purchased other type of items. Table 1 shows the Purchase index.

The data have been collected between the month of August to November in 2016, 2017 etc. The data is categorized based on the amount spent by customers during festival season like Durga Puja, Eid and Diwali etc. Table 2 shows the statistical data used for the analysis and in this case the survey was done for 169 customers with 10 numbers of products.

Figure 1. Block diagram of proposed model for price based (PB) item-to-item clustering and price-based clustering (PBC) for individual items

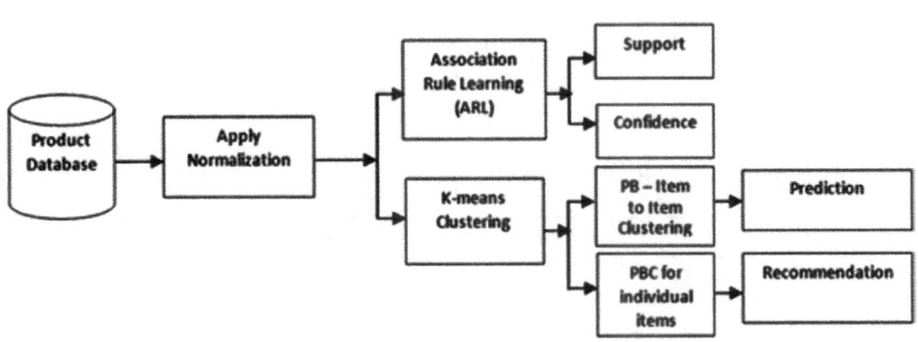

Table 1. Purchase index

Total Money Spent	Category
Above 12000	Diamond
> 9000-12000	Gold
> 6000-9000	Silver
> 3000-6000	Bronze
Below 3000	Copper

Table 2 shows the data set (Customers' Purchase amount) in the form N by M matrices, where N is the rows (# of customers) and M is the column (# of products) purchased by each customer. Table 3, Table 4 and Table 5 indicate cluster size and money spent in average, for k=3, k=4 and k=5 respectively. The model has been applied on the dataset of a shopping mall and data were collected from 225 customers. Each user was asked to fill out his/her records of purchase data. After normalization 169 user's data are considered in the work. The result generated is shown in Table 3, 4 and 5 for k=3, k=4 and k=5 respectively.

Table 2. Statistics of the data used

Data	Total Number of Customers	Total number of Products
Customers purchase amount	169	10

Table 3. Cluster size vs. money spent when k=3

Cluster #	Cluster Size	Money Spent in Average (Rs.)
1	62	3600
2	98	7003.11
3	9	16679.56

Table 4. Cluster size vs. money spent when k=4

Cluster #	Cluster Size	Money Spent in Average (Rs.)
1	39	3030.8
2	102	6125.08
3	21	8919.86
4	7	18477.86

Table 3 shows that data are divided into three clusters and the cluster size of cluster 1, cluster 2 and cluster 3 are 62, 98 and 9 respectively. Table 4 shows that data are divided into four clusters and the cluster size of cluster 1, cluster 2, cluster 3 and cluster 4 are 39, 102, 21 and 7 respectively. Similarly,

Table 5 shows that data are divided into five clusters and the cluster size of cluster 1, cluster 2, cluster 3, cluster 4 and cluster 5 are 18, 61, 75, 8 and 7 respectively.

EXPERIMENTAL RESULTS

The experimental results for the proposed work is shown in Figure 2, Figure 3 and Figure 4 which show the Cluster size vs. Money spent in average when k=3, k=4 and k=5 respectively.

Table 5. Cluster size vs. money spent when k=5

Cluster #	Cluster Size	Money Spent in Average (Rs.)
1	18	2475.17
2	61	4455.75
3	75	7154.33
4	8	9668.38
5	7	18477.86

Figure 2. Cluster size vs. money spent in Rs. when k=3

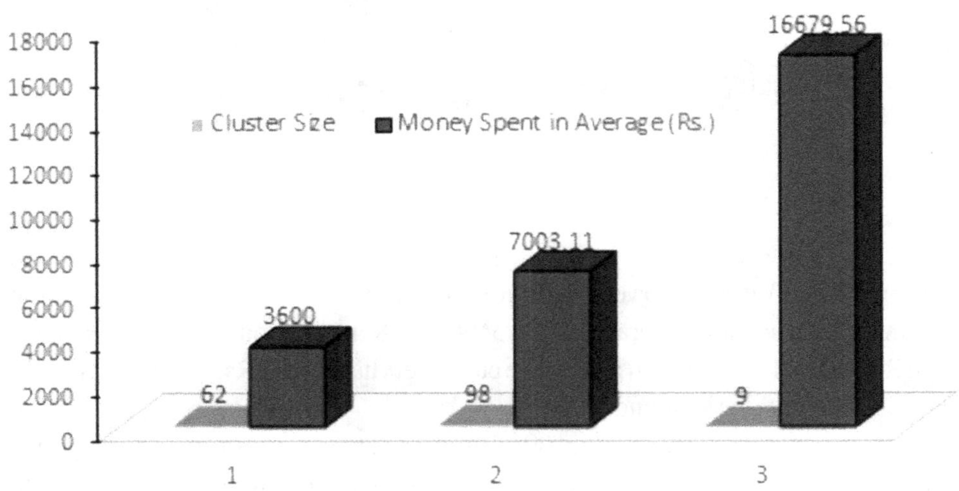

Figure 3. Cluster size vs. Money spent in Rs. when k=4

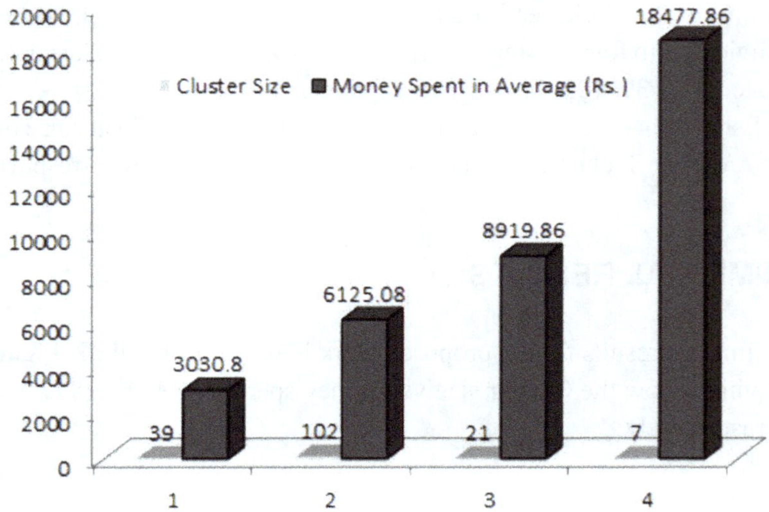

Figure 4. Cluster size vs. Money spent in Rs. when k=5

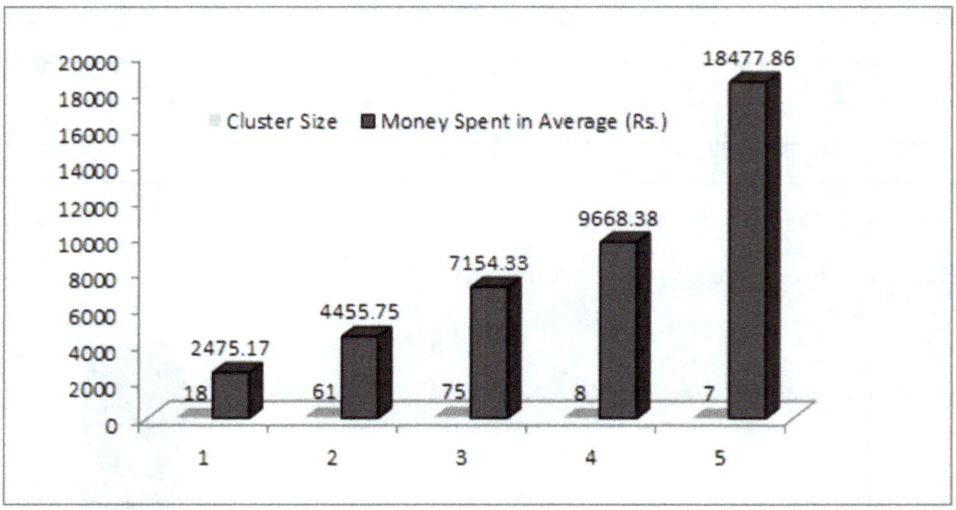

From Table 3 it is observed that out of 169 transaction in 62 transactions the customers have spent an average of Rs. 3600, in 98 transaction the customers have spent Rs.7003.11 and in 9 transaction the customers have spent Rs.16679.56. Similar analysis also hold for k=4 and k=5.

From Figure 2 it is observed that 09 customers out of 169 customers have spent a high amount of money while purchasing in a festival season while 62 customers spent moderately low and 98 customers spent average amount of money for purchasing product for a festival. Similarly, in table 4 the cluster sizes are 39, 102, 21 and 07 for cluster 1, 2, 3 and 4 respectively. From Figure 3 it can be observed that the average money spent by the customers belongs to cluster 4 is Rs. 18477.86 which is considered as high purchasing index. The customers belongs to this category has been considered as Diamond category. In case of cluster 1, cluster 2 and cluster 3 the average purchase amount are Rs. 3030.80, Rs. 6125.08 and Rs. 8919.86 respectively. So, cluster 1 has been considered as Bronze category customers and cluster 2 and cluster 3 have been considered as silver category customers.

Similar analysis also holds for Table 5. In Figure 4, the average purchase amount for cluster size 7 is Rs.18477.86, while the average purchase amount for cluster size 8 is Rs. 9668.38. Cluster size 75 has average purchase amount of Rs. 7154.33, while the cluster size 61 has the average purchase amount Rs. 4455.75 and cluster size 18 has the average purchase amount of Rs. 2475.17. This analysis indicated that out of 169 customers 7 customers crossed over the high purchase index and can be considered as Diamond category customers, 8 belongs to the category of moderately high purchase index and are considered as Gold category, while 75 customers belong to the average purchase index can be considered as Silver category, 61 customers fall into the category of moderately low purchase index and the remaining 18 customers belong to low purchase index category. They are considered as Bronze and Copper category customers. According to these categorization, the retailer can tap the customers (i.e. Silver category), and provide some discount to these customers, thus the profit of the business merchants may increase.

CONCLUSION AND FUTURE WORK

Recommender systems are useful technologies which helps the business firm for extracting additional useful information from its customer databases. This system also helps the customers to purchase products from E-Commerce website. Conversely, they help the business vendor to make profits by generating more sales. As the volume of customer data available on these websites is huge, hence there is an urgent need for new technologies that can improve the scalability of recommender systems. Business enterprise requires doing the financial analysis on regular intervals, which is one of the most important criteria as the actual growth rate of a business is concerned. Our work aims to cluster customer's data and is based on their expenditure in a particular span of time. It will benefit the enterprise to analyze which class of people are their main customers and which products should be kept in stock for their business growth, which helps the business enterprise to attain sustainability.

Enterprises can use our model to predict the stocks available and to retain customers for their business sustainability.

Our experiments suggest the following:

1. In the database, the item set {Shirt, Trouser, Shoe} has a support value of 0.1 since it occurs in 10% of all transactions (17 out of 169 transactions). The rule {Shirt, Trouser} → {Shoe} has a confidence value of 0.1/0.2=0.5 in the database, which means that for 50% of the transactions the rule is correct. The item set{Jeans, T-Shirt, Shoe} has a support value of 0.11 since it occurs in 19 cases out of 169 transactions. The rule {Jeans, T-Shirt} →Shoe has a confidence value of 0.11/0.159=0.69 in the database, which means that for 69% of the transactions the rule is correct. Similarly, in case of women {Top, Jeans} →Shoe has a confidence value of 0.72, which shows that for 72% of the transaction the rule is correct.
2. In this work, we explained the effectiveness of clustering algorithm i.e. bisecting k-means algorithm. In future, better clustering algorithms as well as better prediction generation schemes can be used to improve the prediction quality. Clustering techniques can also be applied as a "first step" for shrinking the candidate set in the nearest neighbour algorithm or for distributing nearest-neighbour computation across several recommender engines. While dividing the population into clusters may hurt the accuracy or recommendations to users near the fringes of their assigned cluster, pre-clustering may be a worthwhile trade-off between accuracy and throughput.

ACKNOWLEDGMENT

The authors acknowledge the support provided by the DST PURSE Scheme, Govt. of India at the University of Kalyani, India.

REFERENCES

Aggarwal, C. C., Wolf, J. L., & Yu, P. S. (1999). A new method for similarity indexing for market data. In *Proceedings of the ACM SIGMOD Conference* (pp. 407-418). Philadelphia: ACM. 10.1145/304182.304218

Agrawal, R., Imielinski, T., & Swami, A. N. (1993). Mining Association Rules between Sets of Items in Large Databases. In *Proc. ACM SIGMOD Int'l Conf. Management of Data* (pp 207-216). Washington, DC: ACM. 10.1145/170035.170072

Agrawal, R., & Srikant, R. (1994). Fast algorithms for mining association rules in large databases. In *VLDB '94: Proceedings of the 20th International Conference on Very Large Data Bases* (pp. 487—499). San Francisco, CA: Academic Press.

Bandyopadhyay, S., Thakur, S. S., & Mandal, J. K. (2018). *A Novel Approach for Product Prediction Using Artificial Neural Networks*. Social Transformation – Digital Way. doi:10.1007/978-981-13-1343-1_56

Celebi, M., Kingravi, H., & Vela, P. (2013). A comparative study of efficient initialization methods for the k-means clustering algorithm. *Expert Systems with Applications*, *40*(1), 200–210. doi:10.1016/j.eswa.2012.07.021

Choi, K., Yoo, D., Kim, G., & Suh, Y. (2012). A hybrid online-product recommendation system: Combining implicit rating-based collaborative filtering and sequential pattern analysis. *Electronic Commerce Research and Applications*, *11*(4), 309–317. doi:10.1016/j.elerap.2012.02.004

Ester, M., Kriegel, H. P., Sander, J., & Xu, X. (1996). A Density-Based Algorithm for Discovering Clusters in Large Spatial Databases with Noise. In *Proceedings of the Second International Conference on Knowledge Discovery and Data Mining* (pp. 226-231). Portland, OR: Academic Press.

Fang, X. (2013). An Improved Apriori Algorithm on the Frequent Item set. *Proc. International Conference on Education Technology and Information System*, 845-848.

Irvin, S. (1994). Using lifetime value analysis for selecting new customers. *Credit World*, *82*(3), 37–39.

Kim, K., & Ahn, H. (2008). A recommender system using GA K-means clustering in an online Shopping market. *Expert Systems with Applications*, *34*(2), 1200–1209. doi:10.1016/j.eswa.2006.12.025

Liu, D. R., & Shih, Y. (2005). Integrating AHP and data mining for product recommendation based on customer lifetime value. *Information & Management*, *42*(3), 387–400. doi:10.1016/j.im.2004.01.008

Mandave, P., Mane, M., & Patil, S. (2013). Data mining using Association Rule Based on Apriori Algorithm and Improved Approach with Illustration. *International Journal of Latest Trends in Engineering and Technology*, *3*(2), 107–113.

Niu, Y. C. (2016). An Analysis of E-Commerce Enterprises Sustainable Development and Financing Structure. *Theoretical Economics Letters*, *6*(5), 1105–1114. doi:10.4236/tel.2016.65106

Oyelade, O. J., Oladipupo, O. O., & Obagbuwa, I. C. (2010). Application of k-Means Clustering algorithm for prediction of Students *Academic Performance. International Journal of Computer Science and Information Security*, *7*(1), 292–295.

Paul, C., Sarkar, D., Chelliah, S., Kalyan, M. C., & Nadkarni, P. P. S. (2017). Recommendation of High Quality Representative Reviews in e-commerce. In: *Proceedings of the Eleventh ACM Conference on Recommender Systems* (pp. 311-315). Como, Italy: ACM. 10.1145/3109859.3109901

Pitsilis, G., Zhang, X., & Wang, W. (2011). Clustering recommenders in collaborative filtering using explicit trust information. In I. Wakeman, E. Gudes, C. D. Jensen, & J. Crampton (Eds.), *Trust Management V. IFIPTM 2011. IFIP Advances in Information and Communication Technology, Advances in Information and Communication Technology* (Vol. 358, pp. 82–97). Berlin: Springer. doi:10.1007/978-3-642-22200-9_9

Sarwar, B., Karypis, G., Konstan, J., & Riedl, J. (2000). Analysis of recommendation algorithms for e-commerce. *Proceedings of the 2nd ACM conference on Electronic commerce* (pp. 158-167). New York: ACM. 10.1145/352871.352887

Sarwar, B. M., Karypis, G., Konstan, A. J., & Riedl, J. T. (2001). Item-based collaborative filtering recommendation algorithms. In *Proc. of the 10th International World Wide Web Conference* (pp. 285-295). Academic Press.

Sumathi, C. P. (2010). Automatic Recommendation of Web Pages in Web Usage Mining. *International Journal on Computer Science and Engineering*, *02*(09), 3046–3052.

Thakur, S. S., Kundu, A., & Sing, J. K. (2011). A Collaborative Filtering Approach: Based on offline customers ratings for Prediction and Recommendation. In *Proc. of the 1st International Conference on Scientific Paradigm Shift in Information Technology & Management* (pp. 215-220). Academic Press.

Thiyagarajan, R., Thangavel, K., & Rathipriya, R. (2014). Recommendation of web pages using weighted K- means clustering. *International Journal of Computers and Applications*, *86*(14), 44–48. doi:10.5120/15057-3517

Chapter 13
Development of an Integrated TOPSIS-Quality Function Deployment Model for Sustainability Assessment of Indian Banks

Rishi Dwivedi
Xavier Institute of Social Service, India

Bhaskar Bhowani
Xavier Institute of Social Service, India

P. Kritee Rao
Xavier Institute of Social Service, India

ABSTRACT

The contribution of banks to sustainable advancement is supreme, considering the vital part they play in funding the economic actions of the human race. But, in the fast-changing banking ecosystem, quantifying the level of sustainability attained by financial institutions is a challenging task due to the need of considering a wide range of economic, environmental, and social dimensions concurrently. In this chapter, a novel method based on the technique for order preference by similarity to ideal solution (TOPSIS) and quality function deployment (QFD) is proposed for the first time to evaluate the sustainable efficiency of banking operations. The QFD technique is employed here to provide due importance to the customers' needs with respect to banks' sustainability, and subsequently calculate the priority weights of the considered criteria of sustainability principles. Then, TOPSIS is employed to rank alternatives based on the extent to which they conform to sustainability principles.

DOI: 10.4018/978-1-5225-8579-4.ch013

INTRODUCTION

Banking industry forms the core of a financial sector, and can be regarded as a backbone to the country's economy. Primarily, banks are the financial institutions which are being entrusted to accept deposits and provide loans. They not only facilitate the flow of funds in our economy but also ensure the financial resources are apportioned efficiently, thereby promoting economic development and growth. Thus, banks are regarded as principal stimuli for financial progress. Besides, they are primarily entrusted for ensuring social stability and sustainable advancement of the economy through channelizing proper supply of money in the market.

The word 'Bank' is said to be derived from French word "Bancus" which means a bench. It is believed that the early bankers, the Jews of Lombardy, transacted their business on benches in the marketplace. The modern banking system began with the opening of Bank of England in 1694. Being originated in the 18th century, Bank of Hindustan was the first bank to be established in India in 1770. Today, Indian banking system consists of 27 public sector banks, 21 private sector banks, 49 foreign banks, 56 regional rural banks, 1562 urban cooperative banks and 94384 rural cooperative banks. Further, based on the reports published by World Bank, 80% of the population in India has bank account ownership as on 2017, compared to 53% in 2014. Besides, according to a report published by India Brand Equity Foundation in 2018, total lending has increased at a compound annual growth rate (CAGR) of 10.94% and total deposits has augmented at a CAGR of 11.66% in financial year 2017-18. Additionally, India's retail credit market is the fourth largest in the emerging countries. It has increased to US$ 281 billion on December 2017 from US$ 181 billion on December 2014. This robust growth of banking industry indicates its bright future. Moreover, with the unremitting rise in the mobile phone users, sustainable penetration of Indian banks is not very far away. But, with exponential development and unswerving opportunity going ahead, banks need to enhance customer's overall experience in order to assure their long-term existence. Sustainability acts as a concord to enhance both current and future potential of the banks to meet human needs and aspirations. In 2005, the World Summit on Social Development identified three core areas that contribute to the philosophy of sustainable development, i.e. economic development, social development and environmental protection. Therefore, it can be said that sustainability in the banking industry is dependent on multiple factors, such as long-term business, profitability, non-performing asset (NPA) etc. which at times are contradictory in nature. Hence, in order to assess the long-term sustenance of banks, a multi-criteria decision-making (MCDM) model needs to be applied. Quality function deployment (QFD) method focuses on improving customers' satisfaction by a systematic analysis of the customers' needs and competitive market pressures with an objective to build a better product or service. QFD aims to identify the customers' requirements, the related technical requirements, and subsequently correlate those customers' requirements into technical requirements, thereby finding the weightage of each. On the other hand, technique for order preference by similarity to ideal solution (TOPSIS) focuses on identifying the best alternative based on assorted criteria considered. It discusses about two aspects, one being positive ideal solution and other being negative ideal solution. The positive ideal solution maximizes the benefit criteria and minimizes the cost criteria, whereas, the negative ideal solution maximizes the cost criteria and minimizes the benefit criteria. Option closest to the positive ideal solution is being regarded as the best candidate on the basis of identified criteria. Thus, this paper proposes to develop an integrated TOPSIS-QFD model for sustainability evaluation of Indian banks. This can help policy maker to better visualize the preeminent parameters leading towards sustainable success of banks while selecting the best suited bank in current Indian market environment.

LITERATURE REVIEW

Banking industry have gradually transformed into an extremely service oriented sector and thereby attracted various researchers and academicians to contribute towards its performance evaluation through various techniques. Therefore, first half of this review of past research work focuses on evaluation of various methods that have been applied to analyze various aspects of banking. Furthermore, diverse MCDM techniques employed for performance evaluation of varied dimensions of different industries are also examined in second part of literature review.

Mercana et al. (2003) presented the financial performance index of a commercial bank. Mavridis (2004) through the study concluded that the best performing banks were the ones that displayed good results in the usage of their intellectual capital. Foehn (2004) identified the criteria for valuing clients and extent of its implementation. Berger and Di-Patti (2006) employed profit efficiency approach to test agency theory in the banking industry. Yiu et al. (2007) explored the implementation of internet banking by retail customers of Hong Kong. Iannotta et al. (2007) studied about the ownership structure, risk and performance in the European banking industry. Kim et al. (2009) concluded that factors, such as structural assurances, relative benefits and propensity to trust have substantial effect on mobile banking penetration. Subramony and Holtom (2012) suggested that employee attrition has significant negative effects on the performance of business units. Paul et al. (2015) analyzed the impact of service quality on customer satisfaction in private and public sector banks. Neema and Nema (2017) studied the proportionality of bank's brand image with brand loyalty and brand equity.

Tan and Pawitra (2001) developed a combined approach involving service quality model (SERVQUAL), Kano technique and QFD methodology to help organizations to evaluate customer satisfaction. Bottani and Rizzi (2006) addressed the issue of how to deploy the QFD model to effectively and efficiently improve logistics processes. Kahraman et al. (2006) proposed an integrated framework based on fuzzy-QFD and a fuzzy optimization models to determine the product technical requirements (PTRs) to be considered in designing a product. Carnevalli and Miguel (2008) presented a review, analysis, classification and codification of the literature on QFD model, published between 2002 and 2006. Bhattacharya et al. (2010) delineated a concurrent engineering approach integrating analytic hierarchy process (AHP) with QFD techniques in combination with cost factor measure (CFM) to rank and subsequently select candidate-suppliers under multiple, conflicting-in-nature criteria environment within a value-chain framework. Ho et al. (2012) designed an integrated methodology, combining QFD model, fuzzy set theory, and AHP approach, to evaluate and select the optimal third-party logistics service providers (3PLs). Dursun and Karsak (2013) applied a fuzzy multi-criteria group decision making approach for supplier selection process. Kannan et al. (2014) proposed a framework using fuzzy TOPSIS method to select green suppliers for a Brazilian electronics company. Biswas et al. (2016) developed a new approach for multi-attribute group decision-making problems through extending the TOPSIS to single-valued neutrosophic environment. Sindhu et al. (2017) aimed to select appropriate site in an Indian case using hybrid AHP and fuzzy TOPSIS model. Dos Santos et al. (2019) proposed a method for the evaluation and selection of green suppliers for the Brazilian furniture industry while employing TOPSIS technique.

It is evident from the above literature review that analyzing banking industry is of great importance in achieving sustainable economic growth for the country, and numerous methods have already been applied to assist decision-making for managers of banks. Further, it is observed that both TOPSIS and QFD models have been implemented independently to evaluate and streamline assorted organization's strategies. Besides, there are substantial published research papers on the integrated implementation of TOPSIS and QFD models to devise competitive long-term management schemes and rules for the enterprises. But, the previous papers have not explored the applicability of integrated TOPSIS-QFD technique to evaluate the performance of banks on sustainable parameters. Hence, this paper intends to first apply the QFD model in banking sector of India to evaluate significance of all the identified criteria for achieving long term sustenance goal of financial stability. Subsequently, TOPSIS technique rank different public sector banks while using the weightage of the chosen criteria as estimated through QFD technique. The detailed implementation procedure of the integrated model in the banking industry is also exhibited. The results derived from the application of the proposed combined TOPSIS-

QFD technique would not only assist the policy makers of the country to enhance the productivity and performance the banks in order to achieve sustainable competitive advantage, but also, help them to determine the comprehensive advancement of the financial institutions on different identified parameters.

QFD METHODOLOGY

These days the consumer has a multitude of options available to select from for similar products and services. Usually, the consumers make their choice on the basis of general discernment of quality or value. But, in order to remain competitive, organizations must determine the driving force behind the customer's perception of value or quality in a product or service. QFD is an effective tool that outlines customers' requirements and deploys them into detailed technical requirements and processes plan to produce the products that satisfy those requirements. The methodology was first developed in Japan by Dr. Yoji Akao in the late 1960s while working for Mitsubishi's shipyard. Dr. Yoji Akao, who is also known as the father of QFD defined it as ''a method for developing a design quality aimed at satisfying the consumer and then translating the consumer's demands into design targets and major quality assurance points to be used throughout the production phase". It is used to translate the needs of customers into measurable design targets and drive them from the assembly level down through the sub-assembly, component and production process levels.

One of the keys to success of any organization is effective communiqué between its various departments The QFD methodology can effectually communicate customers' needs to different departments throughout the organization including design, quality, manufacturing, production, marketing and sales. This allows the entire organization to work together and produce products with high levels of customer perceived value. It focuses on the wants and needs of the customer, not on what the company may believe the customer wants. Moreover, the structured approach followed by QFD provides documentation of decisions made and lessons learned during the entire product development process which can be utilized for projects in future.

QFD methodology utilizes a matrix format called House of Quality (HoQ) matrix to translate the customer wants and needs into product or service design characteristics. The HoQ matrix directly compares how the design or product performs against the other competitors in meeting the requirements. It demonstrates the relationship between the customer wants or "Whats" and the design parameters/ technical specifications or "Hows". The matrix is data intensive and allows the team to capture a large amount of information in one place. The basic structure of HoQ matrix resembles that of a house and a brief description of its different sections are listed below:

1. **Customers' requirements:** The voice of customers, or the wants and needs, of the customer are listed in rows of this matrix. This is usually the first section to be completed;
2. **Technical requirements:** It constitutes the ceiling of matrix and contains the design features/ technical requirements the product would need to meet the customers' requirements;
3. **Interrelationship matrix:** It is the matrix which shows the correlation between customer's requirement and technical requirement/design specification. A set of symbols or values indicate the strength of relationship. A blank box would represent no correlation or influence;
4. **Roof:** This matrix is used to indicate how the design requirements interact with each other. The interrelationships are ratings that range from a strong positive interaction to a strong negative interaction;
5. **Prioritized customers' requirements:** This section visualizes a comparison of the competitor's product regarding fulfilling the needs of customers. This section should be completed using direct feedback from customer surveys or other means of data collection;
6. **Prioritized technical requirements:** It forms the foundation of HoQ matrix. This section lists more specific target values for technical specifications relating to the "Hows" used to satisfy the customer's needs.

DEVELOPMENT OF QFD MODEL FOR ANALYZING SUSTAINABILITY OF BANKS

It has already been noticed that different MCDM techniques can be efficiently applied for resolving multifaceted decision making problems that are characterized by the existence of numerous competing options and an array of contradictory assessment criteria. It assists in focusing on what is imperative, commonsensical and unswerving. In contemporary competitive market environment, banks in order to promote long term growth of financial market should constantly utilize approaches and patterns for examining their functions in organization. Concerning the significance of this dialogue, evaluating banks' performance to achieve sustainable financial stability seems to be of utmost importance. The first step in development of the QFD model for analyzing factors impacting sustenance of banks consists of identifying the stakeholders' expectations while banks perform various actions, like accepting deposits, granting advances, agency functions and utility functions. It is carried out through review of past literature, market survey, questionnaires and customers' feedback forms. The six most important stakeholders' requirements related to achieving long term existence of banks are identified as efficient mobilization of financial funds, encouraging employees to adopt sustainable business practices, sustainable management of monetary resources, resilience to external shocks, encourage and expand access to banking, insurance and financial services, and support productive activities, decent job creation, entrepreneurship, creativity and innovation, which are positioned along the rows of HoQ matrix. After critically analyzing sustainability specifications of banks as desired by stakeholders, 16 technical requirements having their impact on attaining long term sustenance of financial permanence are then recognized. They are located at the top of HoQ matrix along the columns. A brief description of all technical requirements is mentioned as below:

1. **Gross non-performing asset (NPA) (in Rs millions):** Gross NPA in the books of a bank is the total outstanding amount in the borrowal account, other than the interest which has been recorded and not debited to the borrowal account. Gross NPA is the sum of all loan assets that are classified as NPA simply added together;
2. **Net NPA (in Rs millions):** Net NPA is simply the total bad assets (actual) minus the provision left aside. Both Gross and Net NPA has to be kept under control as spiraling figures will eat away profitability and thus affect sustainable growth;
3. **Gross advance (in Rs millions):** It means all outstanding loans and advances including advances for which refinance have been received but excluding rediscounted bills and advances written off at headquarters' level. Growth in advances is always desirable both in terms of geographical diversity as well as client profile and product portfolio, but its implication on asset quality (quantum of stressed assets) has also to be seen for sustainable growth;
4. **Operating profit (in Rs millions):** Operating profit is the profit figure after deducting cost of providing services and operating expenses from net sales. It helps to gauge the overall operating effectiveness and performance of the company;
5. **Net profit (in Rs millions):** Net profit denotes the amount of earnings left with the firm, after deducting all expenses, interest and taxes. It is a key indicator of company's ability to create returns for the owners of the firm. The various

elements leading to profitability are interest spreads, fee based earnings, overhead expense and provisioning levels. All these factors are also required to be examined before commenting on profitability trends and the sustainability of profits;

6. **Total assets (in Rs millions):** Total assets signify the strength of any organization. They are the sum of all current and noncurrent assets. In case of banking companies, total assets comprise of cash balances, investments, advances, fixed and other assets;
7. **Total income (in Rs millions):** The primary components of total income for banking institutions are interest income (from advances, other investments and interest on deposits with RBI and other banks) and other income (through way of commission, exchange and brokerages; sale of investments and assets, exchange transactions, dividend earnings and leasing etc.);
8. **Earning quality (in Rs millions):** The earning profile and the quality of earnings determine the ability of the bank to earn consistently in future. Diversity of income sources is an important input in analyzing the stability of earnings. Diversity in fund-based income is achieved by focusing on different borrower segments, such as industries, trade and retail. Banks also diversify their income streams through non-interest or fee income from guarantees, cash management facility, service charges from retail customer etc. Fee income provides a cushion to profitability, especially in times of pressure on interest spreads. Earnings can be significantly affected due to volatility in interest rates. Thus, the trend in profitability at gross profit levels over the past years is examined to form a view on the sustainability of earnings;
9. **Business (in Rs millions):** Business in banking parlance means summation of total advances and total deposits as per the balance sheet. A sustainable growth in business here means equal attention to both the assets and the liability side;
10. **Capital adequacy ratio:** Like other organizations, adequate capital does provide a bank too with the much required cushion to withstand credit and other risks unique to its business. The absolute size of capital imparts flexibility to a bank to withstand shocks. Therefore, a bank with high absolute capital should be viewed favorably. The proportion of Tier I capital, or core capital, is the primary indicator of the quality of capital the bank has. Although Tier II capital does provide some cushion over short to medium term exposures, such capital needs to be periodically replenished. Also, an analysis of the balance sheet with respect to the identification of hidden reserves (for example, unrealized gains on investment portfolio) and the percentage of the investment portfolio that is marked to market needs to be considered for guaranteeing a sustainable risk shielding ability;

11. **Net interest income (in Rs millions):** Net interest income is the difference between the interest income a bank earns from its lending activities and the interest it pays to depositors. Net interest income can vary from bank to bank due to dissimilarities in the composition and quality of assets and interest-bearing funds and change in yields of interest-earning assets and in interest rates paid on liabilities. If the spread between rate-sensitive assets and rate-sensitive liabilities increases, a rise in interest rate can make interest income rise more than interest expenses. In such a case, net interest income also goes up and vice-versa in case the spread between rate-sensitive assets and rate-sensitive liabilities falls. Net interest income is also greatly impacted by any rise or fall in the quantum of NPAs. For sustainability, endeavor should be towards maximization of interest income because this forms the core business for a banking institution;
12. **Deposit (in Rs millions):** It is a sum of money lodged in a bank for the purpose of earning interest. Deposit mobilization is an integral part of banking activity. Mobilization of savings through intensive deposit collection has been regarded as the major task of banking. It basically comprises of demand and time deposits. Deposits are the primary source of funds for conducting banking operations, but too much funds with banks can lead to a greater interest burden and thus the requirement to deploy them profitably becomes all the more imperative. A sustainable growth in deposits ensures a smooth conduction of business. The cornerstone of an effective deposits management policy lies in the proper monitoring of the credit to deposit ratio (C-D ratio);
13. **Branches:** The branch expansion program in Indian banks were started in the country since the nationalization of banks in 1969. Reach of banking services in rural areas is the precursor of growth. An extensive presence across geographies does bring in an obvious advantage in terms of economies of scale, but mindless expansion, just for the sake of competition impairs sustainability. Therefore, a judicious approach should be adopted towards branch expansion that would be viable;
14. **Complaints:** This parameter refers to quality of services offered by banking institutions. Fair practices and customer centricity is the first step towards bringing about customer satisfaction, customer service and customer experience. In India, the banking ombudsman scheme introduced by the Reserve Bank of India in 1995 set the tone for redressal of customer complaints against deficiency in banking services. Since then it has become an important parameter for determining a bank's service quality as every bank tends to strengthen their internal grievance redressal mechanism and thereby reinforce sustainability;

15. **Business per employee (in Rs millions):** It is a measure of how competently a bank is using its employees. Normally, high business per employee is a positive sign that suggests the bank is finding ways to generate more revenue out of each of its workers;
16. **Profit per employee (in Rs millions):** This parameter is employed by banks to check and improve productivity of employees and thus overall profitability.

The subsequent step comprises of implementation of QFD technique to estimate the priority weights for all the identified technical requirements. This stage principally initiates with development of the related HoQ matrix that can further be separated into three steps. The customers' requirements and technical specifications of the bank's sustainability identified in the preceding stage are placed along the rows and columns of HoQ matrix respectively. These technical specifications can either be beneficial (higher-the-better) or non-beneficial (lower-the-better) in characteristic. They are represented by the value of the corresponding improvement driver, i.e. +1 for beneficial criterion and –1 for non-beneficial criterion. After that, the relative importance to each of the customer requirements is allocated for which a scale of 1-5 is utilized, where 1 - not important, 2 - less important, 3 - moderately important, 4 - important and 5 - most important. Next, the inter-relationship matrix showing the relative strength between customers' requirements and technical requirements is established while critically analyzing the degree of association existing between them. A suitable scale of 1-9, where 1 - very very weak, 2 - very weak, 3 - weaker, 4 - weak, 5 - moderate, 6 - strong, 7 - stronger, 8 - very strong and 9 - very very strong, is set for allocating these correlation indices. The weight for all the technical requirements is subsequently calculated from the filled up HoQ matrix. This priority weight is calculated using the following equation:

$$w_j = ID_j \times \sum_{i=1}^{n} P_{r_i} \times Correlation\,index \tag{1}$$

where w_j is the weight for j^{th} technical requirement, n is the number of customers' requirements, ID_j is the value of improvement driver for j^{th} technical requirement, Pr_i is the priority assigned to i^{th} customer requirement and correlation index is the relative importance of jth technical requirement with respect to i^{th} customer requirement. The derived criteria weights as shown in Table 1 are subsequently employed to compute the performance scores of the different alternative banks in the next section while employing TOPSIS technique.

Table 1. Weightage of technical parameters

Parameter	Weightage
Gross NPA	7.33
Gross advances	7.57
Operating profit	6.38
Net profit	6.70
Total assets	4.89
Total income	7.96
Business	7.64
Earnings quality	6.23
Capital adequacy ratio	8.27
Net NPA	6.93
Net interest income	6.07
Deposit	5.52
Branches	6.07
Complaints	2.99
Business per employee	4.81
Profit per employee	4.65

APPLICATION OF TOPSIS APPROACH FOR RANKING BANKS

TOPSIS is a valuable method in resolving decision problems encompassing multiple criteria (Hwang and Yoon, 1981). It presents a more pragmatic form of modelling than non-compensatory techniques that incorporate or eliminate optimal choice on the basis of rigid cut-offs. TOPSIS is a generally pertinent technique with straightforward mathematical calculations. It assists decision maker to systematize the problem to be deciphered, and accomplish evaluation of the alternatives. TOPSIS methodology employs distance-based approach for examining the performance of alternatives while considering Euclidean distance. The best alternative in this technique is the one which has minimal distance from positive ideal solution, whereas its distance from negative ideal solution is highest. Positive ideal solution makes the most of the beneficial criteria and diminishes the cost criteria. On the other hand, negative ideal solution gets the most out of the cost criteria and reduces the beneficial criteria. The procedural steps involved in TOPSIS methodology are as follows:

Step 1: TOPSIS method initiates with formation of decision matrix. The decision matrix represents the performance measures of alternatives with respect to various criteria.

Step 2: In the next step, different attribute dimensions are converted into non-dimensionless attributes in order to allow comparison across the criteria. Following expression is employed to normalize the decision matrix:

$$r_{ij} = \frac{x_{ij}}{\sqrt{\sum_{i=1}^{m} x^2_{ij}}} \tag{2}$$

Step 3: Subsequently weighted normalized decision matrix is estimated using the formulae given below:

$$t_{ij} = r_{ij} \cdot w_j \text{ i= 1, 2, ...,m; j= 1, 2, ..., n} \tag{3}$$

where w_j denotes weightage of criterion C_j with respect to the ultimate objective of the selection problem.

Step 4: The negative ideal solution and positive ideal solution is calculated employing the two equations mentioned below:

$$S_n = \left\{\max(t_{ij} | i = 1, 2, \ldots .m) | j \in J\_, \min(t_{ij} | i = 1, 2, \ldots ., m) | j \in J_+\right\} \equiv \left\{t_{wj} | j = 1, 2, \ldots .n\right\} \tag{4}$$

$$S_p = \left\{\min(t_{ij} | i = 1, 2, \ldots .m) | j \in J\_, \max(t_{ij} | i = 1, 2, \ldots ., m) | j \in J_+\right\} \equiv \left\{t_{bj} | j = 1, 2, \ldots .n\right\} \tag{5}$$

where, $J\_ = \left\{J = 1, 2, \ldots ., n | j\right\}$ associated with the criteria having a negative impact, and $J_+ = \left\{J = 1, 2, \ldots ., n | j\right\}$ associated with the criteria having a positive impact.

Step 5: Next, separation measure from the negative ideal solution and positive ideal solution is computed while employing following expressions:

$$d_{iw} = \sqrt{\sum_{j=1}^{n} \left(t_{ij} - t_{wj}\right)^2} \text{ i= 1,2,......,m} \tag{6}$$

$$d_{ib} = \sqrt{\sum_{j=1}^{n} \left(t_{ij} - t_{bj}\right)^2} \text{ i= 1,2,......,m} \tag{7}$$

Step 6: Estimation of the relative closeness to ideal solution, using the formula given below, is the consequent step:

$$\delta_{iw} = \frac{d_{iw}}{\left(d_{iw} + d_{ib}\right)} \text{ i= 1,2,......,m} \tag{8}$$

where, $0 \leq \delta_{iw} \leq 1$; $\delta_{iw} = 1$, if the alternative solution has the paramount condition and $\delta_{iw} = 0$, if the alternative option has the worst condition.

Step 7: Finally, rank preorder of different alternatives is derived while arranging the options in decreasing order of δ_{iw}.

With a view to prove the applicability, robustness and solution exactness of the integrated TOPSIS-QFD model in selecting best bank on sustainability factor, a list of 21 public sector banks in India are evaluated. Some of the public sector banks are not considered in this research work owing to unavailability of their information on identified criteria on public domain. Then, following the steps of the considered hybrid approach, at first decision matrix is developed for the identified 21 public sector bank with respect to 16 chosen technical parameters. The 16 criteria important for achieving sustainability on which banks are going to be evaluated are already discussed in the preceding section. After the decision matrix has been developed, it is normalized to lessen data idleness and enhance data integrity. Next, normalized decision matrix is multiplied with the weights of diverse selected criteria to obtain weighted normalized matrix. The weights for identified attributes for evaluation of banks have been derived previously while employing QFD technique. Subsequently, positive ideal solution and negative ideal solutions are estimated using Equations 5 and 4 respectively. Separation measure from the negative ideal solution and positive ideal solution is computed in next step employing Equation 6 and Equation 7 in that order. Table 2 shows the relative closeness coefficient of 21 banks with respect to positive ideal solution on selected sustainability criteria along with their ranks.

Here, Canara Bank is identified as the best bank in public sector category on the basis of stakeholders' sustainability requirements and Bank of India is the second-best choice, as exhibited in Table 2. It is further noticed that State Bank of India is worst choice. Additionally, it is obvious from the closer view at the properties of Canara Bank that it can effectively satisfy all the sustenance criteria values. This unquestionably rationalizes its choice as the best bank in public sector based on selected stakeholders' needs.

Table 2. Ranking of banks

Bank	Relative Closeness Coefficient	Rank
Allahabad Bank	0.243407	10
Andhra Bank	0.241499	13
Bank of Baroda	0.267791	4
Bank of India	0.279783	2
Bank of Maharashtra	0.236390	15
Canara Bank	0.289043	1
Central Bank of India	0.254015	7
Corporation Bank	0.242695	11
Dena Bank	0.232496	17
Indian Bank	0.239009	14
Indian Overseas Bank	0.247917	9
Oriental Bank of Commerce	0.227578	20
Punjab and Sind Bank	0.228997	19
Punjab National Bank	0.265622	5
Syndicate Bank	0.251153	8
UCO Bank	0.242357	12
Union Bank of India	0.271569	3
United Bank of India	0.229342	18
Vijaya Bank	0.233340	16
State Bank of India	0.215405	21
IDBI	0.257424	6

FUTURE RESEARCH DIRECTIONS

The future scope of this current research work may include implementation of this integrated TOPSIS-QFD method for the assessment all the banks in India with respect to sustainability parameters in order to identify the superior performing bank of entire nation. This will also assist the policy makers of banks to understand the significance of various parameters in terms of the sustainable growth which in turn will further help them to achieve long term sustainability. Moreover, the growth of an economy or a society can also be estimated through analyzing the sustainability assessment of banks in future research work.

CONCLUSION

India is seventh largest country by area and the second most populous country. It is one of the fastest growing economies in the world According to a report published by India Brand Equity Foundation, GDP of India is estimated to have increased by 6.6% in 2017-18 and expected to grow 7.3% in 2018-19. Despite of its vast economic achievements, it is the country in the world with the largest number of people living in absolute poverty. This exploding population of India is leading towards shrinking of resources. Economy of India is growing; growth in purchasing power parity is visible; industries, business, transport etc. are blooming but huge portion of Indian populace are still living in extreme poverty. This problem of poverty, extinction of natural resources and others can be resolved only through sustainable development. Sustainable progress plays an imminent function in current ethnicities that understand the worth of sufficing the demands of current population without compromising capability of upcoming generations to convene their own needs. Encouraging continuous development requires applying tangible actions, projects, programs, plans and policies that engage the concurrent quest and satisfaction of economic, environmental, and social goals. In this milieu, the exigency for banks to operate as responsible business populace in the society, particularly in an emergent economy, like India, required to be barely overstated. The progress of industries, development of infrastructure in villages, success of social programs etc. are all proportional to the growth of banks. Thus, for the sustainable growth of an economy, banking sector needs to perform well. Hence, this paper demonstrates the application of an integrated TOPSIS-QFD technique of selecting the best bank to fulfill a set of specified sustainability requirements. QFD is fundamentally the adaptation of tools employed in total quality management, that facilitates the formulation of business problems and probable solutions through efficient communiqué among the members of a cross functional team. TOPSIS is one of the most eminent distance-based techniques for MCDM. It ranks the options with respect to their geometric distance from the positive and negative ideal solutions. This method is classified as one of the MCDM methods in which value judgments of criteria are articulated through crisp values. The applicability of the integrated TOPSIS-QFD technique is illustrated through performing a sustainability assessment of public sector banking institutions operating in India. In addition to screening the alternatives based on sustainability principles, the proposed technique provides improvement targets for the least sustainable banking institutions for corrective actions, thus helping the management of those institutes to improve the efficiency of technologies, products and policies.

REFERENCES

Berger, A. N., & Di-Patti, E. B. (2006). Capital structure and firm performance: A new approach to testing agency theory and an application to the banking industry. *Journal of Banking & Finance*, *30*(4), 1065–1102. doi:10.1016/j.jbankfin.2005.05.015

Bhattacharya, A., Geraghty, J., & Young, P. (2010). Supplier selection paradigm: An integrated hierarchical QFD methodology under multiple-criteria environment. *Applied Soft Computing*, *10*(4), 1013–1027. doi:10.1016/j.asoc.2010.05.025

Biswas, P., Pramanik, S., & Giri, B. C. (2016). TOPSIS method for multi-attribute group decision-making under single-valued neutrosophic environment. *Neural Computing & Applications*, *27*(3), 727–737. doi:10.100700521-015-1891-2

Bottani, E., & Rizzi, A. (2006). Strategic management of logistics service: A fuzzy QFD approach. *International Journal of Production Economics*, *103*(2), 585–599. doi:10.1016/j.ijpe.2005.11.006

Carnevalli, J. A., & Miguel, P. C. (2008). Review, analysis and classification of the literature on QFD - Types of research, difficulties and benefits. *International Journal of Production Economics*, *114*(2), 737–754. doi:10.1016/j.ijpe.2008.03.006

Dos Santos, B. M., Godoy, L. P., & Campos, L. M. (2019). Performance evaluation of green suppliers using entropy-TOPSIS-F. *Journal of Cleaner Production*, *207*, 498–509. doi:10.1016/j.jclepro.2018.09.235

Dursun, M., & Karsak, E. E. (2013). A QFD-based fuzzy MCDM approach for supplier selection. *Applied Mathematical Modelling*, *37*(8), 5864–5875. doi:10.1016/j.apm.2012.11.014

Foehn, P. (2004). Client valuation in private banking: Results of a case study in Switzerland. *Managing Service Quality: An International Journal*, *14*(2/3), 195–204. doi:10.1108/09604520410528617

Ho, W., He, T., Lee, C. K. M., & Emrouznejad, A. (2012). Strategic logistics outsourcing: An integrated QFD and fuzzy AHP approach. *Expert Systems with Applications*, *39*(12), 10841–10850. doi:10.1016/j.eswa.2012.03.009

Hwang, C. L., & Yoon, K. (1981). Methods for multiple attribute decision making. In *Multiple attribute decision making* (pp. 58–191). Berlin: Springer. doi:10.1007/978-3-642-48318-9_3

Iannotta, G., Nocera, G., & Sironi, A. (2007). Ownership structure, risk and performance in the European banking industry. *Journal of Banking & Finance*, *31*(7), 2127–2149. doi:10.1016/j.jbankfin.2006.07.013

Kahraman, C., Ertay, T., & Büyüközkan, G. (2006). A fuzzy optimization model for QFD planning process using analytic network approach. *European Journal of Operational Research*, *171*(2), 390–411. doi:10.1016/j.ejor.2004.09.016

Kannan, D., De Sousa Jabbour, A. B. L., & Jabbour, C. J. C. (2014). Selecting green suppliers based on GSCM practices: Using fuzzy TOPSIS applied to a Brazilian electronics company. *European Journal of Operational Research*, *233*(2), 432–447. doi:10.1016/j.ejor.2013.07.023

Kim, G., Shin, B., & Lee, H. G. (2009). Understanding dynamics between initial trust and usage intentions of mobile banking. *Information Systems Journal*, *19*(3), 283–311. doi:10.1111/j.1365-2575.2007.00269.x

Mavridis, D. G. (2004). The intellectual capital performance of the Japanese banking sector. *Journal of Intellectual Capital*, *5*(1), 92–115. doi:10.1108/14691930410512941

Mercan, M., Reisman, A., Yolalan, R., & Emel, A. B. (2003). The effect of scale and mode of ownership on the financial performance of the Turkish banking sector: Results of a DEA-based analysis. *Socio-Economic Planning Sciences*, *37*(3), 185–202. doi:10.1016/S0038-0121(02)00045-9

Neema, A., & Nema, G. (2017). Impact of brand image on brand loyalty and brand equity in banking sector: A case of State Bank of India. *Pacific Business Review International*, *10*(3), 109–112.

Paul, J., Mittal, A., & Srivastav, G. (2015). Impact of service quality on customer satisfaction in private and public sector banks. *International Journal of Bank Marketing*, *34*(5), 606–622. doi:10.1108/IJBM-03-2015-0030

Sindhu, S., Nehra, V., & Luthra, S. (2017). Investigation of feasibility study of solar farms deployment using hybrid AHP-TOPSIS analysis: Case study of India. *Renewable & Sustainable Energy Reviews*, *73*, 496–511. doi:10.1016/j.rser.2017.01.135

Subramony, M., & Holtom, B. C. (2012). The long-term influence of service employee attrition on customer outcomes and profits. *Journal of Service Research*, *15*(4), 460–473. doi:10.1177/1094670512452792

Tan, K. C., & Pawitra, T. A. (2001). Integrating SERVQUAL and Kano's model into QFD for service excellence development. *Managing Service Quality: An International Journal*, *11*(6), 418–430. doi:10.1108/EUM0000000006520

Yiu, C. S., Grant, K., & Edgar, K. (2007). Factors affecting the adoption of Internet Banking in Hong Kong - Implications for the banking sector. *International Journal of Information Management*, *27*(5), 336–351. doi:10.1016/j.ijinfomgt.2007.03.002

ADDITIONAL READING

Bai, C., & Sarkis, J. (2018). Integrating sustainability into supplier selection: A grey-based TOPSIS analysis. *Technological and Economic Development of Economy*, *24*(6), 2202–2224. doi:10.3846/tede.2018.5582

Barathi Kamath, G. (2007). The intellectual capital performance of the Indian banking sector. *Journal of Intellectual Capital*, *8*(1), 96–123. doi:10.1108/14691930710715088

Bikker, J. A., & Haaf, K. (2002). Competition, concentration and their relationship: An empirical analysis of the banking industry. *Journal of Banking & Finance*, *26*(11), 2191–2214. doi:10.1016/S0378-4266(02)00205-4

Claessens, S., & Laeven, L. (2005). Financial dependence, banking sector competition, and economic growth. *Journal of the European Economic Association*, *3*(1), 179–207. doi:10.1162/1542476053295322

Dikmen, I., Birgonul, M. T., & Kiziltas, S. (2005). Strategic use of quality function deployment (QFD) in the construction industry. *Building and Environment*, *40*(2), 245–255. doi:10.1016/j.buildenv.2004.07.001

Kuhlman, T., & Farrington, J. (2010). What is sustainability? *Sustainability*, *2*(11), 3436–3448. doi:10.3390u2113436

Kumar, K., & Garg, H. (2018). TOPSIS method based on the connection number of set pair analysis under interval-valued intuitionistic fuzzy set environment. *Computational & Applied Mathematics*, *37*(2), 1319–1329. doi:10.100740314-016-0402-0

Patil, C. K., Husain, M., & Halegowda, N. V. (2018). Study of quality function deployment model based on artificial neural network with optimization techniques. *Journal of Advanced Manufacturing Systems*, *17*(01), 119–136. doi:10.1142/S0219686718500087

Režek Jambrak, A., Šimunek, M., Grbeš, F., Mandura, A., & Djekic, I. (2018). Analysis of apple beverages treated with high-power ultrasound: A quality function deployment approach. *Journal of the Science of Food and Agriculture*, *98*(6), 2258–2266. doi:10.1002/jsfa.8714 PMID:28981162

KEY TERMS AND DEFINITIONS

Agency Function of Bank: It includes various activities, like collection of checks, accepting bills, remittance facilities, collecting payments, etc.

Balance Sheet: It is a financial statement showing the finance of a company on a particular date.

Banking Ombudsman: It is a quasi-judicial authority which is being formed with intent to resolve the grievances of the customers of a bank.

Credit to Deposit Ratio: It is the ratio of assets and liabilities of the banks.

Portfolio: It is a group of financial assets which are being held by the investors and are managed by financial professionals or fund managers.

Tier 1 Capital: It measures the financial strength or capital adequacy of a bank.

Utility Function of Banks: It comprises of actions, such as undertaking foreign exchange transactions, providing safe custody of valuables, issuing letter of credit to promote the foreign trade, etc.

Compilation of References

Adar, E., Ince, M., Karatop, B., & Bilgili, M. S. (2017). The risk analysis by failure mode and effect analysis (FMEA) and Fuzzy FMEA of supercritical water gasification system used in the sewage sludge treatment. *Journal of Environmental Chemical Engineering*, *5*(1), 1261–1268. doi:10.1016/j.jece.2017.02.006

Agarwal, A. K., & Das, L. M. (2000). Biodiesel development and characterization for use as a fuel in compression ignition engines. *Journal of Engineering for Gas Turbines and Power*, *123*(2), 440–447. doi:10.1115/1.1364522

Agarwal, A., Shankar, R., & Tiwari, M. K. (2006). Modeling agility of supply chain. *Industrial Marketing Management*, *36*(4), 443–457. doi:10.1016/j.indmarman.2005.12.004

Aggarwal, C. C., Wolf, J. L., & Yu, P. S. (1999). A new method for similarity indexing for market data. In *Proceedings of the ACM SIGMOD Conference* (pp. 407-418). Philadelphia: ACM. 10.1145/304182.304218

Agrawal, R., Imielinski, T., & Swami, A. N. (1993). Mining Association Rules between Sets of Items in Large Databases. In *Proc. ACM SIGMOD Int'l Conf. Management of Data* (pp 207-216). Washington, DC: ACM. 10.1145/170035.170072

Agrawal, R., & Srikant, R. (1994). Fast algorithms for mining association rules in large databases. In *VLDB '94: Proceedings of the 20th International Conference on Very Large Data Bases* (pp. 487—499). San Francisco, CA: Academic Press.

Ahmadi, H. B., Kusi-Sarpong, S., & Rezaei, J. (2017). Assessing the social sustainability of supply chains using best worst method. *Resources, Conservation and Recycling*, *126*, 99–106. doi:10.1016/j.resconrec.2017.07.020

Ahmad, M., Zafar, M., Khan, M. A., & Sultana, S. (2009). Biodiesel from Pongamia pinnata L. oil: A promising alternative bioenergy source. Energy Sources. *Part A*, *31*(16), 1436–1442.

Ajanovic, A., & Haas, R. (2016). Dissemination of electric vehicles in urban areas: Major factors for success. *Energy*, *115*(Part 2), 1451–1458. doi:10.1016/j.energy.2016.05.040

Alexander, A. J., & Holownia, B. P. (1978). Wind tunnel test on a Savonius rotor. *Journal of Industrial Aerdynamics*, *3*(4), 343–351. doi:10.1016/0167-6105(78)90037-5

Ali, N. H., Sabri, I. A. A., Noor, N. M. M., & Ismail, F. (2012). Rating and ranking criteria for selected Islands using fuzzy analytic hierarchy process (FAHP). *International Journal of Applied Mathematics and Informatics*, *6*(1), 57–65.

Ali, O. M., Mamat, R., Abdullah, N. R., & Abdullah, A. A. (2016). Analysis of blended fuel properties and engine performance with palm biodiesel-diesel blended fuel. *Renewable Energy*, *86*, 59–67. doi:10.1016/j.renene.2015.07.103

Alptekin, E., & Canakci, M. (2009). Characterization of the key fuel properties of methyl ester–diesel fuel blends. *Fuel*, *88*(1), 75–80. doi:10.1016/j.fuel.2008.05.023

Altan, B. D., & Atılgan, M. (2008). An experimental and numerical study on the improvement of the performance of Savonius wind rotor. *Energy Conversion and Management*, *49*(12), 3425–3432. doi:10.1016/j.enconman.2008.08.021

Altan, B. D., Atılgan, M., & Özdamar, A. (2008). An experimental study on improvement of a Savonius rotor performance with curtaining. *Experimental Thermal and Fluid Science*, *32*(8), 1673–1678. doi:10.1016/j.expthermflusci.2008.06.006

Alves, C. M., Ferreira, C. M. H., Soares, E. V., & Soares, M. V. M. (2017). A multi-metal risk assessment strategy for natural freshwater ecosystems based on the additive inhibitory free metal ion concentration index. *Environmental Pollution*, *223*, 517–523. doi:10.1016/j.envpol.2017.01.053 PMID:28159400

Amy, H. I. L., Hsing, H. C., & He-Yau, K. (2009). Multi-criteria decision making on strategic selection of wind farms. *Renewable Energy*, *34*(1), 120–126. doi:10.1016/j.renene.2008.04.013

Amy, H. I. L., Meng-Chan, H., He-Yau, K., & Pearn, W. L. (2012). A wind turbine evaluation model under a multi-criteria decision making environment. *Energy Conversion and Management*, *64*, 289–300. doi:10.1016/j.enconman.2012.03.029

Anand, A., Rufuss, D. D. W., Rajkumar, V., & Suganthi, L. (2017). Evaluation of sustainability Indicators in smart cities for India using MCDM approach. *Energy Procedia*, *141*, 211–215. doi:10.1016/j.egypro.2017.11.094

Annalee, Y., Karen, L., Ray, C., Mickey, K., & Marc Corbiere, E. (2007). Determinants of Healthcare Workers' Compliance with Infection Control Procedures. *Healthcare Quarterly*, *10*(1), 44–52. doi:10.12927/hcq.2007.18648 PMID:17326369

Anthopolous, L., & Giannakidis, G. (2017). Policy Making in Smart City: Standardizing City's Energy Efficiency with Task-Based Modelling. *Journal of ICT*, *4*(2), 111–146.

Araújo, R. M. (2015). Multicriteria classification with TODIM-FSE. *Procedia Computer Science*, *55*, 559–565. doi:10.1016/j.procs.2015.07.043

Arbab, M. I., Masjuki, H. H., Varman, M., Kalam, M. A., Imtenan, S., & Sajjad, H. (2013). Fuel properties, engine performance and emission characteristic of common biodiesels as a renewable and sustainable source of fuel. *Renewable & Sustainable Energy Reviews*, *22*, 133–147. doi:10.1016/j.rser.2013.01.046

Arkesteijn, K., & Oerlemans, L. (2005). The early adoption of green power by Dutch households: An empirical exploration of factors influencing the early adoption of green electricity for domestic purposes. *Energy Policy*, *33*(2), 183–196. doi:10.1016/S0301-4215(03)00209-X

Askenazy, P. (2006). *Some determinants of reporting workplace accidents in France: The role of labour contract*. Paris: Centre Pour La Recherche Economique Et ses Applications.

Atanassov, K. T. (1986). Intuitionistic fuzzy sets. *Fuzzy Sets and Systems*, *20*(1), 87–96. doi:10.1016/S0165-0114(86)80034-3

Athanasios, K., Varvara, M., Estivaliz, L. M., & Konstantinos, S. (2016). A Comparative Study of Multiple-Criteria Decision-Making Methods under Stochastic Inputs. *Energies*, *9*(566), 1–21.

Athawale, V. M., Chatterjee, P., & Chakraborty, S. (2012). Selection of industrial robots using compromise ranking method. *International Journal of Industrial and Systems Engineering*, *11*(1/2), 3–15. doi:10.1504/IJISE.2012.046651

Aydin, H., & Bayindir, H. (2010). Performance and emission analysis of cottonseed oil methyl ester in a diesel engine. *Renewable Energy*, *35*(3), 588–592. doi:10.1016/j.renene.2009.08.009

Azadeh, A., Arani, H. V., & Dashti, H. (2014). A stochastic programming approach towards optimization of biofuel supply chain. *Energy*, *76*, 513–525. doi:10.1016/j.energy.2014.08.048

Bach, G. (1931). *Investigation Concerning Savonius rotors and related Machines*. Quebec, Canada: Brace Research Institute.

Badi, I., Abdulshahed, A., & Shetwan, A. (2018). A case study of supplier selection for steel making company in Libya by using combinative distance-based assessment (CODAS) model. *Decision Making: Applications in Management and Engineering*, *1*(1), 1–12.

Banapurmath, N. R., Tewari, P. G., & Hosmath, R. S. (2008). Performance and emission characteristics of a DI compression ignition engine operated on Honge, Jatropha and sesame oil methyl esters. *Renewable Energy*, *33*(9), 1982–1988. doi:10.1016/j.renene.2007.11.012

Bandyopadhyay, S., Thakur, S. S., & Mandal, J. K. (2018). *A Novel Approach for Product Prediction Using Artificial Neural Networks*. Social Transformation – Digital Way. doi:10.1007/978-981-13-1343-1_56

Banerjee, S., & Majumder, M. (2016, April). Application of AHP for selection of source for Microgrid system. In *Energy Efficient Technologies for Sustainability (ICEETS), 2016 International Conference on* (pp. 112-115). IEEE.

Banthia, J. K. (2004). *Census of India 2001 - Primary Census Abstracts*. Registrar General & Census Commissioner, Govt. of India.

Banwet, D. K., & Majumdar, A. (2014). Comparative analysis of AHP TOPSIS and GA-TOPSIS methods for selection of raw materials in textile industries. *Proceedings of the International Conference on Industrial Engineering and Operations Management*, 2071-2080.

Behçet, R. (2011). Performance and emission study of waste anchovy fish biodiesel in a diesel engine. *Fuel Processing Technology*, *92*(6), 1187–1194. doi:10.1016/j.fuproc.2011.01.012

Behnam, V., Zandieh, M., & Moghaddam, R. T. (2011). Two novel FMCDM methods for alternative-fuel buses selection. *Applied Mathematical Modelling*, *35*(3), 1396–1412. doi:10.1016/j.apm.2010.09.018

Berger, A. N., & Di-Patti, E. B. (2006). Capital structure and firm performance: A new approach to testing agency theory and an application to the banking industry. *Journal of Banking & Finance*, *30*(4), 1065–1102. doi:10.1016/j.jbankfin.2005.05.015

Bhattacharya, A., Geraghty, J., & Young, P. (2010). Supplier selection paradigm: An integrated hierarchical QFD methodology under multiple-criteria environment. *Applied Soft Computing*, *10*(4), 1013–1027. doi:10.1016/j.asoc.2010.05.025

Bhattacharya, A., Sarkar, B., & Mukherjee, S. K. (2005). Integrating AHP with QFD for robot selection under requirement perspective. *International Journal of Production Research*, *43*(17), 3671–3685. doi:10.1080/00207540500137217

Bhowmik, C., Baruah, A., Bhowmik, S., & Ray, A. (2018). Green energy sources selection for sustainable energy planning using multi-criteria decision-making approach. In IOP Conference Series: Materials Science and Engineering (Vol. 377, No. 1, p. 012029). IOP Publishing. doi:10.1088/1757-899X/377/1/012029

Bhowmik, C., Bhowmik, S., & Ray, A. (2018). Social acceptance of green energy determinants using principal component analysis. *Energy*, *160*, 1030–1046. doi:10.1016/j.energy.2018.07.093

Bhowmik, C., Bhowmik, S., Ray, A., & Pandey, K. M. (2017). Optimal green energy planning for sustainable development: A review. *Renewable & Sustainable Energy Reviews*, *71*, 796–813. doi:10.1016/j.rser.2016.12.105

Biswas, A., Gupta, R., & Sharma, K. K. (2007). Experimental Investigation of Overlap and Blockage Effects on Three-Bucket Savonius Rotors. *Wind Engineering*, *31*(5), 363–368. doi:10.1260/030952407783418702

Biswas, P., Pramanik, S., & Giri, B. C. (2016). TOPSIS method for multi-attribute group decision-making under single-valued neutrosophic environment. *Neural Computing & Applications*, *27*(3), 727–737. doi:10.100700521-015-1891-2

Biswas, T., & Das, M. (2018). Selection of hybrid vehicle for green environment using multi-attributive border approximation area comparison method. *Management Science Letters*, *8*(2), 121–130. doi:10.5267/j.msl.2017.11.004

Blackwell B.F. (1978) *Wind tunnel performance data for two- & three- bucket Savonius rotor*. Sandia Laboratories, Sand 76-0131 under act AT/29-11-789.

Boral, S., & Chakraborty, S. (2016). A case-based reasoning approach for non-traditional machining processes selection. *Advances in Production Engineering & Management*, *11*(4), 311–323. doi:10.14743/apem2016.4.229

Borjalilu, N., & Ghambari, M. (2018). Optimal maintenance strategy selection based on a fuzzy analytical network process: A case study on a 5-MW powerhouse. *International Journal of Engineering Business Management*, *10*, 1–10. doi:10.1177/1847979018776172

Bottani, E., & Rizzi, A. (2006). Strategic management of logistics service: A fuzzy QFD approach. *International Journal of Production Economics*, *103*(2), 585–599. doi:10.1016/j.ijpe.2005.11.006

Brandl, C., Mertens, A., & Schlick, C.M. (2017a). Ergonomic analysis of working postures using OWAS in semi-trailer assembly, applying an individual sampling strategy. *Int. J. Occup. Saf. Ergon, JOSE,* 1-8. . doi:10.1080/10803548.2016.1191224

Brandl, C., Mertens, A., & Schlick, C. M. (2017b). Effect of sampling interval on the reliability of ergonomic analysis using the Ovako working posture analysing system (OWAS). *International Journal of Industrial Ergonomics*, *57*, 68–73. doi:10.1016/j.ergon.2016.11.013

Brans, J. P., & Mareschal, B. (1994). The PROMCALC and GAIA decision-support system for multi criteria decision aid. *Decision Support Systems*, *12*(4-5), 297–310. doi:10.1016/0167-9236(94)90048-5

Brans, J. P., & Vincke, P. (1985). Preference ranking organization method: The PROMETHEE method for MCDM. *Management Science*, *31*(6), 647–656. doi:10.1287/mnsc.31.6.647

Brans, J. P., Vincke, P., & Mareschal, B. (1986). How to select and how to rank project: The PROMETHEE method. *European Journal of Operational Research*, *24*(2), 228–238. doi:10.1016/0377-2217(86)90044-5

Brauers, W. K. M., & Zavadskas, E. K. (2006). The MOORA method and its application to privatization in a transition economy. *Control and Cybernetics*, *35*(2), 445–469.

Bühler, F., Cocron, P., Neumann, I., Franke, T., & Krems, J. F. (2014). Is EV experience related to EV acceptance? Results from a German field study. *Transportation Research Part F: Traffic Psychology*, *25*, 85–90. doi:10.1016/j.trf.2014.05.002

Burillo, P., & Bustince, H. (1986). Construction theorems for intuitionistic fuzzy sets. *Fuzzy Sets and Systems*, *84*(3), 271–281. doi:10.1016/0165-0114(95)00313-4

Cakır, S., & Percin, S. (2013). Çok kriterli karar verme teknikleriyle lojistik firmalarinda performans ölçümü. *EgeAkademik Bakış*, *13*(4), 449–459.

Caliskan, H., Kursuncu, B., Kurbanoglu, C., & Guven, S. Y. (2013). Material selection for the tool holder working under hard milling conditions using different multi criteria decision making methods. *Materials & Design*, *45*, 473–479. doi:10.1016/j.matdes.2012.09.042

Caragliu, A., Del Bo, C., & Nijkamp, P. (2011). Smart cities in Europe. *Journal of Urban Technology*, *18*(2), 65–82. doi:10.1080/10630732.2011.601117

Carnevalli, J. A., & Miguel, P. C. (2008). Review, analysis and classification of the literature on QFD - Types of research, difficulties and benefits. *International Journal of Production Economics*, *114*(2), 737–754. doi:10.1016/j.ijpe.2008.03.006

Celebi, M., Kingravi, H., & Vela, P. (2013). A comparative study of efficient initialization methods for the k-means clustering algorithm. *Expert Systems with Applications*, *40*(1), 200–210. doi:10.1016/j.eswa.2012.07.021

Chaghooshi, A. J., Janatifar, H., & Dehghan, M. (2014). An application of AHP and similarity-based approach to personnel selection. *International Journal of Business Management and Economics*, *1*(1), 24–32.

Chakraborty, S., & Prasad, K. (2018). A quality function deployment-based expert system for cotton fiber selection. *Journal of the Institution of Engineers (India): Series E, 99*(1), 43-53.

Chakraborty, S., Chatterjee, P., & Prasad, K. (2018). An Integrated DEMATEL-VIKOR Method-based approach for cotton fiber selection and evaluation. *Journal of the Institution of Engineers (India): Series E, 99*(1), 63-73.

Chakraborty, S., Chatterjee, P., & Prasad, K. (2018). An Integrated DEMATEL-VIKOR Method-Based Approach for Cotton Fibre Selection and Evaluation. Journal of the Institution of Engineers. doi:10.100740034-018-0113-8

Chakraborty, S., Das, P. P., & Kumar, V. (2017). A grey fuzzy logic approach for cotton fiber selection. *Journal of the Institution of Engineers (India): Series E, 98*(1), 1-9.

Chakraborty, S., & Bandhopadyay, S. (2017). Cotton fiber selection and grading-A PROMETHEE-GAIA-based approach. *International Journal of Clothing Science and Technology*, *29*(5), 646–660. doi:10.1108/IJCST-12-2016-0137

Chakraborty, S., & Boral, S. (2017). A developed case-based reasoning system for machine tool selection. *Benchmarking: An International Journal*, *24*(5), 1364–1385. doi:10.1108/BIJ-07-2016-0103

Chakraborty, S., & Prasad, K. (2016). A QFD-based expert system for industrial truck selection in manufacturing organizations. *Journal of Manufacturing Technology Management*, *27*(6), 800–817. doi:10.1108/JMTM-02-2016-0020

Chakraborty, S., Ranjan, R., & Mondal, P. (2014). A state-wise performance appraisal of the Indian roads using PROMETHEE-GIS approach. *Benchmarking: An International Journal*, *9*(25), 1–19.

Chamanehpour, E., & Ahmadizadeh, A. (2017). Site selection of wind power plant using multi-criteria decision-making methods in GIS: A case study. *Computational Ecology and Software*, *7*(2), 49–64.

Chatterjee, P., Athawale, V. M., & Chakraborty, S. (2009). Selection of materials using compromise ranking and outranking methods. *Materials & Design*, *30*(10), 4043–4053. doi:10.1016/j.matdes.2009.05.016

Chatterjee, P., Athawale, V. M., & Chakraborty, S. (2010). Selection of industrial robots using compromise ranking and outranking methods. *Robotics and Computer-integrated Manufacturing*, *26*(5), 483–489. doi:10.1016/j.rcim.2010.03.007

Chatterjee, P., & Chakraborty, S. (2013). Selection of materials using multi-criteria decision-making methods with minimum data. *Decision Science Letters*, *2*(3), 135–148.

Chatterjee, P., & Chakraborty, S. (2016). A comparative analysis of VIKOR method and its variants. *Decision Science Letters*, *5*(4), 469–486. doi:10.5267/j.dsl.2016.5.004

Chatterjee, P., & Chakraborty, S. (2017). Development of a metamodel for the determination of technological value of cotton fiber using design of experiments and the TOPSIS method. *Journal of Natural Fibers*, *15*(6), 882–895. doi:10.1080/15440478.2017.1376303

Chatterjee, P., Mondal, S., Boral, S., Banerjee, A., & Chakraborty, S. (2017). A novel hybrid method for non-traditional machining process selection using factor relationship and multi-attributive border approximation method. *Facta Universitatis Series: Mechanical Engineering*, *15*(3), 439–456. doi:10.22190/FUME170508024C

Chatterjee, P., Mukherjee, P., & Chakraborty, S. (2011). Supplier selection using compromise ranking and outranking methods. *Journal of Industrial Engineering International*, *7*(14), 61–73.

Choi, K., Yoo, D., Kim, G., & Suh, Y. (2012). A hybrid online-product recommendation system: Combining implicit rating-based collaborative filtering and sequential pattern analysis. *Electronic Commerce Research and Applications*, *11*(4), 309–317. doi:10.1016/j.elerap.2012.02.004

Chowdhury, S., Mazumder, M. A. J., Al-Attas, O., & Husain, T. (2016). Heavy metals in drinking water: Occurrences, implications, and future needs in developing countries. *The Science of the Total Environment*, *569–570*, 476–488. doi:10.1016/j.scitotenv.2016.06.166 PMID:27355520

CNEV. (2017). *A survey of investment and capacity planning of new energy vehicles in China*. Retrieved from www.chinanev.net/news/newscontent/id/9582

Cobb, J. (2015b). *How Nissan and Renault are Dominating the Electric Car Game*. Retrieved from www.hybridcars.com/how-nissan-and-renault-are-dominating-the-electric-car-game/

Cocron, P., Bühler, F., Franke, T., Neumann, I., Dielmann, B., & Krems, J. F. (2013). Energy recapture through deceleration – regenerative braking in electric vehicles from a user perspective. *Ergonomics*, *56*(8), 1203–1215. doi:10.1080/00140139.2013.803160 PMID:23767823

Cooper, S. P., Burau, K. E., Frankowski, R., Shipp, E., Deljunco, D., Whitworth, R., ... Hanis, C. (2006). A cohort study of injuries in migrant farm worker families in South Texas. *Annals of Epidemiology*, *16*(4), 313–320. doi:10.1016/j.annepidem.2005.04.004 PMID:15994097

Crouch, G. I., & Brent Ritchie, J. R. (1999). Tourism, Competitiveness, and Societal Prosperity. *Journal of Business Research*, *44*(3), 137–152. doi:10.1016/S0148-2963(97)00196-3

Curl, C. L., Fenske, R. A., Kissel, J. C., Shirai, J. H., Moate, T. F., Griffith, W., ... Thompson, B. (2002). Evaluation of take-home organophosphorus pesticide exposure among agricultural workers and their children. *Environmental Health Perspectives*, *110*(12), A787–A792. doi:10.1289/ehp.021100787 PMID:12460819

Daimler. (2016a). *Interview Prof. Dr. Thomas Weber: Electric mobility at Daimler will be in the six figures by 2020*. Retrieved from http://media.daimler.com/marsMediaSite/en

Debbarma, B., Chakraborti, P., Bose, P. K., Deb, M., & Banerjee, R. (2017). Exploration of PROMETHEE II and VIKOR methodology in a MCDM approach for ascertaining the optimal performance-emission trade-off vantage in a hydrogen-biohol dual fuel endeavour. *Fuel*, *210*, 922–935. doi:10.1016/j.fuel.2017.08.016

Debbarma, S., Das, B., Giri, A., Majumder, S., & Bose, P. K. (2011). Back propagation artificial neural network (BPANN) based performance analysis of diesel engine using biodiesel. *Journal of Renewable and Sustainable Energy*, *3*(1), 013101–013112. doi:10.1063/1.3517229

Debbarma, S., & Misra, R. D. (2017). Effects of Iron Nanoparticles Blended Biodiesel on the Performance and Emission Characteristics of a Diesel Engine. *Journal of Energy Resources Technology*, *139*(042212), 1–12.

Debbarma, S., & Misra, R. D. (2018). Effects of Iron NanoparticleFuel Additive on thePerformance and ExhaustEmissions of a CompressionIgnition Engine Fueled With Diesel and Biodiesel. *Journal of Thermal Science and Engineering Applications*, *10*(4), 041002-1–12. doi:10.1115/1.4038708

Dehnavi, A., Aghdam, I. N., Pradhan, B., & Varzandeh, M. H. M. (2015). A new hybrid model using step-wise weight assessment ratio analysis (SWARA) technique and adaptive neuro-fuzzy inference system (ANFIS) for regional landslide hazard assessment in Iran. *Catena*, *135*, 122–148. doi:10.1016/j.catena.2015.07.020

Delucchi, M., Wang, Q., & Sperling, D. (1989). Electric vehicles: Performance, life-cycle costs, emissions and recharging requirements. Transpn. *Res A*, *23A*(3), 255–278.

Demirbas, A. (2007). Alternatives to petroleum diesel fuel. Energy Sources. *Part B*, *2*(4), 343–351.

DeMuri, G. P., & Purschwitz, M. A. (2000). Fatal injuries in children: A review. *WMJ: Official Publication of the State Medical Society of Wisconsin*, *99*(9), 51–55. PMID:11220197

Deng, H. (2007). A similarity-based approach to ranking multicriteria alternatives. *International Conference on Intelligent Computing, Lecture Notes in Artificial Intelligence*, *4682*, 253-262. 10.1007/978-3-540-74205-0_28

Dey, A. R., & Misra, R. D. (2016). Effect of infiltration of bio-lubricant on the performance of a compression ignition engine fuelled with biodiesel blends. *Clean Technologies and Environmental Policy*, 1–11.

Diakoulaki, D., Mavrotas, G., & Papayannakis, L. (1995). Determining objective weights in multiple criteria problems: The CRITIC method. *Computers Ops Res.*, *22*(7), 763–770. doi:10.1016/0305-0548(94)00059-H

Dijk, M., Orsato, R. J., & Kemp, R. (2013). The emergence of an electric mobility trajectory. *Energy Policy*, *52*, 135–145. doi:10.1016/j.enpol.2012.04.024

Dinesh, K., Tamilvanan, A., Vaishnavi, S., Gopinath, M., & Rajmohan, K. S. (2016). () Biodiesel production using Calophyllum inophyllum (Tamanu) seed oil and its compatibility test in a CI engine. *Biofuels*, 1–7. doi:10.1080/17597269.2016.1187543

Dos Santos, B. M., Godoy, L. P., & Campos, L. M. (2019). Performance evaluation of green suppliers using entropy-TOPSIS-F. *Journal of Cleaner Production*, *207*, 498–509. doi:10.1016/j.jclepro.2018.09.235

Doyen, L., & Gaudoin, O. (2004). Classes of imperfect repair models based on reduction of failure intensity or virtual age. *Reliability Engineering & System Safety*, *84*(1), 45–56. doi:10.1016/S0951-8320(03)00173-X

Durairaj, S., Sathiyasekar, K., & Ilangkumaran, M. (2016). Selection of alternate fuel for electrical power generator using hybrid multi criteria decision making technique. *University Politehnica of Bucharest Scientific Bulletin Series C-Electrical Engineering and Computer Science, 78*(1), 247-258.

Durairaj, S., Sekar, K. S., Ilangkumaran, M., RamManohar, M., Thyalan, B., Yuvaraj, E., & Ramesh, S. (2014). Multi-Criteria Decision Model for Biodiesel Selection in an Electrical Power Generator Based on Fahp-Gra-Topsis. *Int. J. Res. Eng. Technol*, 2319-2363.

Dursun, M., & Karsak, E. E. (2013). A QFD-based fuzzy MCDM approach for supplier selection. *Applied Mathematical Modelling*, *37*(8), 5864–5875. doi:10.1016/j.apm.2012.11.014

Earle, W. C. (1930). Malaria in Puerto Rico and its relation to the cultivation of Sugar Cane. *Southern Medical Journal*, *23*(5), 499–452. doi:10.1097/00007611-193005000-00021

Ebeling, C. (2000). An Introduction to Reliability and Maintainability Engineering. Tata McGraw-Hill Company Ltd.

Edmund, E. E. (2015). Analysis of Occupational Hazards and Safety of Workers in Selected Working Environments within Enugu Metropolis. *Journal of Environmental & Analytical Toxicology*, *5*(6), 1000337. doi:10.4172/2161-0525.1000337

Elsayed, E. A. (2012). *Reliability engineering* (Vol. 88). Hoboken, NJ: John Wiley & Sons.

Emma, M., Kieran, S., & Vida, M. (2013). An assessment of sustainable housing affordability using a multiple criteria decision making method. *Omega, 41*, 270-279.

Emma, M., Kieran, S., & Vida, M. (2013). An assessment of sustainable housing affordability using a multiple criteria decision making method. *Omega*, *41*(2), 270–279. doi:10.1016/j.omega.2012.05.002

Emovon, I., Norman, R. A., & Murphy, A. J. (2018). Hybrid MCDM based methodology for selecting the optimum maintenance strategy for ship machinery systems. *Journal of Intelligent Manufacturing*, *29*(3), 519–531. doi:10.100710845-015-1133-6

EPA. (2008). The provision and quality of drinking water in Ireland: A report for the year. Environmental Protection Agency, 102.

Ester, M., Kriegel, H. P., Sander, J., & Xu, X. (1996). A Density-Based Algorithm for Discovering Clusters in Large Spatial Databases with Noise. In *Proceedings of the Second International Conference on Knowledge Discovery and Data Mining* (pp. 226-231). Portland, OR: Academic Press.

Etghani, M. M., Shojaeefard, M. H., Khalkhali, A., & Akbari, M. (2013). A hybrid method of modified NSGA-II and TOPSIS to optimize performance and emissions of a diesel engine using biodiesel. *Applied Thermal Engineering*, *59*(1–2), 309–315. doi:10.1016/j.applthermaleng.2013.05.041

Fang, X. (2013). An Improved Apriori Algorithm on the Frequent Item set. *Proc. International Conference on Education Technology and Information System*, 845-848.

Fattah, I. M. R., Masjuki, H. H., Kalam, M. A., Mofijur, M., & Abedin, M. J. (2014). Effect of antioxidant on the performance and emission characteristics of a diesel engine fueled with palm biodiesel blends. *Energy Conversion and Management*, *79*, 265–272. doi:10.1016/j.enconman.2013.12.024

Foehn, P. (2004). Client valuation in private banking: Results of a case study in Switzerland. *Managing Service Quality: An International Journal*, *14*(2/3), 195–204. doi:10.1108/09604520410528617

Ford. (2017). *MediaFord*. Retrieved from https://media.ford.com/content/fordmediamobile/fna/us/en/news/2017/01/03/ford-addingelectrified-f-150-mustang-transit-by-2020.html

Formica, S. (2000). *Dissertation Thesis, Destination Attractiveness as a Function of Supply and Demand Interaction,* Virginia Tech.

Franke, T., Rauh, N., & Krems, J. F. (2015). Individual differences in BEV drivers' range stress during first encounter of a critical range situation. *Applied Ergonomics*. PMID:26456746

Freedman, D. B. (2015). Towards better test utilization–strategies to improve physician ordering and their impact on patient outcomes. *The Electronic Journal of the International Federation of Clinical Chemistry and Laboratory Medicine, 26*(1), 15–30. Retrieved from https://edition.cnn.com/2016/03/04/us/flint-water-crisis-fast-facts/index.html https://timesofindia.indiatimes.com/city/kolkata/Lead-poisoning-affects-20-Kolkata-kids/articleshow/21062303.cms https://www.dawn.com/news/1157836

Gan, S., & Ng, H. K. (2010). Effects of antioxidant additives on pollutant formation from the combustion of palm oil methyl ester blends with diesel in a non-pressurised burner. *Energy Conversion and Management*, *51*(7), 1536–1546. doi:10.1016/j.enconman.2010.02.012

Garg, A., & Deshmukh, S. G. (2009). Flexibility in maintenance: A Framework. *Global Journal of Flexible Systems Management*, *10*(2), 21–33. doi:10.1007/BF03396559

Gargama, H., & Chaturvedi, S. K. (2011). Criticality assessment models for failure mode effects and criticality analysis using fuzzy logic. *IEEE Transactions on Reliability*, *60*(1), 102–110. doi:10.1109/TR.2010.2103672

Ghadge, S. V., & Raheman, H. (2005). Biodiesel production from mahua (Madhuca indica) oil having high free fatty acids. *Biomass and Bioenergy*, *28*(6), 601–605. doi:10.1016/j.biombioe.2004.11.009

Ginevičius, R. (2011). A new determining method for the criteria weights in multicriteria evaluation. *International Journal of Information Technology & Decision Making*, *10*(06), 1067–1095. doi:10.1142/S0219622011004713

Gite, L. P., & Kot, L. S. (2003). Accidents in Indian Agriculture. *Technical Bulletin No. CIAE/2003/103*. Coordinating Cell, All India Coordinated Research Project on Human Engineering and Safety in Agriculture, Central Institute of Agricultural Engineering, Nabibagh, Bhopal.

Godiganur, S., Murthy, C. S., & Reddy, R. P. (2010). Performance and emission characteristics of a Kirloskar HA394 diesel engine operated on fish oil methyl esters. *Renewable Energy*, *35*(2), 355–359. doi:10.1016/j.renene.2009.07.007

GOI. (2002). *India Vision 2020. Planning commission*. New Delhi: Govt. of India.

GOI. (2006). *Population projections for India and States 2001-2006, Report of the technical group on population projections constituted by the National Commission on Population*. New Delhi: Govt. of India.

Gomes, L. F. A. M., Rangel, L. A. D., & Maranhão, F. J. C. (2009). Multicriteria analysis of natural gas destination in Brazil: An application of the TODIM method. *Mathematical and Computer Modelling*, *50*(1-2), 92–100. doi:10.1016/j.mcm.2009.02.013

Goonetilleke, A., Yigitcanlar, T., Ayoko, G., & Egodawatta, P. (2014). *Sustainable urban water environment: Climate, pollution and adaptation*. Cheltenham, UK: Edward Elgar. doi:10.4337/9781781004647

Gooroochurn, N., & Sugiyarto, G. (2004). *Discussion Paper, Measuring Competitiveness, TTRI*. Nottingham University Business School, University of Nottingham.

Gui, M. M., Lee, K. T., & Bhatia, S. (2008). Feasibility of edible oil vs. non-edible oil vs. waste edible oil as biodiesel feedstock. *Energy*, *33*(11), 1646–1653. doi:10.1016/j.energy.2008.06.002

Guo, C., Wang, Y., & Jiang, W. (2013). An empirical study of evaluation index system and measure method on city's soft power: 17 cities in Shandong province. *Cross-Cultural Communication*, *9*(6), 27–31.

Gupta, H., & Barua, M. K. (2017). Supplier selection among SMEs on the basis of their green innovation ability using BWM and fuzzy TOPSIS. *Journal of Cleaner Production*, *152*, 242–258. doi:10.1016/j.jclepro.2017.03.125

Gupta, R., Biswas, A., & Sharma, K. K. (2008). Comparative study of a three-bucket Savonius rotor with a combined three-bucket Savonius – three-bladed Darrieus rotor. *Renewable Energy*, *33*(9), 1974–1981. doi:10.1016/j.renene.2007.12.008

Habibullah, N. S., & Rajnarayan, R. T. (2004). Occupational Health Research in India. *Industrial Health*, *42*(2), 141–148. doi:10.2486/indhealth.42.141 PMID:15128163

Hall, D., Moultak, M., & Lusey, N. (2017). *Electric Vehicle Capitals of the World: Demonstrating the Path to Electric Drive*. Washington, DC: International Council on Clean Transportation. Retrieved from www.theicct.org/sites/default/files/publications/Global-EV-Capitals_WhitePaper_06032017_vF.pdf

Hazar, H. (2010). Cotton methyl ester usage in a diesel engine equipped with insulated combustion chamber. *Applied Energy*, *87*(1), 134–140. doi:10.1016/j.apenergy.2009.07.024

Hekmatpanah, M., Shahin, A., & Ravichandran, N. (2011). The application of FMEA in the oil industry in Iran: The case of four litre oil canning process of Sepahan Oil Company. *African Journal of Business Management*, *5*(7), 3019.

Helkamp, J., & Lundstrom, W. (2002). Tractor Related Deaths Among West Virginia Farmers. *Annals of Epidemiology*, *12*(7), 510. doi:10.1016/S1047-2797(02)00344-7

Hellig, T., Mertens, A., & Brandl, C. (2018). The interaction effect of working postures on muscle activity and subjective discomfort during static working postures and its correlation with OWAS. *International Journal of Industrial Ergonomics*, *68*, 25–33. doi:10.1016/j.ergon.2018.06.006

Helmbrecht, M., Bengler, K., & Vilimek, R. (2013). Strategies for Efficient Driving with Electric Vehicles. In E. Brandenburg, L. Doria, A. Gross, T. Günzler, & H. Smieszek (Eds.), Grundlagen and Anwendungen der Mensch-Maschine-Interaktion - Foundations and applications of human-machine interaction. 10. Berliner Werkstatt Mensch-Maschine-Systeme10-12. ZMMS, Technische Universität Berlin, prometeigraduiertenkolleg (pp. 83–91). Academic Press.

Hemmati, N., Rahiminezhad Galankashi, M., Imani, D. M., & Farughi, H. (2018). Maintenance policy selection: A fuzzy-ANP approach. *Journal of Manufacturing Technology Management*, *29*(7), 1253–1268. doi:10.1108/JMTM-06-2017-0109

Heracles, P., & Stefan I. (2014). Micro-siting/positioning of wind turbines: introducing a multi-criteria decision analysis framework. *Proceedings of Engineering and Technological Innovation*, 23-27.

He-Yau, K., Meng-Chan, H., Pearn, W. L., Amy, H. I. L., & Mei-Sung, K. (2011). An Integrated Multi-Criteria Decision Making Model for Evaluating Wind Farm Performance. *Energies*, *4*(11), 2002–2026. doi:10.3390/en4112002

Hignett, S., & McAtamney, L. (2000). Rapid entire body assessment (REBA). *Applied Ergonomics*, *31*(2), 201–205. doi:10.1016/S0003-6870(99)00039-3 PMID:10711982

Honda. (2016). *Honda adds all-new dedicated hybrid model to growing electrified lineup*. Retrieved from http://news.honda.com/newsandviews/article.aspx?id=9463-en

Hope, A., Kelleher, C., Holmes, L., & Hennessy, T. (1999). Health and safety practices among farmers and other workers: A needs assessment. *Occupational Medicine (Philadelphia, Pa.)*, *49*, 231–235. PMID:10474914

Ho, W., He, T., Lee, C. K. M., & Emrouznejad, A. (2012). Strategic logistics outsourcing: An integrated QFD and fuzzy AHP approach. *Expert Systems with Applications, 39*(12), 10841–10850. doi:10.1016/j.eswa.2012.03.009

Huiyun, X., Zengzhen, W., Lorann, S., Thomas, J. K., Xuzhen, H., & Xianghua, F. (2000). Agricultural work-related injuries among farmers in Hubei, People's Republic of China. *American Journal of Public Health, 90*(8), 1269–1276. doi:10.2105/AJPH.90.8.1269 PMID:10937008

Hwang, C. L., & Yoon, K. (1981). Methods for multiple attribute decision making. In *Multiple attribute decision making* (pp. 58–191). Berlin: Springer. doi:10.1007/978-3-642-48318-9_3

Iannotta, G., Nocera, G., & Sironi, A. (2007). Ownership structure, risk and performance in the European banking industry. *Journal of Banking & Finance, 31*(7), 2127–2149. doi:10.1016/j.jbankfin.2006.07.013

Ighravwe, D. E., & Ayoola Oke, S. (2017). Ranking maintenance strategies for sustainable maintenance plan in manufacturing systems using fuzzy axiomatic design principle and fuzzy-TOPSIS. *Journal of Manufacturing Technology Management, 28*(7), 961–992. doi:10.1108/JMTM-01-2017-0007

International Labour Organisation. (2013). *ILO calls for urgent global action to fight occupational diseases*. Available from: http://www.ilo.org/global/about-the-ilo/media-centre/pressreleases/WCMS_211627/lang--en/index.htm

Irvin, S. (1994). Using lifetime value analysis for selecting new customers. *Credit World, 82*(3), 37–39.

Jaishankar, M., Tseten, T., Anbalagan, N., Mathew, B. B., & Beeregowda, K. N. (2014). Toxicity, mechanism and health effects of some heavy metals. *Interdisciplinary Toxicology, 7*(2), 60–72. doi:10.2478/intox-2014-0009 PMID:26109881

Jena, J., & Misra, R. D. (2014). Effect of fuel oxygen on the energetic and exergetic efficiency of a compression ignition engine fuelled separately with palm and karanja biodiesels. *Energy, 68*, 411–419. doi:10.1016/j.energy.2014.02.079

Jindal, S., & Goyal, K. (2012). Evaluation of performance and emission of *Hibwascus cannabinus* (Ambadi) seed oil biodiesel. *Clean Technologies and Environmental Policy, 14*(4), 633–639. doi:10.100710098-011-0426-5

Jong, M., Joss, S., Schraven, D., Zhan, C., & Weijnen, M. (2015). Sustainable–smart–resilient–low carbon–eco–knowledge cities; making sense of a multitude of concepts promoting sustainable urbanization. *Journal of Cleaner Production, 109*, 25–38. doi:10.1016/j.jclepro.2015.02.004

Joshua, J., Mathew, S. G., & Harikrishnan, A. R. (2016). Selection of an optimum maintenance strategy for improving the production efficiency in a casting unit. *International Journal of Science Technology & Engineering, 3*(2), 138–141.

Kaergaard, A., & Andersen, J. H. (2000). Musculoskeletal disorders of the neck and shoulders in female sewing machine operators: Prevalence, incidence, and prognosis. *Occupational and Environmental Medicine*, *57*(8), 528–534. doi:10.1136/oem.57.8.528 PMID:10896959

Kahraman, C., Ertay, T., & Büyüközkan, G. (2006). A fuzzy optimization model for QFD planning process using analytic network approach. *European Journal of Operational Research*, *171*(2), 390–411. doi:10.1016/j.ejor.2004.09.016

Kalam, M. A., & Masjuki, H. H. (2008). Testing palm biodiesel and NPAA additives to control NOx and CO while improving efficiency in diesel engines. *Biomass and Bioenergy*, *32*(12), 1116–1122. doi:10.1016/j.biombioe.2008.02.009

Kalligeros, S., Zannikos, F., Stournas, S., Lois, E., Anastopoulos, G., Teas, C., & Sakellaropoulos, F. (2003). An investigation of using biodiesel/marine diesel blends on the performance of a stationary diesel engine. *Biomass and Bioenergy*, *24*(2), 141–149. doi:10.1016/S0961-9534(02)00092-2

Kamoji, M. A., Kedare, S. B., & Prabhu, S. V. (2008). Experimental investigations on single stage modified Savonius rotor. *Applied Energy*, *86*(7-8), 1064–1073. doi:10.1016/j.apenergy.2008.09.019

Kannan, D., De Sousa Jabbour, A. B. L., & Jabbour, C. J. C. (2014). Selecting green suppliers based on GSCM practices: Using fuzzy TOPSIS applied to a Brazilian electronics company. *European Journal of Operational Research*, *233*(2), 432–447. doi:10.1016/j.ejor.2013.07.023

Kannan, G. R., & Anand, R. (2012). Effect of injection pressure and injection timing on DI diesel engine fuelled with biodiesel from waste cooking oil. *Biomass and Bioenergy*, *46*, 343–352. doi:10.1016/j.biombioe.2012.08.006

Karabasevic, D., Zavadskas, E. K., Turskis, Z., & Stanujkic, D. (2016). The framework for the selection of personnel based on the SWARA and ARAS methods under uncertainties. *Informatica*, *27*(1), 49–65. doi:10.15388/Informatica.2016.76

Karande, P., & Chakraborty, S. (2012). Application of multi-objective optimization on the basis of ratio analysis (MOORA) method for materials selection. *Materials & Design*, *37*(1), 317–324. doi:10.1016/j.matdes.2012.01.013

Karhu, O., Härkönen, R., Sorvali, P., & Vepsäläinen, P. (1981). Observing workin postures in industry: Examples of OWAS application. *Applied Ergonomics*, *12*(1), 13–17. doi:10.1016/0003-6870(81)90088-0 PMID:15676393

Karhu, O., Kansi, P., & Kuorinka, I. (1977). Correcting working postures in industry: A practical method for analysis. *Applied Ergonomics*, *8*(4), 199–201. doi:10.1016/0003-6870(77)90164-8 PMID:15677243

Karthikeyan, C., Veeraragavathatham, D., Karpagam, D., & Ayisha, F. S. (2009). Traditional tools in agricultural practices. *Indian Journal of Traditional Knowledge*, *8*(2), 212–217.

Kaur, I., Mandiyal, D., Singh, B. P., Kumar, R., & Chawla, J. (2016). Amino-functionalized mesoporous MCM-41: An efficient adsorbent for the removal of chromium (III) ions from aqueous solution. *Journal of Water Supply: Research & Technology - Aqua*, *65*(6), 480–493. doi:10.2166/aqua.2016.118

Kazan, H., & Ozdemir, O. (2014). Financial performance assessment of large scale conglomerates via TOPSIS and CRITIC methods. *International Journal of Management and Sustainability*, *3*(4), 203–224.

Kee, D., & Karwowski, W. (2001). LUBA: An assessment technique for postural loading on the upper body based on joint motion discomfort and maximum holding time. *Applied Ergonomics*, *32*(4), 357–366. doi:10.1016/S0003-6870(01)00006-0 PMID:11461037

Keršuliene, V., Zavadskas, E. K., & Turskis, Z. (2010). Selection of rational dispute resolution method by applying new step-wise weight assessment ratio analysis (SWARA). *Journal of business economics and management*, *11*(2), 243–258. doi:10.3846/jbem.2010.12

Keshavarz Ghorabaee, M., Zavadskas, E. K., Olfat, L., & Turskis, Z. (2015). Multi-criteria inventory classification using a new method of evaluation based on distance from average solution (EDAS). *Informatica*, *26*(3), 435–451. doi:10.15388/Informatica.2015.57

Khan, M. H. (1975). *Improvement of Savonius Rotor-windmill* (M.S. thesis). University of the Phillipines, Lasbonas.

Khandekar, A. V., & Chakraborty, S. (2017). Application of fuzzy axiomatic design principles for cotton fiber selection. *Journal of the Textile Institute*, *109*(6), 730–739. doi:10.1080/00405000.2017.1366700

Khan, M. H. (1988). Model & prototype performance characteristics of S-rotor wind mills. *Journal of Wind Engineering*, *12*, 59–75.

Khan, M. R., & Sudha, S. (2012). Evaluation of genotoxicity in automobile mechanics occupationally exposed to polycyclic aromatic hydrocarbons using micronuclei and other nuclear abnormalities. *Iranian Journal of Cancer Prevention*, *5*(2), 87–92. PMID:25628826

Khatibi, V., & Gholam Ali, M. (2009). Intuitionistic fuzzy set vs. fuzzy set application in medical pattern recognition. *Artificial Intelligence in Medicine*, *47*(1), 43–52. doi:10.1016/j.artmed.2009.03.002 PMID:19375899

Kibret, S., Alemu, Y., Boelee, E., Tekie, H., Alemu, D., & Petros, B. (2010). The impact of a small-scale irrigation scheme on malaria transmission in Ziway area, Central Ethiopia. *Tropical Medicine & International Health*, *15*, 41–50. PMID:19917039

Kim, G., Shin, B., & Lee, H. G. (2009). Understanding dynamics between initial trust and usage intentions of mobile banking. *Information Systems Journal*, *19*(3), 283–311. doi:10.1111/j.1365-2575.2007.00269.x

Kim, K., & Ahn, H. (2008). A recommender system using GA K-means clustering in an online Shopping market. *Expert Systems with Applications*, *34*(2), 1200–1209. doi:10.1016/j.eswa.2006.12.025

Kirubakaran, B., & Ilangkumaran, M. (2015). The selection of optimum maintenance strategy based on ANP integrated with GRA-TOPSIS. *Journal for Global Business Advancement*, *8*(2), 190–215. doi:10.1504/JGBA.2015.069530

Kirubakaran, B., & Ilangkumaran, M. (2016). Selection of optimum maintenance strategy based on FAHP integrated with GRA–TOPSIS. *Annals of Operations Research*, *245*(1-2), 285–313. doi:10.100710479-014-1775-3

Knapp, L. W. (1966). Occupational and Rural Accidents. *Archives of Environmental Health*, *13*(4), 501–506. doi:10.1080/00039896.1966.10664604 PMID:5922020

Knapp, L. W. Jr. (1965). Agricultural Injuries Prevention. *Journal of Occupational Medicine.*, *7*(11), 553–745. doi:10.1097/00043764-196511000-00001 PMID:5831719

Knezevic, J. (1997). Systems Maintainability Analysis, Engineering and Management. London: Chapman & Hall.

Knezevic, J. (1993). *Reliability, maintainability, and supportability: a probabilistic approach.* McGraw-Hill Companies.

Knittel, C. R. (2012). Reducing petroleum consumption from transportation. *The Journal of Economic Perspectives*, *26*(1), 93–118. doi:10.1257/jep.26.1.93

Kolodner, J. L. (1992). An introduction to case-based reasoning. *Artificial Intelligence Review*, *6*(1), 3–34. doi:10.1007/BF00155578

Kozak, M., & Rimmington, M. (1999). Measuring tourist destination competitiveness: Conceptual considerations and empirical findings. *International Journal of Hospitality Management*, *18*(3), 273–283. doi:10.1016/S0278-4319(99)00034-1

Krylovas, A., Zavadskas, E. K., Kosareva, N., & Dadelo, S. (2014). New KEMIRA method for determining criteria priority and weights in solving MCDM problem. *International Journal of Information Technology & Decision Making*, *13*(06), 1119–1133. doi:10.1142/S0219622014500825

Kumar, R., & Ray, A. (2015). Optimal selection of material: an eclectic decision. *Journal of the Institution of Engineers (India): Series C, 96*(1), 29-33.

Kumar, A., Singh, J. K., Mohan, D., & Varghese, M. (2008). Farm hand tools injuries: A case study from northern India. *Safety Science*, *46*(1), 54–65. doi:10.1016/j.ssci.2007.03.003

Kumar, A., Varghese, M., Pradhan, C. K., Goswami, A., Ghosh, S. N., & Nag, P. K. (1986). Evaluation of working with spade in agriculture. *The Indian Journal of Medical Research*, *84*, 424–429. PMID:3781600

Kumar, R., & Chawla, J. (2014). Removal of cadmium metal from water/wastewater by nano-metal oxide: A review. *Water Quality, Exposure, and Health*, *5*(4), 215–226. doi:10.100712403-013-0100-8

Kumar, R., Chawla, J., & Kaur, I. (2015). Removal of cadmium metal from water by carbon based nanosorbents: A review. *Journal of Water and Health*, *13*(1), 18–33. doi:10.2166/wh.2014.024 PMID:25719462

Kumar, V., Arora, H. D., & Pal, K. (2016). Intuitionistic Trapezoidal Fuzzy Prioritized Weighted Average Operators: An Algorithm for the Suitable Treatment for Lung Cancer. *Journal of Applied Probability and Statistics*, *11*(1), 91–104.

Kumar, V., Arora, H., & Pal, K. (2016). Intuitionistic trapezoidal fuzzy prioritized weighted geometric operator: An algorithm for the selection of suitable treatment for lung cancer. *Proceedings of Fifth International Conference on Soft Computing for Problem Solving, 436*, 605-615.

Kumar, V., Bharti, I., & Sharma, Y. K. (2012). Fuzzy diagnosis procedure of the types of glaucoma. *International Journal of Applied Information Systems*, *1*(6), 42–45. doi:10.5120/ijais12-450191

Kumar, V., & Vashist, D. (2013). Fuzzy procedure for the selection of car among various brands. *International Journal of Engineering Research and Technology*, *6*(3), 337–342.

Kumru, M., & Kumru, P. Y. (2013). Fuzzy FMEA application to improve purchase process in public hospital. *Applied Soft Computing*, *13*(1), 721–733. doi:10.1016/j.asoc.2012.08.007

Lahu, P. M., Gandigude, A. U., & Kale, S. A. (2014). Selection of Material for Wind Turbine Blade by Analytic Hierarchy Process (AHP) Method. *Applied Mechanics and Materials*, *612*, 145–150. doi:10.4028/www.scientific.net/AMM.612.145

Lambert, F. (2017b). *BMW explains its electric vehicle plans in more details and it's going in the right direction*. Retrieved from https://electrek.co/2017/03/22/bmw-electric-vehicle-plans

Lazakis, I., & Ölçer, A. (2016). Selection of the best maintenance approach in the maritime industry under fuzzy multiple attributive group decision-making environment. *Proceedings of the Institution of Mechanical Engineers, Part M: Journal of Engineering for the Maritime Environment*, *230*(2), 297–309.

Lee, J. H., Hancock, M. G., & Hu, M. C. (2014). Towards an effective framework for building smart cities: Lessons from Seoul and San Francisco. *Technological Forecasting and Social Change*, *89*, 80–99. doi:10.1016/j.techfore.2013.08.033

Lee, S. H., Han, J. H., Leem, Y. T., & Yigitcanlar, T. (2008). Towards Ubiquitous City: Concept, Planning, and Experiences. *Igi Global*, *2*, 148–169.

Liaquat, A. M., Masjuki, H. H., Kalam, M. A., Fattah, I. M. R., Hazrat, M. A., Varman, M., ... Shahabuddin, M. (2013). Effect of coconut biodiesel blended fuels on engine performance and emission characteristics. *Procedia Engineering*, *56*, 583–590. doi:10.1016/j.proeng.2013.03.163

Lidouh, K., Smet De, Y., & Zimányi, E. (2011). An adaptation of the ´GAIA visualization method for cartography. *Procs. of the IEEE Symposium on Computational Intelligence in Multicriteria Decision-Making*, 29-35.

Li, M. Y., & Cao, P. P. (2018). Extended TODIM method for multi-attribute risk decision making problems in emergency response. *Computers & Industrial Engineering*. doi:10.1016/j.cie.2018.06.027

Lin, L., Cunshan, Z., Vittayapadung, S., Xiangqian, S., & Mingdong, D. (2011). Opportunities and challenges for biodiesel fuel. *Applied Energy*, *88*(4), 1020–1031. doi:10.1016/j.apenergy.2010.09.029

Liu, D. R., & Shih, Y. (2005). Integrating AHP and data mining for product recommendation based on customer lifetime value. *Information & Management*, *42*(3), 387–400. doi:10.1016/j.im.2004.01.008

Llamazares, B. (2018). An analysis of the generalized TODIM method. *European Journal of Operational Research*, *269*(3), 1041–1049. doi:10.1016/j.ejor.2018.02.054

Lombardi, M., & Fargnoli, M. (2018). Prioritization of hazards by means of a QFD- based procedure. *International Journal of Safety and Security Engineering.*, *8*(2), 342–353. doi:10.2495/SAFE-V8-N2-342-353

Lourenzutti, R., & Krohling, R. A. (2013). A study of TODIM in a intuitionistic fuzzy and random environment. *Expert Systems with Applications*, *40*(16), 6459–6468. doi:10.1016/j.eswa.2013.05.070

Loveday, E. (2017). *LG Chem expects Chevrolet Bolt sales to exceed 30,000 in 2017*. Retrieved from http://insideevs.com/lg-chem-expects-chevrolet-bolt-sales-exceed-30000-2017/

Lucas, P. F. J., & van der Gaag, L. C. (1991). *Principles of Expert Systems*. Amsterdam: Centre for Mathematics and Computer Science.

Lutsey, N. (2015). *Global climate change mitigation potential from a transition to electric vehicles*. Academic Press.

MacCrimmon, K. R. (1968). *Decision Making among Multiple-Attribute Alternatives: A Survey and Consolidated Approach*. Santa Monica, CA: RAND Co.

Macpherson, R. B. (1972). *Design, Development and testing of Low Head High Efficiency Kinetic Energy Machine* (MSc thesis). University of Massachusetts, Amherst, MA.

Madic, M., & Radovanovic, M. (2015). Ranking of some most commonly used nontraditional machining processes using ROV and CRITIC methods. U.P.B. Sci. *Bull., Series D*, *77*(2), 193–204.

Madiwale, S., Karthikeyan, A., & Bhojwani, V. (2018). Properties investigation and performance analysis of a diesel engine fuelled with Jatropha, Soybean, Palm and Cottonseed biodiesel using Ethanol as an additive. *Materials Today: Proceedings*, *5*(1), 657–664. doi:10.1016/j.matpr.2017.11.130

Mahamarakkalage, F. (1980). *On the Performance and Wake Aerodynamics of the Savonius Wind Turbine* (PhD thesis). University of Peradenyia, Sri Lanka.

Ma, J., Wang, N., & Kong, D. (2009). Market forecasting modeling study for new energy vehicle based on AHP and logit regression. *Journal of Tongji University*, *37*(8), 1079–1084.

Majumdar, A. (2010). Selection of raw materials in textile spinning industry using fuzzy multi-criteria decision making approach. *Fibers and Polymers*, *11*(1), 121–127. doi:10.100712221-010-0121-x

Majumdar, A., Majumdar, P. K., & Sarkar, B. (2005b). Determination of the technological value of cotton fiber: A comparative study of the traditional and multi-criteria decision-making approaches. *AUTEX Research Journal*, *5*(2), 71–80.

Majumdar, A., Mangla, R., & Gupta, A. (2010). Developing a decision support system software for cotton fiber grading and selection. *Indian Journal of Fibre and Textile Research*, *35*(3), 195–200.

Majumdar, A., Sarkar, B., & Majumdar, P. K. (2004). Application of analytic hierarchy process for the selection of cotton fibers. *Fibers and Polymers*, *5*(4), 297–302. doi:10.1007/BF02875528

Majumdar, A., Sarkar, B., & Majumdar, P. K. (2005a). Determination of quality value of cotton fiber using hybrid AHP-TOPSIS method of multi-criteria decision-making. *Journal of the Textile Institute*, *96*(5), 303–309. doi:10.1533/joti.2005.0013

Majumdar, A., Sarkar, B., & Majumdar, P. K. (2006). A new algorithm of cotton fiber selection and laydown using TOPSIS method of multi-criteria decision making. *Indian Journal of Fibre and Textile Research*, *31*(2), 248–255.

Majumdar, A., & Singh, S. P. (2014). A new approach to determine the quality value of cotton fibers using multi-criteria decision making and genetic algorithm. *Fibers and Polymers*, *15*(12), 2658–2664. doi:10.100712221-014-2658-6

Mandave, P., Mane, M., & Patil, S. (2013). Data mining using Association Rule Based on Apriori Algorithm and Improved Approach with Illustration. *International Journal of Latest Trends in Engineering and Technology*, *3*(2), 107–113.

Manh, C. D., Beebe, N. W., Van, V. N., Quang, T. L., Lein, C. T., Nguyen, D. V., ... Cooper, R. D. (2010). Vectors and malaria transmission in deforested, rural communities in north-central Vietnam. *Malaria Journal*, *9*(1), 259. doi:10.1186/1475-2875-9-259 PMID:20846447

Mardani, A., Nilashi, M., Zakuan, N., Loganathan, N., Soheilirad, S., Saman, M. Z. M., & Ibrahim, O. (2017). A systematic review and meta-Analysis of SWARA and WASPAS methods: Theory and applications with recent fuzzy developments. *Applied Soft Computing*, *57*, 265–292. doi:10.1016/j.asoc.2017.03.045

Mavridis, D. G. (2004). The intellectual capital performance of the Japanese banking sector. *Journal of Intellectual Capital*, *5*(1), 92–115. doi:10.1108/14691930410512941

McAtamney, L., & Corlett, N. E. (1993). RULA: A survey method for the investigation of work related upper limb disorders. *Applied Ergonomics*, *24*(2), 91–99. doi:10.1016/0003-6870(93)90080-S PMID:15676903

Melián-González, A., & García-Falcón, J. M. (2003). Competitive potential of tourism in destinations. *Annals of Tourism Research*, *30*(3), 720–740. doi:10.1016/S0160-7383(03)00047-1

Meng Tay, K., & Peng Lim, C. (2006). Fuzzy FMEA with a guided rules reduction system for prioritization of failures. *International Journal of Quality & Reliability Management*, *23*(8), 1047–1066. doi:10.1108/02656710610688202

Mercan, M., Reisman, A., Yolalan, R., & Emel, A. B. (2003). The effect of scale and mode of ownership on the financial performance of the Turkish banking sector: Results of a DEA-based analysis. *Socio-Economic Planning Sciences*, *37*(3), 185–202. doi:10.1016/S0038-0121(02)00045-9

Mierlo, J. V., Maggetto, G., & Lataire, P. (2006). Which energy source for road transport in the future? A comparison of battery, hybrid and fuel cell vehicles. *Energy Conversion and Management*, *47*(17), 2748–2760. doi:10.1016/j.enconman.2006.02.004

Mishra, D., & Satapathy, S. (2018). Drudgery Reduction of Farm Women of Odisha by Improved Serrated Sickle. *International Journal of Mechanical Engineering and Technology*, *9*(2), 53–61.

Mittal, V. K., Bhatia, B. S., & Ahuja, S. S. (1996). *A Study of the Magnitude, Causes, and Profile of Victims of Injuries with Selected Farm Machines in Punjab. Final Report of ICAR adhoc Research Project*. Ludhiana, India: Department of Farm Machinery and Power Engineering, Punjab Agricultural University.

Mobley, R. K. (2002). An introduction to predictive maintenance (2nd ed.). Elsevier Science.

Modi, V. J., & Fernando, M. S. U. K. (1989). On the performance of the Savonius wind turbine. *ASME Journal of Solar Engineering*, *11*(1), 71–76. doi:10.1115/1.3268289

Modi, V. J., Roth, M. N. J., & Fernando, S. U. K. (1984). optimal configuration studies and prototype design of a wind energy operated irrigation system. *Journal of Wind Engineering and Industrial Aerodynamics*, *16*(1), 85–96. doi:10.1016/0167-6105(84)90050-3

Mohamed, F., & Santawy, E. (2012). A VIKOR Method for Solving Personnel Training Selection Problem. *International Journal of Computational Science*, *1*(2), 9–12.

Mohan, D. (2000). Equipment-related injuries in agriculture: An international perspective. *Injury Control and Safety Promotion*, *7*(3), 1–12.

Mohan, D., & Patel, R. (1992). Design of safer agricultural equipment: Application of ergonomics and epidemiology. *International Journal of Industrial Ergonomics*, *10*(4), 301–309. doi:10.1016/0169-8141(92)90097-J

Mojola, O. (1985). Aerodynamic design of the Savonius windmill rotor. *Journal of Wind Engineering and Industrial Aerodynamics*, *21*(2), 223–231. doi:10.1016/0167-6105(85)90005-4

Montano, D. (2014). Chemical and biological work-related risks across occupations in Europe: A review. *Journal of Occupational Medicine and Toxicology (London, England)*, *9*(28). Retrieved from http://www.occup-med.com/content/9/1/28 PMID:25071862

Moore, H. (2007). You Can't Be a Meat-Eating Environmentalist. *American Chronicle*. Retrieved from http://www.americanchronicle.com/articles/view/24825

Moradi, M., & Ebrahimi, E. (2014). Applying fuzzy AHP and similarity-based approach for economic evaluating companies based on corporate governance measures. *Global Journal of Management Studies and Researches*, *1*(1), 10–20.

Mouli, G. C., Bauer, P., & Zeman, M. (2016). System design for a solar powered electric vehicle charging station for workplaces. *Applied Energy*, *168*(Suppl C), 434–443. doi:10.1016/j.apenergy.2016.01.110

Mukhametzyanov, I., & Pamucar, D. (2018). A sensitivity analysis in MCDM problems: A statistical approach. *Decision Making: Applications in Management and Engineering*, *1*(2), 51–80.

Murphy, D. J. (1992). *Safety and Health for Production Agriculture*. St. Joseph, MI: American Society of Agricultural Engineers.

Muthukumar, G. P., & Krishnan, S. S. (2016). A similarity measure of intuitionistic fuzzy soft sets and its application in medical diagnosis. *Applied Soft Computing*, *41*, 148–156. doi:10.1016/j.asoc.2015.12.002

Naeini, H. S., Karuppiah, K., Tamrin, S. B., & Dalal, K. (2014). Ergonomics in agriculture: An approach in prevention of work-related musculoskeletal disorders (WMSDs). *J Agriculture and Environmental Sciences*, *3*, 33–51.

Nag, P. K., & Nag, A. (2004). Drudgery, accidents and injuries in Indian agriculture. *Industrial Health*, *42*(4), 149–162. doi:10.2486/indhealth.42.149 PMID:15128164

Ndayishimiye, P., & Tazerout, M. (2011). Use of palm oil based biofuel in the internal combustion engines. *Energy*, *36*(3), 1790–1796. doi:10.1016/j.energy.2010.12.046

Neema, A., & Nema, G. (2017). Impact of brand image on brand loyalty and brand equity in banking sector: A case of State Bank of India. *Pacific Business Review International*, *10*(3), 109–112.

Newman, B. G. (1974). Measurement on a Savonius rotor with variable gap. *Proceeding of Sherbrook University Symposium on wind energy*, 116-127.

Nezami, F. G., & Yildirim, M. B. (2013). A sustainability approach for selecting maintenance strategy. *International Journal of Sustainable Engineering*, *6*(4), 332–343. doi:10.1080/19397038.2013.765928

Ng, H. K., & Gan, S. (2010). Combustion performance and exhaust emissions from the non-pressurised combustion of palm oil biodiesel blends. *Applied Thermal Engineering*, *30*(16), 2476–2484. doi:10.1016/j.applthermaleng.2010.06.020

Ng, J. H., Ng, H. K., & Gan, S. (2012). Characterisation of engine-out responses from a light duty diesel engine fuelled with palm methyl ester (PME). *Applied Energy*, *90*(1), 58–67. doi:10.1016/j.apenergy.2011.01.028

Nilsson, K., Pinzke, S., & Lundqvist, P. (2010). Occupational Injuries to Senior Farmers in Sweden. *Journal of Agricultural Safety and Health*, *16*(1), 19–29. doi:10.13031/2013.29246 PMID:20222268

Niu, Y. C. (2016). An Analysis of E-Commerce Enterprises Sustainable Development and Financing Structure. *Theoretical Economics Letters*, *6*(5), 1105–1114. doi:10.4236/tel.2016.65106

Nobuyuki, F. (1992). On the torque mechanism of Savonius rotors. *Journal of Wind Engineering and Industrial Aerodynamics*, *40*(3), 277–292. doi:10.1016/0167-6105(92)90380-S

Nwafor, O. M. I. (2004). Emission characteristics of diesel engine operating on rapeseed methyl ester. *Renewable Energy*, *29*(1), 119–129. doi:10.1016/S0960-1481(03)00133-2

Nwankwo, C. M., Karanja, S., & Vasanthakaalam, H. (2018). The occurrence of occupational health hazards in districts health facilities in Kigali, Rwanda. *Int J Community Med Public Health*, *5*(1), 21–29. doi:10.18203/2394-6040.ijcmph20175509

Oluoch, I., Njogu, P., & Ndeda, J.O.H. (2017). Effects of Occupational Safety and Health Hazards' Exposure on Work Environment in the Water Service Industry within Kisumu County - Kenya. *IOSR Journal of Environmental Science, Toxicology and Food Technology*, *11*(5-1), 46-51. doi:10.9790/2402-1105014651

Ong, H. C., Silitonga, A. S., Masjuki, H. H., Mahlia, T. M. I., Chong, W. T., & Boosroh, M. H. (2013). Production and comparative fuel properties of biodiesel from non-edible oils: Jatropha curcas, Sterculia foetida and Ceiba pentandra. *Energy Conversion and Management*, *73*, 245–255. doi:10.1016/j.enconman.2013.04.011

Oyelade, O. J., Oladipupo, O. O., & Obagbuwa, I. C. (2010). Application of k-Means Clustering algorithm for prediction of Students.*Academic Performance. International Journal of Computer Science and Information Security*, *7*(1), 292–295.

Özcan, E. C., Ünlüsoy, S., & Eren, T. (2017). A combined goal programming–AHP approach supported with TOPSIS for maintenance strategy selection in hydroelectric power plants. *Renewable & Sustainable Energy Reviews*, *78*, 1410–1423. doi:10.1016/j.rser.2017.04.039

Ozcanli, M., Gungor, C., & Aydin, K. (2013). Biodiesel fuel specifications: A review. *Energy Sources. Part A, Recovery, Utilization, and Environmental Effects*, *35*(7), 635–647. doi:10.1080/15567036.2010.503229

Özcanli, M., Keskin, A., & Aydın, K. (2011). Biodiesel production from terebinth (Pwastacia terebinthus) oil and its usage in diesel engine. *International Journal of Green Energy*, *8*(5), 518–528. doi:10.1080/15435075.2011.588766

Pal, S. K., & Shiu, S. C. (2004). *Foundations of soft case-based reasoning* (Vol. 8). Hoboken, NJ: John Wiley & Sons. doi:10.1002/0471644676

Pamučar, D., & Ćirović, G. (2015). The selection of transport and handling resources in logistics centers using Multi-Attributive Border Approximation area Comparison (MABAC). *Expert Systems with Applications*, *42*(6), 3016–3028. doi:10.1016/j.eswa.2014.11.057

Pamučar, D., Sremac, S., Stević, Ž., Ćirović, G., & Tomić, D. (2019). New multi-criteria LNN WASPAS model for evaluating the work of advisors in the transport of hazardous goods. *Neural Computing & Applications*, 1–24.

Panchal, Chattergi, Shukla, Choudhury, & Temostallin. (2017). Integrated fuzzy AHP-CODAS framework for maintenance decision in urea fertilizer industry. *Econ Comput Econ Cyb., 51*(3), 179-196.

Panchal, D., & Kumar, D. (2016). Integrated framework for behavior analysis in a process plant. *Journal of Loss Prevention in the Process Industries*, *40*, 147–161. doi:10.1016/j.jlp.2015.12.021

Panchal, D., & Kumar, D. (2016). Stochastic behaviour analysis of power generating unit in thermal power plant using Fuzzy methodology. *Opsearch.*, *53*(1), 16–40. doi:10.100712597-015-0219-4

Panchal, D., & Kumar, D. (2017). Risk analysis of compressor house unit in a thermal power plant using integrated fuzzy FMEA and GRA approach. *Int. J. Industrial and Systems Engineering.*, *25*(2), 228–250. doi:10.1504/IJISE.2017.081519

Panchal, D., & Kumar, D. (2017). *Stochastic behaviour analysis of real industrial system. Int J Syst Assur Eng Manag*. doi:10.100713198-017-0579-7

Panchal, D., Mangla, S. K., Tyagi, M., & Ram, M. (2018). Risk analysis for clean and sustainable production in a urea fertilizer industry. *International Journal of Quality & Reliability Management*, *35*(7), 1459–1476. doi:10.1108/IJQRM-03-2017-0038

Panchal, D., Singh, A. K., Chatterjee, P., Zavadkas, E. K., & Ghorabaee, M. K. (2018). A new fuzzy methodology- based structured framework for RAM and risk analysis. *Applied Soft Computing*, *74*, 242–254. doi:10.1016/j.asoc.2018.10.033

Panchal, D., & Srivastava, P. (2018). *Qualitative analysis of CNG dispensing system using fuzzy FMEA–GRA integrated approach. Int J Syst Assur Eng Manag*. doi:10.100713198-018-0750-9

Patel, S. K., Varma, M. R., & Kumar, A. (2010). Agricultural injuries in Etawah district of Uttar Pradesh in India. *Safety Science*, *48*(2), 222–229. doi:10.1016/j.ssci.2009.08.003

Paul, C., Sarkar, D., Chelliah, S., Kalyan, M. C., & Nadkarni, P. P. S. (2017). Recommendation of High Quality Representative Reviews in e-commerce. In: *Proceedings of the Eleventh ACM Conference on Recommender Systems* (pp. 311-315). Como, Italy: ACM. 10.1145/3109859.3109901

Paul, J., Mittal, A., & Srivastav, G. (2015). Impact of service quality on customer satisfaction in private and public sector banks. *International Journal of Bank Marketing*, *34*(5), 606–622. doi:10.1108/IJBM-03-2015-0030

Perimenis, A., Walimwipi, H., Zinoviev, S., Müller-Langer, F., & Miertus, S. (2011). Development of a decision support tool for the assessment of biofuels. *Energy Policy*, *39*(3), 1782–1793. doi:10.1016/j.enpol.2011.01.011

Perumal, V., & Ilangkumaran, M. (2017). Experimental analysis of engine performance, combustion and emission using pongamia biodiesel as fuel in CI engine. *Energy*, *129*, 228–236. doi:10.1016/j.energy.2017.04.120

Pietrosemoli, L., & Monroy, C. R. (2013). The impact of sustainable construction and knowledge management on sustainability goals. A review of the Venezuelan renewable energy sector. *Renewable & Sustainable Energy Reviews*, *27*, 683–691. doi:10.1016/j.rser.2013.07.056

Pitsilis, G., Zhang, X., & Wang, W. (2011). Clustering recommenders in collaborative filtering using explicit trust information. In I. Wakeman, E. Gudes, C. D. Jensen, & J. Crampton (Eds.), *Trust Management V. IFIPTM 2011. IFIP Advances in Information and Communication Technology, Advances in Information and Communication Technology* (Vol. 358, pp. 82–97). Berlin: Springer. doi:10.1007/978-3-642-22200-9_9

Plata, V., Gauthier-Maradei, P., Romero-Bohórquez, A. R., Kafarov, V., & Castillo, E. (2015). Characterization of insoluble material isolated from Colombian palm oil biodiesel. *Biomass and Bioenergy*, *74*, 6–14. doi:10.1016/j.biombioe.2014.12.024

Poh, K. L., & Ang, B. W. (1999). Transportation fuels and policy for Singapore: An AHP planning approach. *Computers & Industrial Engineering*, *37*(3), 507–525. doi:10.1016/S0360-8352(00)00020-6

Prasad, K., & Chakraborty, S. (2015). Development of a QFD-based expert system for CNC turning centre selection. *Journal of Industrial Engineering International*, *11*(4), 575–594. doi:10.100740092-015-0122-x

Prasad, K., & Chakraborty, S. (2018). Application of the modified similarity-based method for cutting fluid selection. *Decision Science Letters*, *7*, 273–286. doi:10.5267/j.dsl.2017.8.002

Prasad, K., Mahanty, S., Maity, S. R., & Chakraborty, S. (2014). Development of an expert system for materials selection. *Journal of Materials Education*, *36*(5-6), 117–138.

Prasanna Kumar, G. V., & Dewangan, K. N. (2009). Article. *Safety Science*, *47*, 199–205. doi:10.1016/j.ssci.2008.03.007

Puhan, S., Jegan, R., Balasubbramanian, K., & Nagarajan, G. (2009). Effect of injection pressure on performance, emission and combustion characteristics of high linolenic linseed oil methyl ester in a DI diesel engine. *Renewable Energy*, *34*(5), 1227–1233. doi:10.1016/j.renene.2008.10.001

Puhan, S., Vedaraman, N., Ram, B. V., Sankarnarayanan, G., & Jeychandran, K. (2005). Mahua oil (Madhuca Indica seed oil) methyl ester as biodiesel-preparation and emission characterstics. *Biomass and Bioenergy*, *28*(1), 87–93. doi:10.1016/j.biombioe.2004.06.002

Qi, D. H., Chen, H., Geng, L. M., & Bian, Y. Z. (2010). Experimental studies on the combustion characteristics and performance of a direct injection engine fueled with biodiesel/diesel blends. *Energy Conversion and Management*, *51*(12), 2985–2992. doi:10.1016/j.enconman.2010.06.042

Rahman, S. M. A., Masjuki, H. H., Kalam, M. A., Abedin, M. J., Sanjid, A., & Sajjad, H. (2013). Impact of idling on fuel consumption and exhaust emissions and available idle-reduction technologies for diesel vehicles – a review. *Energy Conversion and Management*, *74*, 171–182. doi:10.1016/j.enconman.2013.05.019

Rai, D., Jha, G. K., Chatterjee, P., & Chakraborty, S. (2013). Material selection in manufacturing environment using compromise ranking and regret theory-based compromise ranking methods: A comparative study. *Universal Journal of Materials Science*, *1*(2), 69–77.

Rai, R. N., & Bolia, N. (2014). Optimal decision support for air power potential. *IEEE Transactions on Engineering Management*, *61*(2), 310–322. doi:10.1109/TEM.2013.2293420

Raj, T., & Attri, R. (2011). Identification and modeling of barriers in the implementation of TQM. *International Journal of Productivity and Quality Management.*, *28*(2), 153–179. doi:10.1504/IJPQM.2011.041844

Ramadhas, A. S., Jayaraj, S., & Muraleedharan, C. (2005). Biodiesel production from high FFA rubber seed oil. *Fuel*, *84*(4), 335–340. doi:10.1016/j.fuel.2004.09.016

Ramesh, W. V., & Sakthivel, G. (2018). Selection Of Optimum Fuel Blend In An Ic Engine Using FAHP-PROMETHEE. *IOP Conf. Series: Materials Science and Engineering*, *390*(1), 1-12. 10.1088/1757-899X/390/1/012028

Ranjan, R., Chatterjee, P., & Chakraborty, S. (2016). Performance evaluation of Indian railway zones using DEMATEL and VIKOR methods. *Benchmarking: An International Journal, 23*(1), 78-95.

Ranjan, R., & Chakraboty, S. (2015). Performance Evaluation of Indian Technical Institutions Using PROMETHEE-GAIA Approach. *Informatics in Education*, *1*(14), 103–125. doi:10.15388/infedu.2015.07

Ranjan, R., Chatterjee, P., & Chakraborty, S. (2014). Evaluating performance of engineering departments in an Indian University using DEMATEL and compromise ranking methods. *Opsearch*, *2*(52), 307–328.

Ranjan, R., Chatterjee, P., & Chakraborty, S. (2015). Evaluating performance of engineering departments in an Indian University using DEMATEL and compromise ranking methods. *Opsearch*, *52*(2), 307–328. doi:10.100712597-014-0186-1

Ranjan, R., Chatterjee, P., & Chakraborty, S. (2015). Performance evaluation of Indian states in tourism using an integrated PROMETHEE-GAIA approach. *Opsearch*, *1*(53), 63–84.

Rao, G. V., Reddy, C., & Chakravarthy, V. V. K. (2016). Parthasarthy. Performance and emissions of diesel engine fuelled with ethanol and palm oil methyl ester. *Intenational Journal of Engineering and Technology*, *3*, 108–113.

Rao, R. R., & Kalyankar, V. D. (2014). Optimization of Modern Machining Processes Using Advanced Optimization Techniques: A Review. *International Journal of Advanced Manufacturing Technology*, *47*(5-8), 1159–1188. doi:10.100700170-014-5894-4

Rao, R. V. (2007). *Decision making in the manufacturing environment using graph theory and fuzzy multiple attribute decision making methods*. London, UK: Springer-Verlag.

Rashid, U., Anwar, F., Moser, B. R., & Ashraf, S. (2008). Production of sunflower oil methyl esters by optimized alkali-catalyzed methanolysis. *Biomass and Bioenergy*, *32*(12), 1202–1205. doi:10.1016/j.biombioe.2008.03.001

Rassafi, A. A., Vaziri, M., & Azadani, A. N. (2006). Strategies for utilizing alternative fuels by Iranian passenger cars. *International Journal of Environmental Science and Technology*, *3*(1), 59–68. doi:10.1007/BF03325908

Rauh, N., Franke, T., & Krems, J. F. (2015). Understanding the impact of electric vehicle driving experience on range anxiety. *Human Factors*, *57*(1), 177–187. doi:10.1177/0018720814546372 PMID:25790577

Rautiainen, R. H., & Reynolds, S. J. (2002). Mortality and morbidity in agriculture in the United States. *Journal of Agricultural Safety and Health*, *8*(3), 259–276. doi:10.13031/2013.9054 PMID:12363178

Ravi, V., & Shankar, R. (2005). Analysis of interactions among the barriers of reverse logistics. *Technological Forecasting and Social Change*, *72*(8), 1011–1029. doi:10.1016/j.techfore.2004.07.002

Reijnders, L., & Huijbregts, M. A. J. (2008). Palm oil and the emission of carbon based greenhouse gases. *Journal of Cleaner Production*, *16*(4), 477–482. doi:10.1016/j.jclepro.2006.07.054

Rezaei, J. (2015). Best-worst multi-criteria decision-making method. *Omega*, *53*, 49–57. doi:10.1016/j.omega.2014.11.009

Rezaei, J., Papakonstantinou, A., Tavasszy, L., Pesch, U., & Kana, A. (2018). Sustainable product-package design in a food supply chain: A multi-criteria life cycle approach. *Packaging Technology & Science*, 1–17. doi:10.1002/pts.2418

Rigdon, S. E., & Basu, A. P. (2000). *Statistical methods for the reliability of repairable systems*. New York: Wiley.

Rim, K.-T., & Lim, C.-H. (2014). Biologically Hazardous Agents at Work and Efforts to Protect Workers' Health: A Review of Recent Reports. *Safety and Health at Work*, *5*(2), 43–52. doi:10.1016/j.shaw.2014.03.006 PMID:25180133

Robertoes Koekoeh, K. W., & Peeyush, S. (2016). Article. *Agriculture and Agricultural Science Procedia.*, *9*, 323–332. doi:10.1016/j.aaspro.2016.02.142

Roman-Liu, D. (2014). Comparison of concepts in easy-to-use methods for MSD risk assessment. *Applied Ergonomics*, *45*(3), 420–427. doi:10.1016/j.apergo.2013.05.010 PMID:23849897

Rondini, A., Lagorio, A., Pezzotta, G., & Pinto, R. (2016). A multi-criteria decision-making approach for prioritising product-service systems implementation in smart cities. *International Journal of Management and Decision Making*, *17*(4), 415–446. doi:10.1504/IJMDM.2018.095729

Roth, N. J. (1982). *Prototype Design and Performance of Savonius Rotor Based on Irrigation System* (Master's thesis). University of British Columbia.

Saaty, L. T. (1980). *The Analytic Hierarchy Process*. New York: McGraw Hill Company.

Saaty, L. T., & Vargas, L. G. (2001). *Models, Methods, Concepts &Applications of the Analytical Hierarchy Process*. Boston: Kluwer Academic Publishers; doi:10.1007/978-1-4615-1665-1

Saaty, T. L. (1990). How to make a decision: The analytic hierarchy process. *European Journal of Operational Research*, *48*(1), 9–26. doi:10.1016/0377-2217(90)90057-I

Safari, H., & Ebrahimi, E. (2014). Using modified similarity multiple criteria decision making technique to rank countries in terms of human development index. *Journal of Industrial Engineering and Management*, *7*(1), 254–275.

Safari, H., Khanmohammadi, E., Hafezamini, A., & Ahangari, S. S. (2013). A new technique for multi criteria decision making based on modified similarity method. *Middle East Journal of Scientific Research*, *14*(5), 712–719.

Saha, U. K., Thotla, S., & Maity, D. (2008). Optimum design configuration of Savonius rotor through wind tunnel experiments. *Journal of Wind Engineering and Industrial Aerodynamics*, *96*(8-9), 1359–1375. doi:10.1016/j.jweia.2008.03.005

Sahoo, P. K., & Das, L. M. (2009). Combustion analysis of Jatropha, Karanja and Polanga based biodiesel as fuel in a diesel engine. *Fuel*, *88*(6), 994–999. doi:10.1016/j.fuel.2008.11.012

Sahoo, P. K., Naik, S. N., Babu, M. K. G., & Das, L. M. (2007). Biodiesel development from high acid value polanga seed oil and performance in a CI engine. *Fuel*, *86*(3), 448–454. doi:10.1016/j.fuel.2006.07.025

Sakthivel, G., & Nagarajan, G. (2011). *Experimental studies on the performance and emission characteristics of a diesel engine fuelled with ethyl ester of waste fish oil and its diesel blends* (No. 2011-01-2402). SAE Technical Paper.

Sakthivel, G., & Ilangkumaran, M. (2017). Optimisation of compression ignition engine performance with fishoil biodiesel using Taguchi-Fuzzy approach. *Journal International Journal of Ambient Energy*, *38*(2), 146–160.

Sakthivel, G., Ilangkumaran, M., & Gaikwad, A. (2015). A hybrid multi-criteria decision modeling approach for the best biodiesel blend selection based on ANP-TOPSIS analysis. *Ain Shams Engineering Journal*, *6*(1), 239–256. doi:10.1016/j.asej.2014.08.003

Salimi, N., & Rezaei, J. (2018). Evaluating firms' R&D performance using best worst method. *Evaluation and Program Planning*, *66*, 147–155. doi:10.1016/j.evalprogplan.2017.10.002 PMID:29091790

Salman, A. K. (2016). Multi-criteria Decision-making for Selection of Wind Turbine with Rated Power of 1000 KW using Dubois and Prade Fuzzy Operator. *International Journal of Computing and Network Technology*, *4*(1), 27–32. doi:10.12785/ijcnt/040104

Salvi, B.L., & Jindal, S. (2013). A Comparative Study of engine performance and exhaust emission characteristics of linseed oil biodiesel blends with diesel fuel in a direct injection diesel engine. *Journal of The Institution of Engineers (India): Series C, 94*(1), 1–8.

Sambo, M. N., Idris, S. H., & Shamang, A. (2012). Determinants of occupational health hazards among roadside automobile mechanics in Zaria, Northern Western Nigeria. *BOMJ*, *9*(1), 5–9.

Sarin, A., Arora, R., Singh, N. P., Sarin, R., Malhotra, R. K., & Kundu, K. (2009). Effect of blends of Palm-Jatropha-Pongamia biodiesels on cloud point and pour point. *Energy*, *34*(11), 2016–2021. doi:10.1016/j.energy.2009.08.017

Sarin, R., Sharma, M., Sinharay, S., & Malhotra, R. K. (2007). Jatropha–palm biodiesel blends: An optimum mix for Asia. *Fuel*, *86*(10-11), 1365–1371. doi:10.1016/j.fuel.2006.11.040

Sarwar, B. M., Karypis, G., Konstan, A. J., & Riedl, J. T. (2001). Item-based collaborative filtering recommendation algorithms. In *Proc. of the 10th International World Wide Web Conference* (pp. 285-295). Academic Press.

Sarwar, B., Karypis, G., Konstan, J., & Riedl, J. (2000). Analysis of recommendation algorithms for e-commerce. *Proceedings of the 2nd ACM conference on Electronic commerce* (pp. 158-167). New York: ACM. 10.1145/352871.352887

Satyanarayana, K., Mariano, A., & Vargas, J. (2011). A review on microalgae, a versatile source for sustainable energy and materials. *International Journal of Energy Research*, *35*(4), 291–31. doi:10.1002/er.1695

Savonius, S. J. (1931). The S-rotor and its application. *Mechanical Engineering (New York, N.Y.)*, *53*, 1–6.

Saylers, A. T. (1985). Blade configuration optimization & performance characteristics of a simple Savonius rotor. *Proceedings - Institution of Mechanical Engineers*, *199*(3), 185–191. doi:10.1243/PIME_PROC_1985_199_023_02

Schalkwijk, W. V., & Scrosati, B. (2007). *Advances in lithium-ion batteries*. Springer Science & Business Media.

Schank, R. C. (1982). *Dynamic memory: A theory of reminding and learning in computers and people* (Vol. 240). Cambridge, UK: Cambridge University Press.

Scott, J. A., Ho, W., & Dey, P. K. (2012). A review of multi-criteria decision-making methods for bioenergy systems. *Energy*, *42*(1), 146–156. doi:10.1016/j.energy.2012.03.074

Seiti, H., Hafezalkotob, A., & Fattahi, R. (2018). Extending a pessimistic–optimistic fuzzy information axiom based approach considering acceptable risk: Application in the selection of maintenance strategy. *Applied Soft Computing*, *67*, 895–909. doi:10.1016/j.asoc.2017.11.017

Seiti, H., Tagipour, R., Hafezalkotob, A., & Asgari, F. (2017). Maintenance strategy selection with risky evaluations using RAHP. *Journal of Multi-Criteria Decision Analysis*, *24*(5-6), 257–274. doi:10.1002/mcda.1618

Sen, D. K., Datta, S., & Mahapatra, S. S. (2016). Application of TODIM (Tomada de Decisión Inerativa Multicritero) for industrial robot selection. Benchmarking. *International Journal (Toronto, Ont.)*, *23*(7), 1818–1833.

Senthil, R., & Silambarasan, R. (2015). Annona: A new biodiesel for diesel engine: A comparative experimental investigation. *Journal of the Energy Institute*, *88*(4), 459–469. doi:10.1016/j.joei.2014.09.011

Shafiee, M., Labib, A., Maiti, J., & Starr, A. (2017). Maintenance strategy selection for multi-component systems using a combined analytic network process and cost-risk criticality model. *Proceedings of the Institution of Mechanical Engineers, Part O: Journal of Risk and Reliability*, 1-35.

Shafiee, M. (2015). Maintenance strategy selection problem: An MCDM overview. *Journal of Quality in Maintenance Engineering*, *21*(4), 378–402. doi:10.1108/JQME-09-2013-0063

Shafiqur, R., & Salman, A. K. (2017). Multi-Criteria Wind Turbine Selection using Weighted Sum Approach. *International Journal of Advanced Computer Science and Applications*, *8*(6), 128–132.

Shanmugam, P., Sivakumar, V., Murugesan, A., & Ilangkumaran, M. (2011a). Performance and exhaust emissions of a diesel engine using hybrid fuel with an artificial neural network. *Energy Sources. Part A, Recovery, Utilization, and Environmental Effects*, *33*(15), 1440–1450. doi:10.1080/15567036.2010.539085

Shanmugam, P., Sivakumar, V., Murugesan, A., & Umarani, C. (2011b). Experimental study on diesel engine using hybrid fuel blends. *International Journal of Green Energy*, *8*(6), 655–668. doi:10.1080/15435075.2011.588891

Shannon, C. E. (1948). The Mathematical Theory of Communication. *The Bell System Technical Journal*, *27*(3), 379–423. doi:10.1002/j.1538-7305.1948.tb01338.x

Sharma, A., Muqeem, M., Sherwani, A., & Ahmad, M. (2018). Optimization of diesel engine input parameters running on Polanga biodiesel to improve performance and exhaust emission using MOORA technique with standard deviation. *Energy Sources. Part A, Recovery, Utilization, and Environmental Effects*, *40*(22), 1–18. doi:10.1080/15567036.2018.1511647

Sharma, K. K., Gupta, R., Singh, S. K., & Singh, S. R. (2005). Experimental investigation of the characteristics of a Savonius wind turbine. *Wind Engineering*, *29*(1), 77–82. doi:10.1260/0309524054353674

Sharma, R. K., Kumar, D., & Kumar, P. (2007). Modeling system behaviour for risk and reliability analysis using KBARAM. *Quality and Reliability Engineering International*, *23*(8), 973–998. doi:10.1002/qre.849

Sharma, R. K., & Sharma, P. (2012). Integrated framework to optimize RAM and cost decision in process plant. *Journal of Loss Prevention in the Process Industries*, *25*(6), 883–904. doi:10.1016/j.jlp.2012.04.013

Sharma, R., Kumar, D., & Kumar, P. (2005). Systematic failure mode and effect analysis using fuzzy linguistic modeling. *International Journal of Quality & Reliability Management*, *22*(9), 986–1004. doi:10.1108/02656710510625248

Sharp, B. L., Kleinschmidt, I., Streat, E., Maharaj, R., Barnes, K. I., Durrheim, D. N., ... Grange, L. A. (2007). Seven years of regional malaria control collaboration--Mozambique, South Africa, and Swaziland. *The American Journal of Tropical Medicine and Hygiene*, *76*(1), 42–47. doi:10.4269/ajtmh.2007.76.42 PMID:17255227

Shemshadi, A., Shirazi, H., Toreihi, M., & Tarokh, M. J. (2011). A fuzzy VIKOR method for supplier selection based on entropy measure for objective weighting. *Expert Systems with Applications*, *38*(10), 12160–12167. doi:10.1016/j.eswa.2011.03.027

Sindhu, S., Nehra, V., & Luthra, S. (2017). Investigation of feasibility study of solar farms deployment using hybrid AHP-TOPSIS analysis: Case study of India. *Renewable & Sustainable Energy Reviews*, *73*, 496–511. doi:10.1016/j.rser.2017.01.135

Sivaraja, C. M., & Sakthivel, G. (2017). Compression ignition engine performance modelling using hybrid MCDM techniques for the selection of optimum fish oil biodiesel blend at different injection timings. *Energy*, *139*, 118–141. doi:10.1016/j.energy.2017.07.134

Sivasegaram, S. (1978). Secondary parameters affecting the performance of resistance type vertical axis wind rotors. *Wind Engineering*, *2*, 49–58.

Soheili-Fard, F., Rahbar, A., & Marzban, A. (2017). Ergonomic investigation of workers in tea factories using REBA and OWAS methods -case study: (Langroud region, Guilan, Iran). *Agricultural Engineering International: CIGR Journal*, *19*(3), 112–119.

Sonkar, V., Abhishek, K., Datta, S., & Mahapatra, S. S. (2014). Multi-objective optimization in drilling of GFRP composites: A degree of similarity approach. *Procedia Materials Science*, *6*, 538–543. doi:10.1016/j.mspro.2014.07.068

Splansky, G. L., Corey, D., Yang, Q., Atwood, L. D., Cupples, L. A., Benjamin, E. J., ... Levy, D. (2007). The Third Generation Cohort of the National Heart, Lung, and Blood Institute's Framingham Heart Study: Design, recruitment, and initial examination. *American Journal of Epidemiology*, *165*(11), 1328–1335. doi:10.1093/aje/kwm021 PMID:17372189

Srivastava, A., & Prasad, R. (2000). Triglycerides-based diesel fuels. *Renewable & Sustainable Energy Reviews*, *4*(2), 111–133. doi:10.1016/S1364-0321(99)00013-1

Stanford, A. (1998). Liberalisation of the UK energy market: An opportunity for green energy. *Renewable Energy*, *15*(1-4), 215–217. doi:10.1016/S0960-1481(98)00160-8

Stević, Ž., Pamučar, D., Vasiljević, M., Stojić, G., & Korica, S. (2017). Novel integrated multi-criteria model for supplier selection: Case study construction company. *Symmetry*, *9*(11), 279. doi:10.3390ym9110279

Stojić, G., Stević, Ž., Antuchevičienė, J., Pamučar, D., & Vasiljević, M. (2018). A Novel Rough WASPAS Approach for Supplier Selection in a Company Manufacturing PVC Carpentry Products. *Information*, *9*(5), 121. doi:10.3390/info9050121

Strazdins, L., & Bammer, G. (2004). Women, work and musculoskeletal health. *Social Science & Medicine*, *58*(6), 997–1005. doi:10.1016/S0277-9536(03)00260-0 PMID:14723897

Subramony, M., & Holtom, B. C. (2012). The long-term influence of service employee attrition on customer outcomes and profits. *Journal of Service Research*, *15*(4), 460–473. doi:10.1177/1094670512452792

Sumathi, C. P. (2010). Automatic Recommendation of Web Pages in Web Usage Mining. *International Journal on Computer Science and Engineering*, *02*(09), 3046–3052.

Suresh, K. B., Subbaraju, N. V., Reddy, S. M., & Nageswara, R. D. (2006). The material selection for typical wind turbine blades using a MADM approach & analysis of blades. *Conference on MCDM 2006*, 1-11.

Sureshkumar, K., Velraj, R., & Ganesan, R. (2008). Performance and exhaust emission characteristics of a CI engine fueled with Pongamia pinnata methyl ester (PPME) and its blends with diesel. *Renewable Energy*, *33*(10), 2294–2302. doi:10.1016/j.renene.2008.01.011

Szmidt, E., & Kacprzyk, J. (2000). Distances between intuitionistic fuzzy sets and Systems. *Information and Control*, *114*, 505–518.

Szmidt, E., & Kacprzyk, J. (2001). Entropy for intuitionistic fuzzy sets. *Fuzzy Sets and Systems*, *118*(3), 467–477. doi:10.1016/S0165-0114(98)00402-3

Tadesse, T., & Kumie, A. (2007). Prevalence of factors affecting work-related injury among workers engaged in small and medium-scale industries in Gondar Wereda, North Gondar zone, Amhara Regional state, Ethiopia. *The Ethiopian Journal of Health Development*, *21*(1), 25–34. doi:10.4314/ejhd.v21i1.10028

Tan, K. C., & Pawitra, T. A. (2001). Integrating SERVQUAL and Kano's model into QFD for service excellence development. *Managing Service Quality: An International Journal*, *11*(6), 418–430. doi:10.1108/EUM0000000006520

Tao, W., Chen, G., Zeng, G., Yan, M., & Wang, L. (2016). Influence of silver nanoparticles on heavy metals of pore water in contaminated river sediments. *Chemosphere*, *162*, 117–124. doi:10.1016/j.chemosphere.2016.07.043 PMID:27494311

Tesla. (2017a). *Tesla fourth quarter & full year 2016 update*. Retrieved from http://files.shareholder.com/downloads/ABEA-4CW8X0/4066363333x0x929284/22C29259-6C19-41AC-9CAB-899D148F323D/TSLA_Update_Letter_2016_4Q.pdf

Thakur, S. S., Kundu, A., & Sing, J. K. (2011). A Collaborative Filtering Approach: Based on offline customers ratings for Prediction and Recommendation. In *Proc. of the 1st International Conference on Scientific Paradigm Shift in Information Technology & Management* (pp. 215-220). Academic Press.

Thangaraj, S., & Shireen, N. (2017). Occupational health hazards among automobile mechanics working in an urban area of Bangalore - a cross sectional study. *International Journal of Medical Science and Public Health*, *6*(1), 18. doi:10.5455/ijmsph.2016.31052016540

Thiyagarajan, R., Thangavel, K., & Rathipriya, R. (2014). Recommendation of web pages using weighted K- means clustering. *International Journal of Computers and Applications*, *86*(14), 44–48. doi:10.5120/15057-3517

Tiwari, A., & Mishra, N. (2017). Sustainable and smart city project: An overview of the application of multi criteria decision making techniques and approaches for Indian context. *International Journal of Management and Applied Science*, *3*(7), 1–6.

Triantaphyllou, E. (2000). Multi-criteria decision making methods. In *Multi-Criteria Decision Making Methods: A comparative Study*. Boston, MA: Springer. doi:10.1007/978-1-4757-3157-6_2

Tseng, M. L., Lin, Y. H., Tan, K., Chen, R. H., & Chen, Y. H. (2014). Using TODIM to evaluate green supply chain practices under uncertainty. *Applied Mathematical Modelling*, *38*(11-12), 2983–2995. doi:10.1016/j.apm.2013.11.018

Tsita, K. G., & Pilavachi, P. A. (2012). Evaluation of alternative fuels for the Greek road transport sector using the analytic hierarchy process. *Energy Policy*, *48*, 677–686. doi:10.1016/j.enpol.2012.05.079

Tzeng, G. H., Lin, C. W., & Opricovic, S. (2005). Multi-criteria analysis of alternative-fuel buses for public transportation. *Energy Policy*, *33*(11), 1373–1383. doi:10.1016/j.enpol.2003.12.014

UNFCCC. (2015b). *Adoption of the Paris Agreement*. United Nations Framework Convention on Climate Change. Retrieved from https://unfccc.int/resource/docs/2015/cop21/eng/l09rC1.pdf

United Nations. (1992). *Report of the intergovernmental negotiating committee for a framework convention on climate change on the work of the second part of its fifth session*. United Nations Framework Convention on Climate Change 1992.

United Nations. (2014). *Report of the conference of the parties serving as meeting of the parties to the Kyoto Protocol on its tenth sessio*n. United Nations Framework Convention on Climate Change 2015.

USEPA. (2014). *Inventory of US. Green house gas emissions and Sinks: 1990-2012*. EPA.

Uysal, F., & Tosun, Ö. (2014). Multi criteria analysis of the residential properties in Antalya using TODIM method. *Procedia: Social and Behavioral Sciences*, *109*, 322–326. doi:10.1016/j.sbspro.2013.12.465

van Nieuwenhuyse, A., Somville, P. R., Crombez, G., Burdorf, A., Verbeke, G., Johannik, K., ... Moens, G. F. (2006). The role of physical workload and pain related fear in the development of low back pain in young workers: Evidence from the BelCoBack Study; results after one year of follow up. *Occupational and Environmental Medicine*, *63*(1), 45–52. doi:10.1136/oem.2004.015693 PMID:16361405

Varatharajan, K., & Cheralathan, M. (2013). Effect of aromatic amine antioxidants on NOx emissions from a soybean biodiesel powered DI diesel engine. *Fuel Processing Technology*, *106*, 526–532. doi:10.1016/j.fuproc.2012.09.023

Vedaraman, N., Puhan, S., Nagarajan, G., & Velappan, K. C. (2011). Preparation of palm oil biodiesel and effect of various additives on NOx emission reduction in B20: An experimental study. *International Journal of Green Energy*, *8*(3), 383–397. doi:10.1080/15435075.2011.557847

Velasquez, M., & Hester, P. T. (2013). An analysis of multi-criteria decision making methods. *International Journal of Operations Research*, *10*(2), 56–66.

Vilela, L., Mata, T. M., & Caetano, N. S. (2010, November). Biodiesel production from fish oil with high acidity. In *3rd international symposium on energy from biomass and waste* (pp. 1-14). Academic Press.

Voaklander, K. D., Kelly, D. C., Rowe, B. H., Schopflocher, D. P., Svenson, L., Yiannakoulias, N., & Pickett, W. (2006). Pain, medication, and injury in older farmers. *American Journal of Industrial Medicine*, *49*(5), 374–382. doi:10.1002/ajim.20292 PMID:16526061

Volkswagen. (2016). *New Group strategy adopted: Volkswagen Group to become a world leading provider of sustainable mobility*. Retrieved from www.volkswagenag.com/en/news/2016/6/2025.html

Vyavahare, R. T., & Kallurkar, S. P. (2012). Anthropometric and strength data of Indian agricultural workers for equipment design: A review. *Agricultural Engineering International: CIGR Journal*, *14*(4), 102–114.

Wang, F., Shi, Q., & Hu, Z. (2015). A sustainability approach to maintenance strategy selection based on fuzzy AHP and VIKOR algorithms. *Recent Developments on Reliability, Maintenance and Safety*, *717*. doi:10.2495/QR2MSE140841

Wang, J. Q., & Zhang, Z. H. (2008). Programming method of multi-criteria decision-making based on intuitionistic fuzzy number with incomplete certain information. *Control Decision*, *23*, 1145–1148.

Wang, J. Q., & Zhang, Z. H. (2009). Multi-criteria decision-making method with incomplete certain information based on intuitionistic fuzzy number. *Control Decision*, *24*, 226–230.

Wang, J., Wang, J. Q., & Zhang, H. Y. (2016). A likelihood-based TODIM approach based on multi-hesitant fuzzy linguistic information for evaluation in logistics outsourcing. *Computers & Industrial Engineering*, *99*, 287–299. doi:10.1016/j.cie.2016.07.023

Wang, L., Chu, J., & Wu, J. (2007). Selection of optimum maintenance strategies based on a fuzzy analytic hierarchy process. *International Journal of Production Research*, *107*(1), 151–163. doi:10.1016/j.ijpe.2006.08.005

Wang, Y., Yeo, G. T., & Ng, A. K. (2014). Choosing optimal bunkering ports for liner shipping companies: A hybrid Fuzzy-Delphi–TOPSIS approach. *Transport Policy*, *35*, 358–365. doi:10.1016/j.tranpol.2014.04.009

Wan, S. P., & Dong, J. Y. (2010). Method of trapezoidal intuitionistic fuzzy number for multi-attribute group decision. *Control Decision*, *25*(5), 773–776.

Wei, G. W. (2012). Hesitant fuzzy prioritized operators and their application to multiple attribute decision making. *Knowledge-Based Systems*, *31*, 176–182. doi:10.1016/j.knosys.2012.03.011

WHO. (2008). Guidelines for drinking water quality: Recommendations. World Health Organization.

Wiley, K., Vucinich, N., Miller, J., & Vanzi, M. (2004, November). *Confined Animal Facilities in California*. Retrieved from http://sor.senate.ca.gov/sites/sor.senate.ca.gov/files/%7BD51D1D55-1B1F-4268-80CC-C636EE939A06%7D.pdf

Wu, Y., Wang, J., Hu, Y., Ke, Y., & Li, L. (2018). An extended TODIM-PROMETHEE method for waste-to-energy plant site selection based on sustainability perspective. *Energy*, *156*, 1–16. doi:10.1016/j.energy.2018.05.087

Xu, F., Liu, X., Chen, W., Zhou, C., & Cao, B. (2018). Ontology-Based Method for Fault Diagnosis of Loaders. *Sensors*, *18*(3), 1-22.

Xuewu, X., & Yahui. (2012). Modeling and simulation of brushless dc motor control system for EPS applications. *2012 IEEE International Conference on Computer Science and Automation Engineering (CSAE).*

Yager, R. R. (2004). Modeling prioritized multi-criteria decision making. *IEEE Transactions on Systems, Man, and Cybernetics. Part B, Cybernetics*, *34*(6), 2396–2404. doi:10.1109/TSMCB.2004.837348 PMID:15619938

Yazdani, M., Chatterjee, P., Zavadskas, E. K., & Streimikiene, D. (2018). A novel integrated decision-making approach for evaluation and selection of renewable energy technologies. *Clean Technologies and Environmental Policy*, *20*(2), 403–420. doi:10.100710098-018-1488-4

Yazdani, M., Zarate, P., Kazimieras Zavadskas, E., & Turskis, Z. (2018). A Combined Compromise Solution (CoCoSo) method for multi-criteria decision-making problems. *Management Decision*, MD-05-2017-0458. doi:10.1108/MD-05-2017-0458

Yedla, S., & Shrestha, R. M. (2003). Multi-criteria approach for the selection of alternative options for environmentally sustainable transport system in Delhi. *Transportation Research Part A, Policy and Practice*, *37*(8), 717–729. doi:10.1016/S0965-8564(03)00027-2

Yigitcanlar, T., & Baum, S. (2008). Benchmarking local e-government. In Electronic Government: Concepts, Methodologies, Tools, and Applications. IGI Global. doi:10.4018/978-1-59904-947-2.ch033

Yigitcanlar, T. (2006). Australian local governments' practice and prospects with online planning. *URISA Journal*, *18*(2), 7–17.

Yigitcanlar, T., Dodson, J., Gleeson, B., & Sipe, N. (2007). Travel self-containment in master planned estates: Analysis of recent Australian trends. *Urban Policy and Research*, *25*(1), 129–149. doi:10.1080/08111140701255823

Yigitcanlar, T., & Kamruzzaman, M. (2014). Investigating the interplay between transport, land use and the environment: A review of the literature. *International Journal of Environmental Science and Technology*, *11*(8), 2121–2132. doi:10.100713762-014-0691-z

Yigitcanlar, T., & Kamruzzaman, M. (2015). Planning, development and management of sustainable cities: A commentary from the guest editors. *Sustainability*, *7*(11), 14677–14688. doi:10.3390u71114677

Yılmaz, B., & Harmancıoglu, N. B. (2010). Multi-criteria decision making for water resource management: A case study of the Gediz River Basin, Turkey. *Water S.A.*, *36*(5), 563–576. doi:10.4314/wsa.v36i5.61990

Yılmaz, S., & Selim, H. (2013). A review on the methods for biomass to energy conversion systems design. *Renewable & Sustainable Energy Reviews*, *25*, 420–430. doi:10.1016/j.rser.2013.05.015

Yiu, C. S., Grant, K., & Edgar, K. (2007). Factors affecting the adoption of Internet Banking in Hong Kong - Implications for the banking sector. *International Journal of Information Management*, *27*(5), 336–351. doi:10.1016/j.ijinfomgt.2007.03.002

You, P., Guo, S., Zhao, H., & Zhao, H. (2017). Operation performance evaluation of power grid Enterprise using a hybrid BWM-TOPSIS method. *Sustainability*, *9*(12), 23–29. doi:10.3390u9122329

Zadeh, L. A. (1965). Fuzzy sets. *Information and Control*, *8*(3), 338–353. doi:10.1016/S0019-9958(65)90241-X

Zarbakhshnia, N., Soleimani, H., & Ghaderi, H. (2018). Sustainable third-party reverse logistics provider evaluation and selection using fuzzy SWARA and developed fuzzy COPRAS in the presence of risk criteria. *Applied Soft Computing*, *65*, 307–319. doi:10.1016/j.asoc.2018.01.023

Zavadskas, E. K., Turskis, Z., Antucheviciene, J., & Zakarevicius, A. (2012). Optimization of weighted aggregated sum product assessment. *Elektronika ir Elektrotechnika*, *122*(6), 3–6. doi:10.5755/j01.eee.122.6.1810

Zhang, Y., Dube, M. A., McLean, D. D., & Kates, M. (2003). Biodiesel production from waste cooking oil: 1. Process design and technological assessment. *Bioresource Technology*, *89*(1), 1–16. doi:10.1016/S0960-8524(03)00040-3 PMID:12676496

Zhang, Z. (2014). Intuitionistic trapezoidal fuzzy prioritized operators and their application to multiple attribute group decision making. *Brit. J. Math. Comput. Sci.*, *4*(14), 1951–1998. doi:10.9734/BJMCS/2014/10133

Zhou, C., & Roseman, J. M. (1994). Agricultural injuries among a population based sample of farm operators in Alabama. *American Journal of Industrial Medicine*, *25*(3), 385–402. doi:10.1002/ajim.4700250307 PMID:8160657

Zindani, D., Maity, S. R., Bhowmik, S., & Chakraborty, S. (2017). A material selection approach using the TODIM (TOmada de Decisao Interativa Multicriterio) method and its analysis. *International Journal of Materials Research*, *108*(5), 345–354. doi:10.3139/146.111489

Zolfani, S. H., & Saparauskas, J. (2013). New application of SWARA method in prioritizing sustainability assessment indicators of energy system. *The Engineering Economist*, *24*(5), 408–414.

Zolfani, S. H., Yazdani, M., & Zavadskas, E. K. (2018). An extended stepwise weight assessment ratio analysis (SWARA) method for improving criteria prioritization process. *Soft Computing*, 1–7.

Related References

To continue our tradition of advancing academic research, we have compiled a list of recommended IGI Global readings. These references will provide additional information and guidance to further enrich your knowledge and assist you with your own research and future publications.

Abresch, J. (2003). Geographic Information Systems Research and Data Centers. In A. Hanson & B. Levin (Eds.), *Building a Virtual Library* (pp. 52–64). Hershey, PA: IGI Global. doi:10.4018/978-1-59140-106-3.ch004

Ajiboye, O. E., & Yusuff, O. S. (2017). Foreign Land Acquisition: Food Security and Food Chains – The Nigerian Experience. In I. Management Association (Ed.), Natural Resources Management: Concepts, Methodologies, Tools, and Applications (pp. 1524-1545). Hershey, PA: IGI Global. doi:10.4018/978-1-5225-0803-8.ch072

Alapiki, H. E., & Amadi, L. A. (2018). Sustainable Food Consumption in the Neoliberal Order: Challenges and Policy Implications. In A. Obayelu (Ed.), *Food Systems Sustainability and Environmental Policies in Modern Economies* (pp. 90–123). Hershey, PA: IGI Global. doi:10.4018/978-1-5225-3631-4.ch005

Altaş, A. (2018). Geographical Information System Applications Utilized in Museums in Turkey Within the Scope of the Cultural Heritage Tourism: A Case Study of Mobile Application of Müze Asist. In S. Chaudhuri & N. Ray (Eds.), *GIS Applications in the Tourism and Hospitality Industry* (pp. 42–60). Hershey, PA: IGI Global. doi:10.4018/978-1-5225-5088-4.ch002

Andreea, I. R. (2018). Beyond Macroeconomics of Food and Nutrition Security. *International Journal of Sustainable Economies Management*, *7*(1), 13–22. doi:10.4018/IJSEM.2018010102

Anwar, J. (2017). Reproductive and Mental Health during Natural Disaster: Implications and Issues for Women in Developing Nations – A Case Example. In I. Management Association (Ed.), Gaming and Technology Addiction: Breakthroughs in Research and Practice (pp. 446-472). Hershey, PA: IGI Global. doi:10.4018/978-1-5225-0778-9.ch021

Awadh, H., Aksissou, M., Benhardouze, W., Darasi, F., & Snaiki, J. (2018). Socioeconomic Status of Artisanal Fishers in the West Part of Moroccan Mediterranean. *International Journal of Social Ecology and Sustainable Development*, *9*(1), 40–52. doi:10.4018/IJSESD.2018010104

Aye, G. C., & Haruna, R. F. (2018). Effect of Climate Change on Crop Productivity and Prices in Benue State, Nigeria: Implications for Food Security. In V. Erokhin (Ed.), *Establishing Food Security and Alternatives to International Trade in Emerging Economies* (pp. 244–268). Hershey, PA: IGI Global. doi:10.4018/978-1-5225-2733-6.ch012

Azizan, S. A., & Suki, N. M. (2017). Consumers' Intentions to Purchase Organic Food Products. In T. Esakki (Ed.), *Green Marketing and Environmental Responsibility in Modern Corporations* (pp. 68–82). Hershey, PA: IGI Global. doi:10.4018/978-1-5225-2331-4.ch005

Barakabitze, A. A., Fue, K. G., Kitindi, E. J., & Sanga, C. A. (2017). Developing a Framework for Next Generation Integrated Agro Food-Advisory Systems in Developing Countries. In I. Management Association (Ed.), Agri-Food Supply Chain Management: Breakthroughs in Research and Practice (pp. 47-67). Hershey, PA: IGI Global. doi:10.4018/978-1-5225-1629-3.ch004

Beachcroft-Shaw, H., & Ellis, D. (2018). Using Successful Cases to Promote Environmental Sustainability: A Social Marketing Approach. In I. Management Association (Ed.), Sustainable Development: Concepts, Methodologies, Tools, and Applications (pp. 936-953). Hershey, PA: IGI Global. doi:10.4018/978-1-5225-3817-2.ch042

Behnassi, M., Kahime, K., Boussaa, S., Boumezzough, A., & Messouli, M. (2017). Infectious Diseases and Climate Vulnerability in Morocco: Governance and Adaptation Options. In I. Management Association (Ed.), Public Health and Welfare: Concepts, Methodologies, Tools, and Applications (pp. 91-109). Hershey, PA: IGI Global. doi:10.4018/978-1-5225-1674-3.ch005

Bekele, F., & Bekele, I. (2017). Social and Environmental Impacts on Agricultural Development. In W. Ganpat, R. Dyer, & W. Isaac (Eds.), *Agricultural Development and Food Security in Developing Nations* (pp. 21–56). Hershey, PA: IGI Global. doi:10.4018/978-1-5225-0942-4.ch002

Benaouda, A., & García-Peñalvo, F. J. (2018). Towards an Intelligent System for the Territorial Planning: Agricultural Case. In F. García-Peñalvo (Ed.), *Global Implications of Emerging Technology Trends* (pp. 158–178). Hershey, PA: IGI Global. doi:10.4018/978-1-5225-4944-4.ch010

Bhaskar, A., Rao, G. B., & Vencatesan, J. (2017). Characterization and Management Concerns of Water Resources around Pallikaranai Marsh, South Chennai. In P. Rao & Y. Patil (Eds.), *Reconsidering the Impact of Climate Change on Global Water Supply, Use, and Management* (pp. 102–121). Hershey, PA: IGI Global. doi:10.4018/978-1-5225-1046-8.ch007

Bianchini, E., & Sicurella, S. (2010). GIS: A New Tool for Criminology and Victimology's Studies. In R. Sette (Ed.), *Cases on Technologies for Teaching Criminology and Victimology: Methodologies and Practices* (pp. 87–110). Hershey, PA: IGI Global. doi:10.4018/978-1-60566-872-7.ch006

Bogataj, D., & Drobne, D. (2017). Control of Perishable Goods in Cold Logistic Chains by Bionanosensors. In S. Joo (Ed.), *Applying Nanotechnology for Environmental Sustainability* (pp. 376–402). Hershey, PA: IGI Global. doi:10.4018/978-1-5225-0585-3.ch016

Bogataj, D., & Drobne, D. (2017). Control of Perishable Goods in Cold Logistic Chains by Bionanosensors. In I. Management Association (Ed.), Materials Science and Engineering: Concepts, Methodologies, Tools, and Applications (pp. 471-497). Hershey, PA: IGI Global. doi:10.4018/978-1-5225-1798-6.ch019

Bogueva, D., & Marinova, D. (2018). What Is More Important: Perception of Masculinity or Personal Health and the Environment? In D. Bogueva, D. Marinova, & T. Raphaely (Eds.), *Handbook of Research on Social Marketing and Its Influence on Animal Origin Food Product Consumption* (pp. 148–162). Hershey, PA: IGI Global. doi:10.4018/978-1-5225-4757-0.ch010

Bouzid, M. (2017). Waterborne Diseases and Climate Change: Impact and Implications. In M. Bouzid (Ed.), *Examining the Role of Environmental Change on Emerging Infectious Diseases and Pandemics* (pp. 89–108). Hershey, PA: IGI Global. doi:10.4018/978-1-5225-0553-2.ch004

Bowles, D. C. (2017). Climate Change-Associated Conflict and Infectious Disease. In M. Bouzid (Ed.), *Examining the Role of Environmental Change on Emerging Infectious Diseases and Pandemics* (pp. 68–88). Hershey, PA: IGI Global. doi:10.4018/978-1-5225-0553-2.ch003

Buck, J. J., & Lowry, R. K. (2017). Oceanographic Data Management: Quills and Free Text to the Digital Age and "Big Data". In P. Diviacco, A. Leadbetter, & H. Glaves (Eds.), *Oceanographic and Marine Cross-Domain Data Management for Sustainable Development* (pp. 1–22). Hershey, PA: IGI Global. doi:10.4018/978-1-5225-0700-0.ch001

Buse, C. G. (2017). Are Climate Change Adaptation Policies a Game Changer?: A Case Study of Perspectives from Public Health Officials in Ontario, Canada. In M. Bouzid (Ed.), *Examining the Role of Environmental Change on Emerging Infectious Diseases and Pandemics* (pp. 230–257). Hershey, PA: IGI Global. doi:10.4018/978-1-5225-0553-2.ch010

Calderon, F. A., Giolo, E. G., Frau, C. D., Rengel, M. G., Rodriguez, H., Tornello, M., ... Gallucci, R. (2018). Seismic Microzonation and Site Effects Detection Through Microtremors Measures: A Review. In N. Ceryan (Ed.), *Handbook of Research on Trends and Digital Advances in Engineering Geology* (pp. 326–349). Hershey, PA: IGI Global. doi:10.4018/978-1-5225-2709-1.ch009

Carfì, D., Donato, A., & Panuccio, D. (2018). A Game Theory Coopetitive Perspective for Sustainability of Global Feeding: Agreements Among Vegan and Non-Vegan Food Firms. In I. Management Association (Ed.), Game Theory: Breakthroughs in Research and Practice (pp. 71-104). Hershey, PA: IGI Global. doi:10.4018/978-1-5225-2594-3.ch004

Carr, D. A., & Carr, T. R. (2007). Geographic Information System Applications in the Public Sector. In G. Garson (Ed.), *Modern Public Information Technology Systems: Issues and Challenges* (pp. 293–311). Hershey, PA: IGI Global. doi:10.4018/978-1-59904-051-6.ch015

Castagnolo, V. (2018). Analyzing, Classifying, Safeguarding: Drawing for the Borgo Murattiano Neighbourhood of Bari. In G. Carlone, N. Martinelli, & F. Rotondo (Eds.), *Designing Grid Cities for Optimized Urban Development and Planning* (pp. 93–108). Hershey, PA: IGI Global. doi:10.4018/978-1-5225-3613-0.ch006

Chekima, B. (2018). The Dilemma of Purchase Intention: A Conceptual Framework for Understanding Actual Consumption of Organic Food. *International Journal of Sustainable Economies Management*, 7(2), 1–13. doi:10.4018/IJSEM.2018040101

Chen, Y. (2017). Sustainable Supply Chains and International Soft Landings: A Case of Wetland Entrepreneurship. In B. Christiansen & F. Kasarcı (Eds.), *Corporate Espionage, Geopolitics, and Diplomacy Issues in International Business* (pp. 232–247). Hershey, PA: IGI Global. doi:10.4018/978-1-5225-1031-4.ch013

Çıtak, L., Akel, V., & Ersoy, E. (2018). Investors' Reactions to the Announcement of New Constituents of BIST Sustainability Index: An Analysis by Event Study and Mean-Median Tests. In M. Risso & S. Testarmata (Eds.), *Value Sharing for Sustainable and Inclusive Development* (pp. 270–289). Hershey, PA: IGI Global. doi:10.4018/978-1-5225-3147-0.ch012

Crossland, M. D. (2009). Geographic Information Systems as Decision Tools. In M. Khosrow-Pour, D.B.A. (Ed.), Encyclopedia of Information Science and Technology, Second Edition (pp. 1630-1633). Hershey, PA: IGI Global. doi:10.4018/978-1-60566-026-4.ch257

Crossland, M. D., Herschel, R. T., Perkins, W. C., & Scudder, J. N. (2002). Geographic Information Systems: How Cognitive Style Impacts Decision-Making Effectiveness. In M. Mahmood (Ed.), *Advanced Topics in End User Computing* (Vol. 1, pp. 285–301). Hershey, PA: IGI Global. doi:10.4018/978-1-930708-42-6.ch017

D'Aleo, V., D'Aleo, F., & Bonanno, R. (2018). New Food Industries Toward a New Level of Sustainable Supply: Success Stories, Business Models, and Strategies. In V. Erokhin (Ed.), *Establishing Food Security and Alternatives to International Trade in Emerging Economies* (pp. 74–97). Hershey, PA: IGI Global. doi:10.4018/978-1-5225-2733-6.ch004

Dagevos, H., & Reinders, M. J. (2018). Flexitarianism and Social Marketing: Reflections on Eating Meat in Moderation. In D. Bogueva, D. Marinova, & T. Raphaely (Eds.), *Handbook of Research on Social Marketing and Its Influence on Animal Origin Food Product Consumption* (pp. 105–120). Hershey, PA: IGI Global. doi:10.4018/978-1-5225-4757-0.ch007

Deenapanray, P. N., & Ramma, I. (2017). Adaptations to Climate Change and Climate Variability in the Agriculture Sector in Mauritius: Lessons from a Technical Needs Assessment. In I. Management Association (Ed.), Natural Resources Management: Concepts, Methodologies, Tools, and Applications (pp. 655-680). Hershey, PA: IGI Global. doi:10.4018/978-1-5225-0803-8.ch030

Deenapanray, P. N., & Ramma, I. (2017). Adaptations to Climate Change and Climate Variability in the Agriculture Sector in Mauritius: Lessons from a Technical Needs Assessment. In I. Management Association (Ed.), Natural Resources Management: Concepts, Methodologies, Tools, and Applications (pp. 655-680). Hershey, PA: IGI Global. doi:10.4018/978-1-5225-0803-8.ch030

Deshpande, S., Basu, S. K., Li, X., & Chen, X. (2017). Smart, Innovative and Intelligent Technologies Used in Drug Designing. In I. Management Association (Ed.), Pharmaceutical Sciences: Breakthroughs in Research and Practice (pp. 1175-1191). Hershey, PA: IGI Global. doi:10.4018/978-1-5225-1762-7.ch045

Dlamini, P. N. (2017). Use of Information and Communication Technologies Tools to Capture, Store, and Disseminate Indigenous Knowledge: A Literature Review. In P. Ngulube (Ed.), *Handbook of Research on Theoretical Perspectives on Indigenous Knowledge Systems in Developing Countries* (pp. 225–247). Hershey, PA: IGI Global. doi:10.4018/978-1-5225-0833-5.ch010

Dolejsova, M., & Kera, D. (2017). The Fermentation GutHub Project and the Internet of Microbes. In S. Konomi & G. Roussos (Eds.), *Enriching Urban Spaces with Ambient Computing, the Internet of Things, and Smart City Design* (pp. 25–46). Hershey, PA: IGI Global. doi:10.4018/978-1-5225-0827-4.ch002

Dolunay, O. (2018). A Paradigm Shift: Empowering Farmers to Eliminate the Waste in the Form of Fresh Water and Energy Through the Implementation of 4R+T. In I. Management Association (Ed.), Sustainable Development: Concepts, Methodologies, Tools, and Applications (pp. 882-892). Hershey, PA: IGI Global. doi:10.4018/978-1-5225-3817-2.ch039

Dube, P., Heijman, W. J., Ihle, R., & Ochieng, J. (2018). The Potential of Traditional Leafy Vegetables for Improving Food Security in Africa. In V. Erokhin (Ed.), *Establishing Food Security and Alternatives to International Trade in Emerging Economies* (pp. 220–243). Hershey, PA: IGI Global. doi:10.4018/978-1-5225-2733-6.ch011

Duruji, M. M., & Urenma, D. F. (2017). The Environmentalism and Politics of Climate Change: A Study of the Process of Global Convergence through UNFCCC Conferences. In I. Management Association (Ed.), *Natural Resources Management: Concepts, Methodologies, Tools, and Applications* (pp. 77-108). Hershey, PA: IGI Global. 10.4018/978-1-5225-0803-8.ch004

Dutta, U. (2017). Agro-Geoinformatics, Potato Cultivation, and Climate Change. In S. Londhe (Ed.), *Sustainable Potato Production and the Impact of Climate Change* (pp. 247–271). Hershey, PA: IGI Global. doi:10.4018/978-1-5225-1715-3.ch012

Edirisinghe, R., Stranieri, A., & Wickramasinghe, N. (2017). A Taxonomy for mHealth. In N. Wickramasinghe (Ed.), *Handbook of Research on Healthcare Administration and Management* (pp. 596–615). Hershey, PA: IGI Global. doi:10.4018/978-1-5225-0920-2.ch036

Ekpeni, N. M., & Ayeni, A. O. (2018). Global Natural Hazard and Disaster Vulnerability Management. In A. Eneanya (Ed.), *Handbook of Research on Environmental Policies for Emergency Management and Public Safety* (pp. 83–104). Hershey, PA: IGI Global. doi:10.4018/978-1-5225-3194-4.ch005

Ene, C., Voica, M. C., & Panait, M. (2017). Green Investments and Food Security: Opportunities and Future Directions in the Context of Sustainable Development. In M. Mieila (Ed.), *Measuring Sustainable Development and Green Investments in Contemporary Economies* (pp. 163–200). Hershey, PA: IGI Global. doi:10.4018/978-1-5225-2081-8.ch007

Escamilla, I., Ruíz, M. T., Ibarra, M. M., Soto, V. L., Quintero, R., & Guzmán, G. (2018). Geocoding Tweets Based on Semantic Web and Ontologies. In M. Lytras, N. Aljohani, E. Damiani, & K. Chui (Eds.), *Innovations, Developments, and Applications of Semantic Web and Information Systems* (pp. 372–392). Hershey, PA: IGI Global. doi:10.4018/978-1-5225-5042-6.ch014

Escribano, A. J. (2018). Marketing Strategies for Trendy Animal Products: Sustainability as a Core. In F. Quoquab, R. Thurasamy, & J. Mohammad (Eds.), *Driving Green Consumerism Through Strategic Sustainability Marketing* (pp. 169–203). Hershey, PA: IGI Global. doi:10.4018/978-1-5225-2912-5.ch010

Eudoxie, G., & Roopnarine, R. (2017). Climate Change Adaptation and Disaster Risk Management in the Caribbean. In W. Ganpat & W. Isaac (Eds.), *Environmental Sustainability and Climate Change Adaptation Strategies* (pp. 97–125). Hershey, PA: IGI Global. doi:10.4018/978-1-5225-1607-1.ch004

Farmer, L. S. (2017). Data Analytics for Strategic Management: Getting the Right Data. In V. Wang (Ed.), *Encyclopedia of Strategic Leadership and Management* (pp. 810–822). Hershey, PA: IGI Global. doi:10.4018/978-1-5225-1049-9.ch056

Fattal, L. R. (2017). Catastrophe: An Uncanny Catalyst for Creativity. In R. Shin (Ed.), *Convergence of Contemporary Art, Visual Culture, and Global Civic Engagement* (pp. 244–262). Hershey, PA: IGI Global. doi:10.4018/978-1-5225-1665-1.ch014

Forti, I. (2017). A Cross Reading of Landscape through Digital Landscape Models: The Case of Southern Garda. In A. Ippolito (Ed.), *Handbook of Research on Emerging Technologies for Architectural and Archaeological Heritage* (pp. 532–561). Hershey, PA: IGI Global. doi:10.4018/978-1-5225-0675-1.ch018

Gharbi, A., De Runz, C., & Akdag, H. (2017). Urban Development Modelling: A Survey. In S. Faiz & K. Mahmoudi (Eds.), *Handbook of Research on Geographic Information Systems Applications and Advancements* (pp. 96–124). Hershey, PA: IGI Global. doi:10.4018/978-1-5225-0937-0.ch004

Ghosh, I., & Ghoshal, I. (2018). Implications of Trade Liberalization for Food Security Under the ASEAN-India Strategic Partnership: A Gravity Model Approach. In V. Erokhin (Ed.), *Establishing Food Security and Alternatives to International Trade in Emerging Economies* (pp. 98–118). Hershey, PA: IGI Global. doi:10.4018/978-1-5225-2733-6.ch005

Glaves, H. M. (2017). Developing a Common Global Framework for Marine Data Management. In P. Diviacco, A. Leadbetter, & H. Glaves (Eds.), *Oceanographic and Marine Cross-Domain Data Management for Sustainable Development* (pp. 47–68). Hershey, PA: IGI Global. doi:10.4018/978-1-5225-0700-0.ch003

Godulla, A., & Wolf, C. (2018). Future of Food: Transmedia Strategies of National Geographic. In R. Gambarato & G. Alzamora (Eds.), *Exploring Transmedia Journalism in the Digital Age* (pp. 162–182). Hershey, PA: IGI Global. doi:10.4018/978-1-5225-3781-6.ch010

Gomes, P. P. (2018). Food and Environment: A Review on the Sustainability of Six Different Dietary Patterns. In A. Obayelu (Ed.), *Food Systems Sustainability and Environmental Policies in Modern Economies* (pp. 15–31). Hershey, PA: IGI Global. doi:10.4018/978-1-5225-3631-4.ch002

Gonzalez-Feliu, J. (2018). Sustainability Evaluation of Green Urban Logistics Systems: Literature Overview and Proposed Framework. In A. Paul, D. Bhattacharyya, & S. Anand (Eds.), *Green Initiatives for Business Sustainability and Value Creation* (pp. 103–134). Hershey, PA: IGI Global. doi:10.4018/978-1-5225-2662-9.ch005

Goodland, R. (2017). A Fresh Look at Livestock Greenhouse Gas Emissions and Mitigation. In I. Management Association (Ed.), Natural Resources Management: Concepts, Methodologies, Tools, and Applications (pp. 124-139). Hershey, PA: IGI Global. doi:10.4018/978-1-5225-0803-8.ch006

Goundar, S., & Appana, S. (2018). Mainstreaming Development Policies for Climate Change in Fiji: A Policy Gap Analysis and the Role of ICTs. In I. Management Association (Ed.), Sustainable Development: Concepts, Methodologies, Tools, and Applications (pp. 402-432). Hershey, PA: IGI Global. doi:10.4018/978-1-5225-3817-2.ch020

Granell-Canut, C., & Aguilar-Moreno, E. (2018). Geospatial Influence in Science Mapping. In M. Khosrow-Pour, D.B.A. (Ed.), Encyclopedia of Information Science and Technology, Fourth Edition (pp. 3473-3483). Hershey, PA: IGI Global. doi:10.4018/978-1-5225-2255-3.ch302

Grigelis, A., Blažauskas, N., Gelumbauskaitė, L. Ž., Gulbinskas, S., Suzdalev, S., & Ferrarin, C. (2017). Marine Environment Data Management Related to the Human Activity in the South-Eastern Baltic Sea (The Lithuanian Sector). In P. Diviacco, A. Leadbetter, & H. Glaves (Eds.), *Oceanographic and Marine Cross-Domain Data Management for Sustainable Development* (pp. 282–302). Hershey, PA: IGI Global. doi:10.4018/978-1-5225-0700-0.ch012

Guma, I. P., Rwashana, A. S., & Oyo, B. (2018). Food Security Policy Analysis Using System Dynamics: The Case of Uganda. *International Journal of Information Technologies and Systems Approach*, *11*(1), 72–90. doi:10.4018/IJITSA.2018010104

Guma, I. P., Rwashana, A. S., & Oyo, B. (2018). Food Security Indicators for Subsistence Farmers Sustainability: A System Dynamics Approach. *International Journal of System Dynamics Applications*, *7*(1), 45–64. doi:10.4018/IJSDA.2018010103

Gupta, P., & Goyal, S. (2017). Wildlife Habitat Evaluation. In A. Santra & S. Mitra (Eds.), *Remote Sensing Techniques and GIS Applications in Earth and Environmental Studies* (pp. 258–264). Hershey, PA: IGI Global. doi:10.4018/978-1-5225-1814-3.ch013

Hanson, T., & Hildebrand, E. (2018). GPS Travel Diaries in Rural Transportation Research: A Focus on Older Drivers. In I. Management Association (Ed.), Intelligent Transportation and Planning: Breakthroughs in Research and Practice (pp. 609-625). Hershey, PA: IGI Global. doi:10.4018/978-1-5225-5210-9.ch027

Hartman, M. B. (2017). Research-Based Climate Change Public Education Programs. In I. Management Association (Ed.), Natural Resources Management: Concepts, Methodologies, Tools, and Applications (pp. 992-1003). Hershey, PA: IGI Global. doi:10.4018/978-1-5225-0803-8.ch046

Hashim, N. (2017). Zanzibari Seaweed: Global Climate Change and the Promise of Adaptation. In I. Management Association (Ed.), Natural Resources Management: Concepts, Methodologies, Tools, and Applications (pp. 365-391). Hershey, PA: IGI Global. doi:10.4018/978-1-5225-0803-8.ch019

Herrera, J. E., Argüello, L. V., Gonzalez-Feliu, J., & Jaimes, W. A. (2017). Decision Support System Design Requirements, Information Management, and Urban Logistics Efficiency: Case Study of Bogotá, Colombia. In G. Jamil, A. Soares, & C. Pessoa (Eds.), *Handbook of Research on Information Management for Effective Logistics and Supply Chains* (pp. 223–238). Hershey, PA: IGI Global. doi:10.4018/978-1-5225-0973-8.ch012

Huizinga, T., Ayanso, A., Smoor, M., & Wronski, T. (2017). Exploring Insurance and Natural Disaster Tweets Using Text Analytics. *International Journal of Business Analytics*, *4*(1), 1–17. doi:10.4018/IJBAN.2017010101

Hung, K., Kalantari, M., & Rajabifard, A. (2017). An Integrated Method for Assessing the Text Content Quality of Volunteered Geographic Information in Disaster Management. *International Journal of Information Systems for Crisis Response and Management*, *9*(2), 1–17. doi:10.4018/IJISCRAM.2017040101

Husnain, A., & Avdic, A. (2018). Identifying the Contemporary Status of E-Service Sustainability Research. In I. Management Association (Ed.), Sustainable Development: Concepts, Methodologies, Tools, and Applications (pp. 467-485). Hershey, PA: IGI Global. doi:10.4018/978-1-5225-3817-2.ch022

Iarossi, M. P., & Ferro, L. (2017). "The Past is Never Dead. It's Not Even Past": Virtual Archaeological Promenade. In A. Ippolito & M. Cigola (Eds.), *Handbook of Research on Emerging Technologies for Digital Preservation and Information Modeling* (pp. 228–255). Hershey, PA: IGI Global. doi:10.4018/978-1-5225-0680-5.ch010

Ignjatijević, S., & Cvijanović, D. (2018). Analysis of Serbian Honey Production and Exports. In *Exploring the Global Competitiveness of Agri-Food Sectors and Serbia's Dominant Presence: Emerging Research and Opportunities* (pp. 109–139). Hershey, PA: IGI Global. doi:10.4018/978-1-5225-2762-6.ch005

Jana, S. K., & Karmakar, A. K. (2017). Globalization, Governance, and Food Security: The Case of BRICS. In I. Management Association (Ed.), Natural Resources Management: Concepts, Methodologies, Tools, and Applications (pp. 692-712). Hershey, PA: IGI Global. doi:10.4018/978-1-5225-0803-8.ch032

Jana, S. K., & Karmakar, A. K. (2017). Food Security in Asia: Is There Convergence? In I. Management Association (Ed.), Natural Resources Management: Concepts, Methodologies, Tools, and Applications (pp. 109-123). Hershey, PA: IGI Global. doi:10.4018/978-1-5225-0803-8.ch005

John, J., & Kumar, S. (2018). A Locational Decision Making Framework for Shipbreaking Under Multiple Criteria. In I. Management Association (Ed.), Operations and Service Management: Concepts, Methodologies, Tools, and Applications (pp. 504-527). Hershey, PA: IGI Global. doi:10.4018/978-1-5225-3909-4.ch024

John, J., & Srivastava, R. K. (2018). Decision Insights for Shipbreaking using Environmental Impact Assessment: Review and Perspectives. *International Journal of Strategic Decision Sciences*, *9*(1), 45–62. doi:10.4018/IJSDS.2018010104

Joshi, Y., & Rahman, Z. (2018). Determinants of Sustainable Consumption Behaviour: Review and Conceptual Framework. In A. Paul, D. Bhattacharyya, & S. Anand (Eds.), *Green Initiatives for Business Sustainability and Value Creation* (pp. 239–262). Hershey, PA: IGI Global. doi:10.4018/978-1-5225-2662-9.ch011

Juma, D. W., Reuben, M., Wang, H., & Li, F. (2018). Adaptive Coevolution: Realigning the Water Governance Regime to the Changing Climate. In I. Management Association (Ed.), Hydrology and Water Resource Management: Breakthroughs in Research and Practice (pp. 346-357). Hershey, PA: IGI Global. doi:10.4018/978-1-5225-3427-3.ch014

K., S., & Tripathy, B. K. (2018). Neighborhood Rough-Sets-Based Spatial Data Analytics. In M. Khosrow-Pour, D.B.A. (Ed.), *Encyclopedia of Information Science and Technology, Fourth Edition* (pp. 1835-1844). Hershey, PA: IGI Global. doi:10.4018/978-1-5225-2255-3.ch160

Kabir, F. (2018). Towards a More Gender-Inclusive Climate Change Policy. In N. Mahtab, T. Haque, I. Khan, M. Islam, & I. Wahid (Eds.), *Handbook of Research on Women's Issues and Rights in the Developing World* (pp. 354–369). Hershey, PA: IGI Global. doi:10.4018/978-1-5225-3018-3.ch022

Kabir, F. (2018). Towards a More Gender-Inclusive Climate Change Policy. In I. Management Association (Ed.), Climate Change and Environmental Concerns: Breakthroughs in Research and Practice (pp. 525-540). Hershey, PA: IGI Global. doi:10.4018/978-1-5225-5487-5.ch027

Kanyamuka, J. S., Jumbe, C. B., & Ricker-Gilbert, J. (2018). Making Agricultural Input Subsidies More Effective and Profitable in Africa: The Role of Complementary Interventions. In A. Obayelu (Ed.), *Food Systems Sustainability and Environmental Policies in Modern Economies* (pp. 172–187). Hershey, PA: IGI Global. doi:10.4018/978-1-5225-3631-4.ch008

Karimi, H., & Gholamrezafahimi, F. (2017). Study of Integrated Coastal Zone Management and Its Environmental Effects: A Case of Iran. In R. Singh, A. Singh, & V. Srivastava (Eds.), *Environmental Issues Surrounding Human Overpopulation* (pp. 64–88). Hershey, PA: IGI Global. doi:10.4018/978-1-5225-1683-5.ch004

Kaya, I. R., Hutabarat, J., & Bambang, A. N. (2018). "Sasi": A New Path to Sustain Seaweed Farming From Up-Stream to Down-Stream in Kotania Bay, Molucass. *International Journal of Social Ecology and Sustainable Development*, *9*(2), 28–36. doi:10.4018/IJSESD.2018040103

Khader, V. (2018). Technologies for Food, Health, Livelihood, and Nutrition Security. In I. Management Association (Ed.), Food Science and Nutrition: Breakthroughs in Research and Practice (pp. 94-112). Hershey, PA: IGI Global. doi:10.4018/978-1-5225-5207-9.ch005

Kocadağlı, A. Y. (2017). The Temporal and Spatial Development of Organic Agriculture in Turkey. In W. Ganpat, R. Dyer, & W. Isaac (Eds.), *Agricultural Development and Food Security in Developing Nations* (pp. 130–156). Hershey, PA: IGI Global. doi:10.4018/978-1-5225-0942-4.ch006

Koundouri, P., Giannouli, A., & Souliotis, I. (2017). An Integrated Approach for Sustainable Environmental and Socio-Economic Development Using Offshore Infrastructure. In I. Management Association (Ed.), Renewable and Alternative Energy: Concepts, Methodologies, Tools, and Applications (pp. 1581-1601). Hershey, PA: IGI Global. doi:10.4018/978-1-5225-1671-2.ch056

Kumar, A., & Dash, M. K. (2017). Sustainability and Future Generation Infrastructure on Digital Platform: A Study of Generation Y. In N. Ray (Ed.), *Business Infrastructure for Sustainability in Developing Economies* (pp. 124–142). Hershey, PA: IGI Global. doi:10.4018/978-1-5225-2041-2.ch007

Kumar, A., Mukherjee, A. B., & Krishna, A. P. (2017). Application of Conventional Data Mining Techniques and Web Mining to Aid Disaster Management. In A. Kumar (Ed.), *Web Usage Mining Techniques and Applications Across Industries* (pp. 138–167). Hershey, PA: IGI Global. doi:10.4018/978-1-5225-0613-3.ch006

Kumar, C. P. (2017). Impact of Climate Change on Groundwater Resources. In I. Management Association (Ed.), Natural Resources Management: Concepts, Methodologies, Tools, and Applications (pp. 1094-1120). Hershey, PA: IGI Global. doi:10.4018/978-1-5225-0803-8.ch052

Kumari, S., & Patil, Y. (2017). Achieving Climate Smart Agriculture with a Sustainable Use of Water: A Conceptual Framework for Sustaining the Use of Water for Agriculture in the Era of Climate Change. In P. Rao & Y. Patil (Eds.), *Reconsidering the Impact of Climate Change on Global Water Supply, Use, and Management* (pp. 122–143). Hershey, PA: IGI Global. doi:10.4018/978-1-5225-1046-8.ch008

Kumari, S., & Patil, Y. (2018). Achieving Climate Smart Agriculture With a Sustainable Use of Water: A Conceptual Framework for Sustaining the Use of Water for Agriculture in the Era of Climate Change. In I. Management Association (Ed.), Climate Change and Environmental Concerns: Breakthroughs in Research and Practice (pp. 111-133). Hershey, PA: IGI Global. doi:10.4018/978-1-5225-5487-5.ch006

Kursah, M. B. (2017). Least-Cost Pipeline using Geographic Information System: The Limit to Technicalities. *International Journal of Applied Geospatial Research*, *8*(3), 1–15. doi:10.4018/ijagr.2017070101

Lahiri, S., Ghosh, D., & Bhakta, J. N. (2017). Role of Microbes in Eco-Remediation of Perturbed Aquatic Ecosystem. In J. Bhakta (Ed.), *Handbook of Research on Inventive Bioremediation Techniques* (pp. 70–107). Hershey, PA: IGI Global. doi:10.4018/978-1-5225-2325-3.ch004

Lallo, C. H., Smalling, S., Facey, A., & Hughes, M. (2018). The Impact of Climate Change on Small Ruminant Performance in Caribbean Communities. In I. Management Association (Ed.), Climate Change and Environmental Concerns: Breakthroughs in Research and Practice (pp. 193-218). Hershey, PA: IGI Global. doi:10.4018/978-1-5225-5487-5.ch010

Laurini, R. (2017). Nature of Geographic Knowledge Bases. In S. Faiz & K. Mahmoudi (Eds.), *Handbook of Research on Geographic Information Systems Applications and Advancements* (pp. 29–60). Hershey, PA: IGI Global. doi:10.4018/978-1-5225-0937-0.ch002

Lawrence, J., Simpson, L., & Piggott, A. (2017). Protected Agriculture: A Climate Change Adaptation for Food and Nutrition Security. In I. Management Association (Ed.), Natural Resources Management: Concepts, Methodologies, Tools, and Applications (pp. 140-158). Hershey, PA: IGI Global. doi:10.4018/978-1-5225-0803-8.ch007

Leadbetter, A., Cheatham, M., Shepherd, A., & Thomas, R. (2017). Linked Ocean Data 2.0. In P. Diviacco, A. Leadbetter, & H. Glaves (Eds.), *Oceanographic and Marine Cross-Domain Data Management for Sustainable Development* (pp. 69–99). Hershey, PA: IGI Global. doi:10.4018/978-1-5225-0700-0.ch004

Leipnik, M. R., & Mehta, S. S. (2005). Geographic Information Systems (GIS) in E-Marketing. In I. Clarke III & T. Flaherty (Eds.), *Advances in Electronic Marketing* (pp. 193–210). Hershey, PA: IGI Global. doi:10.4018/978-1-59140-321-0.ch011

Lucas, M. R., Rego, C., Vieira, C., & Vieira, I. (2017). Proximity and Cooperation for Innovative Regional Development: The Case of the Science and Technology Park of Alentejo. In L. Carvalho (Ed.), *Handbook of Research on Entrepreneurial Development and Innovation Within Smart Cities* (pp. 199–228). Hershey, PA: IGI Global. doi:10.4018/978-1-5225-1978-2.ch010

Ma, X., Beaulieu, S. E., Fu, L., Fox, P., Di Stefano, M., & West, P. (2017). Documenting Provenance for Reproducible Marine Ecosystem Assessment in Open Science. In P. Diviacco, A. Leadbetter, & H. Glaves (Eds.), *Oceanographic and Marine Cross-Domain Data Management for Sustainable Development* (pp. 100–126). Hershey, PA: IGI Global. doi:10.4018/978-1-5225-0700-0.ch005

Mabe, L. K., & Oladele, O. I. (2017). Application of Information Communication Technologies for Agricultural Development through Extension Services: A Review. In T. Tossy (Ed.), *Information Technology Integration for Socio-Economic Development* (pp. 52–101). Hershey, PA: IGI Global. doi:10.4018/978-1-5225-0539-6.ch003

Malomo, B. I. (2018). A Review of Psychological Resilience as a Response to Natural Hazards in Nigeria. In A. Eneanya (Ed.), *Handbook of Research on Environmental Policies for Emergency Management and Public Safety* (pp. 147–165). Hershey, PA: IGI Global. doi:10.4018/978-1-5225-3194-4.ch008

Manchiraju, S. (2018). Predicting Behavioral Intentions Toward Sustainable Fashion Consumption: A Comparison of Attitude-Behavior and Value-Behavior Consistency Models. In I. Management Association (Ed.), Fashion and Textiles: Breakthroughs in Research and Practice (pp. 1-21). Hershey, PA: IGI Global. doi:10.4018/978-1-5225-3432-7.ch001

Manzella, G. M., Bartolini, R., Bustaffa, F., D'Angelo, P., De Mattei, M., Frontini, F., ... Spada, A. (2017). Semantic Search Engine for Data Management and Sustainable Development: Marine Planning Service Platform. In P. Diviacco, A. Leadbetter, & H. Glaves (Eds.), *Oceanographic and Marine Cross-Domain Data Management for Sustainable Development* (pp. 127–154). Hershey, PA: IGI Global. doi:10.4018/978-1-5225-0700-0.ch006

Mbonigaba, J. (2018). Comparing the Effects of Unsustainable Production and Consumption of Food on Health and Policy Across Developed and Less Developed Countries. In A. Obayelu (Ed.), *Food Systems Sustainability and Environmental Policies in Modern Economies* (pp. 124–158). Hershey, PA: IGI Global. doi:10.4018/978-1-5225-3631-4.ch006

McKeown, A. E. (2017). Nurses, Healthcare, and Environmental Pollution and Solutions: Breaking the Cycle of Harm. In I. Management Association (Ed.), Natural Resources Management: Concepts, Methodologies, Tools, and Applications (pp. 392-415). Hershey, PA: IGI Global. doi:10.4018/978-1-5225-0803-8.ch020

Mennecke, B. E., & West, L. A. Jr. (2001). Geographic Information Systems in Developing Countries: Issues in Data Collection, Implementation and Management. *Journal of Global Information Management*, *9*(4), 44–54. doi:10.4018/jgim.2001100103

Mili, B., Barua, A., & Katyaini, S. (2017). Climate Change and Adaptation through the Lens of Capability Approach: A Case Study from Darjeeling, Eastern Himalaya. In I. Management Association (Ed.), Natural Resources Management: Concepts, Methodologies, Tools, and Applications (pp. 1351-1365). Hershey, PA: IGI Global. doi:10.4018/978-1-5225-0803-8.ch064

Mir, S. A., Shah, M. A., Mir, M. M., & Iqbal, U. (2017). New Horizons of Nanotechnology in Agriculture and Food Processing Industry. In B. Nayak, A. Nanda, & M. Bhat (Eds.), *Integrating Biologically-Inspired Nanotechnology into Medical Practice* (pp. 230–258). Hershey, PA: IGI Global. doi:10.4018/978-1-5225-0610-2.ch009

Mujere, N., & Moyce, W. (2017). Climate Change Impacts on Surface Water Quality. In W. Ganpat & W. Isaac (Eds.), *Environmental Sustainability and Climate Change Adaptation Strategies* (pp. 322–340). Hershey, PA: IGI Global. doi:10.4018/978-1-5225-1607-1.ch012

Mukherjee, A. B., Krishna, A. P., & Patel, N. (2018). Geospatial Technology for Urban Sciences. In *Geospatial Technologies in Urban System Development: Emerging Research and Opportunities* (pp. 99–120). Hershey, PA: IGI Global. doi:10.4018/978-1-5225-3683-3.ch005

Nagarajan, S. K., & Sangaiah, A. K. (2017). Vegetation Index: Ideas, Methods, Influences, and Trends. In N. Kumar, A. Sangaiah, M. Arun, & S. Anand (Eds.), *Advanced Image Processing Techniques and Applications* (pp. 347–386). Hershey, PA: IGI Global. doi:10.4018/978-1-5225-2053-5.ch016

Naraine, L., & Meehan, K. (2017). Strengthening Food Security with Sustainable Practices by Smallholder Farmers in Lesser Developed Economies. In W. Ganpat, R. Dyer, & W. Isaac (Eds.), *Agricultural Development and Food Security in Developing Nations* (pp. 57–81). Hershey, PA: IGI Global. doi:10.4018/978-1-5225-0942-4.ch003

Naraine, L., & Meehan, K. (2017). Strengthening Food Security with Sustainable Practices by Smallholder Farmers in Lesser Developed Economies. In W. Ganpat, R. Dyer, & W. Isaac (Eds.), *Agricultural Development and Food Security in Developing Nations* (pp. 57–81). Hershey, PA: IGI Global. doi:10.4018/978-1-5225-0942-4.ch003

Naraine, L., & Meehan, K. (2017). Strengthening Food Security with Sustainable Practices by Smallholder Farmers in Lesser Developed Economies. In W. Ganpat, R. Dyer, & W. Isaac (Eds.), *Agricultural Development and Food Security in Developing Nations* (pp. 57–81). Hershey, PA: IGI Global. doi:10.4018/978-1-5225-0942-4.ch003

Neufeld, D. J., & Griffith, S. (2006). Isobord's Geographic Information System (GIS) Solution. In M. Khosrow-Pour, D.B.A. (Ed.), Cases on Database Technologies and Applications (pp. 64-80). Hershey, PA: IGI Global. doi:10.4018/978-1-59904-399-9.ch004

Nikolaou, K., Tsakiridou, E., Anastasiadis, F., & Mattas, K. (2017). Exploring Alternative Distribution Channels of Agricultural Products. *International Journal of Food and Beverage Manufacturing and Business Models*, *2*(2), 36–66. doi:10.4018/IJFBMBM.2017070103

Nishat, K. J., & Rahman, M. S. (2018). Disaster, Vulnerability, and Violence Against Women: Global Findings and a Research Agenda for Bangladesh. In N. Mahtab, T. Haque, I. Khan, M. Islam, & I. Wahid (Eds.), *Handbook of Research on Women's Issues and Rights in the Developing World* (pp. 235–250). Hershey, PA: IGI Global. doi:10.4018/978-1-5225-3018-3.ch014

O'Hara, S., Jones, D., & Trobman, H. B. (2018). Building an Urban Food System Through UDC Food Hubs. In A. Burtin, J. Fleming, & P. Hampton-Garland (Eds.), *Changing Urban Landscapes Through Public Higher Education* (pp. 116–143). Hershey, PA: IGI Global. doi:10.4018/978-1-5225-3454-9.ch006

Obayelu, A. E. (2018). Integrating Environment, Food Systems, and Sustainability in Feeding the Growing Population in Developing Countries. In A. Obayelu (Ed.), *Food Systems Sustainability and Environmental Policies in Modern Economies* (pp. 1–14). Hershey, PA: IGI Global. doi:10.4018/978-1-5225-3631-4.ch001

Othman, R., Nath, N., & Laswad, F. (2018). Environmental Reporting and Accounting: Sustainability Hybridisation. In G. Azevedo, J. da Silva Oliveira, R. Marques, & A. Ferreira (Eds.), *Handbook of Research on Modernization and Accountability in Public Sector Management* (pp. 130–158). Hershey, PA: IGI Global. doi:10.4018/978-1-5225-3731-1.ch007

Padigala, B. S. (2018). Traditional Water Management System for Climate Change Adaptation in Mountain Ecosystems. In I. Management Association (Ed.), Climate Change and Environmental Concerns: Breakthroughs in Research and Practice (pp. 630-655). Hershey, PA: IGI Global. doi:10.4018/978-1-5225-5487-5.ch033

Panda, C. K. (2018). Mobile Phone Usage in Agricultural Extension in India: The Current and Future Perspective. In F. Mtenzi, G. Oreku, D. Lupiana, & J. Yonazi (Eds.), *Mobile Technologies and Socio-Economic Development in Emerging Nations* (pp. 1–21). Hershey, PA: IGI Global. doi:10.4018/978-1-5225-4029-8.ch001

Pandian, S. L., Yarrakula, K., & Chaudhury, P. (2018). GIS-Based Decision Support System for Village Level: A Case Study in Andhra Pradesh. In S. Chaudhuri & N. Ray (Eds.), *GIS Applications in the Tourism and Hospitality Industry* (pp. 275–295). Hershey, PA: IGI Global. doi:10.4018/978-1-5225-5088-4.ch012

Quaranta, G., & Salvia, R. (2017). Social-Based Product Innovation and Governance in The Milk Sector: The Case of Carciocacio and Innonatura. In T. Tarnanidis, M. Vlachopoulou, & J. Papathanasiou (Eds.), *Driving Agribusiness With Technology Innovations* (pp. 293–310). Hershey, PA: IGI Global. doi:10.4018/978-1-5225-2107-5.ch015

Rahman, M. K., Schmidlin, T. W., Munro-Stasiuk, M. J., & Curtis, A. (2017). Geospatial Analysis of Land Loss, Land Cover Change, and Landuse Patterns of Kutubdia Island, Bangladesh. *International Journal of Applied Geospatial Research, 8*(2), 45–60. doi:10.4018/IJAGR.2017040104

Rajack-Talley, T. A. (2017). Agriculture, Trade Liberalization and Poverty in the ACP Countries. In W. Ganpat, R. Dyer, & W. Isaac (Eds.), *Agricultural Development and Food Security in Developing Nations* (pp. 1–20). Hershey, PA: IGI Global. doi:10.4018/978-1-5225-0942-4.ch001

Rajamanickam, S. (2018). Exploring Landscapes in Regional Convergence: Environment and Sustainable Development in South Asia. In I. Management Association (Ed.), Sustainable Development: Concepts, Methodologies, Tools, and Applications (pp. 1051-1087). Hershey, PA: IGI Global. doi:10.4018/978-1-5225-3817-2.ch047

Rizvi, S. M., & Dearden, A. (2018). KHETI: ICT Solution for Agriculture Extension and Its Replication in Open and Distance Learning. In U. Pandey & V. Indrakanti (Eds.), *Open and Distance Learning Initiatives for Sustainable Development* (pp. 163–174). Hershey, PA: IGI Global. doi:10.4018/978-1-5225-2621-6.ch008

Roşu, L., & Macarov, L. I. (2017). Management of Drought and Floods in the Dobrogea Region. In I. Management Association (Ed.), Agri-Food Supply Chain Management: Breakthroughs in Research and Practice (pp. 372-403). Hershey, PA: IGI Global. doi:10.4018/978-1-5225-1629-3.ch016

Rouzbehani, K., & Rouzbehani, S. (2018). Mapping Women's World: GIS and the Case of Breast Cancer in the US. *International Journal of Public Health Management and Ethics*, *3*(1), 14–25. doi:10.4018/IJPHME.2018010102

Roy, D. (2018). Success Factors of Adoption of Mobile Applications in Rural India: Effect of Service Characteristics on Conceptual Model. In M. Khosrow-Pour, D.B.A. (Ed.), Green Computing Strategies for Competitive Advantage and Business Sustainability (pp. 211-238). Hershey, PA: IGI Global. doi:10.4018/978-1-5225-5017-4.ch010

Sajeva, M., Lemon, M., & Sahota, P. S. (2017). Governance for Food Security: A Framework for Social Learning and Scenario Building. *International Journal of Food and Beverage Manufacturing and Business Models*, *2*(2), 67–84. doi:10.4018/IJFBMBM.2017070104

Sallam, G. A., Youssef, T., Embaby, M. E., & Shaltot, F. (2011). Using Geographic Information System to Infollow the Fertilizers Pollution Migration. In I. Management Association (Ed.), Green Technologies: Concepts, Methodologies, Tools and Applications (pp. 564-586). Hershey, PA: IGI Global. doi:10.4018/978-1-60960-472-1.ch312

Sambhanthan, A., & Potdar, V. (2017). A Study of the Parameters Impacting Sustainability in Information Technology Organizations. *International Journal of Knowledge-Based Organizations*, *7*(3), 27–39. doi:10.4018/IJKBO.2017070103

Sanga, C., Kalungwizi, V. J., & Msuya, C. P. (2017). Bridging Gender Gaps in Provision of Agricultural Extension Service Using ICT: Experiences from Sokoine University of Agriculture (SUA) Farmer Voice Radio (FVR) Project in Tanzania. In I. Management Association (Ed.), Discrimination and Diversity: Concepts, Methodologies, Tools, and Applications (pp. 682-697). Hershey, PA: IGI Global. doi:10.4018/978-1-5225-1933-1.ch031

Santra, A., & Mitra, D. (2017). Role of Remote Sensing in Potential Fishing Zone Forecast. In A. Santra & S. Mitra (Eds.), *Remote Sensing Techniques and GIS Applications in Earth and Environmental Studies* (pp. 243–257). Hershey, PA: IGI Global. doi:10.4018/978-1-5225-1814-3.ch012

Schaap, D. (2017). SeaDataNet: Towards a Pan-European Infrastructure for Marine and Ocean Data Management. In P. Diviacco, A. Leadbetter, & H. Glaves (Eds.), *Oceanographic and Marine Cross-Domain Data Management for Sustainable Development* (pp. 155–177). Hershey, PA: IGI Global. doi:10.4018/978-1-5225-0700-0.ch007

Seckin-Celik, T. (2017). Sustainability Reporting and Sustainability in the Turkish Business Context. In U. Akkucuk (Ed.), *Ethics and Sustainability in Global Supply Chain Management* (pp. 115–132). Hershey, PA: IGI Global. doi:10.4018/978-1-5225-2036-8.ch006

Segbefia, A. Y., Barnes, V. R., Akpalu, L. A., & Mensah, M. (2018). Environmental Location Assessment for Seaweed Cultivation in Ghana: A Spatial Multi-Criteria Approach. *International Journal of Applied Geospatial Research*, *9*(1), 51–64. doi:10.4018/IJAGR.2018010104

Sen, Y. (2018). How to Manage Sustainability: A Framework for Corporate Sustainability Tools. In I. Management Association (Ed.), Sustainable Development: Concepts, Methodologies, Tools, and Applications (pp. 568-589). Hershey, PA: IGI Global. doi:10.4018/978-1-5225-3817-2.ch026

Shamshiry, E., Abdulai, A. M., Mokhtar, M. B., & Komoo, I. (2015). Regional Landfill Site Selection with GIS and Analytical Hierarchy Process Techniques: A Case Study of Langkawi Island, Malaysia. In P. Thomas, M. Srihari, & S. Kaur (Eds.), *Handbook of Research on Cultural and Economic Impacts of the Information Society* (pp. 248–282). Hershey, PA: IGI Global. doi:10.4018/978-1-4666-8598-7.ch011

Sharma, Y. K., Mangla, S. K., Patil, P. P., & Uniyal, S. (2018). Analyzing Sustainable Food Supply Chain Management Challenges in India. In M. Ram & J. Davim (Eds.), *Soft Computing Techniques and Applications in Mechanical Engineering* (pp. 162–180). Hershey, PA: IGI Global. doi:10.4018/978-1-5225-3035-0.ch008

Silvestrelli, P. (2018). The Impact of Events: To Which Extent Are They Sustainable for Tourist Destinations? Some Evidences From Expo Milano 2015. In M. Risso & S. Testarmata (Eds.), *Value Sharing for Sustainable and Inclusive Development* (pp. 185–204). Hershey, PA: IGI Global. doi:10.4018/978-1-5225-3147-0.ch008

Silvius, G. (2018). Sustainability Evaluation of IT/IS Projects. In I. Management Association (Ed.), Sustainable Development: Concepts, Methodologies, Tools, and Applications (pp. 26-40). Hershey, PA: IGI Global. doi:10.4018/978-1-5225-3817-2.ch002

Singh, R., Srivastava, P., Singh, P., Upadhyay, S., & Raghubanshi, A. S. (2017). Human Overpopulation and Food Security: Challenges for the Agriculture Sustainability. In R. Singh, A. Singh, & V. Srivastava (Eds.), *Environmental Issues Surrounding Human Overpopulation* (pp. 12–39). Hershey, PA: IGI Global. doi:10.4018/978-1-5225-1683-5.ch002

Singh, R., Srivastava, P., Singh, P., Upadhyay, S., & Raghubanshi, A. S. (2017). Human Overpopulation and Food Security: Challenges for the Agriculture Sustainability. In R. Singh, A. Singh, & V. Srivastava (Eds.), *Environmental Issues Surrounding Human Overpopulation* (pp. 12–39). Hershey, PA: IGI Global. doi:10.4018/978-1-5225-1683-5.ch002

Srivastava, N. (2017). Climate Change Mitigation: Collective Efforts and Responsibly. In I. Management Association (Ed.), Natural Resources Management: Concepts, Methodologies, Tools, and Applications (pp. 64-76). Hershey, PA: IGI Global. doi:10.4018/978-1-5225-0803-8.ch003

Stanganelli, M., & Gerundo, C. (2017). Understanding the Role of Urban Morphology and Green Areas Configuration During Heat Waves. *International Journal of Agricultural and Environmental Information Systems*, *8*(2), 50–64. doi:10.4018/IJAEIS.2017040104

Stewart, M. K., Hagood, D., & Ching, C. C. (2017). Virtual Games and Real-World Communities: Environments that Constrain and Enable Physical Activity in Games for Health. *International Journal of Game-Based Learning*, *7*(1), 1–19. doi:10.4018/IJGBL.2017010101

Stone, R. J. (2017). Modelling the Frequency of Tropical Cyclones in the Lower Caribbean Region. In W. Ganpat & W. Isaac (Eds.), *Environmental Sustainability and Climate Change Adaptation Strategies* (pp. 341–349). Hershey, PA: IGI Global. doi:10.4018/978-1-5225-1607-1.ch013

Syed, A., & Jabeen, U. A. (2018). Climate Change Impact on Agriculture and Food Security. In A. Eneanya (Ed.), *Handbook of Research on Environmental Policies for Emergency Management and Public Safety* (pp. 223–237). Hershey, PA: IGI Global. doi:10.4018/978-1-5225-3194-4.ch012

Tam, G. C. (2017). The Global View of Sustainability. In *Managerial Strategies and Green Solutions for Project Sustainability* (pp. 1–24). Hershey, PA: IGI Global. doi:10.4018/978-1-5225-2371-0.ch001

Tam, G. C. (2017). Understanding Project Sustainability. In *Managerial Strategies and Green Solutions for Project Sustainability* (pp. 110–139). Hershey, PA: IGI Global. doi:10.4018/978-1-5225-2371-0.ch005

Tam, G. C. (2017). Perspectives on Sustainability. In *Managerial Strategies and Green Solutions for Project Sustainability* (pp. 53–76). Hershey, PA: IGI Global. doi:10.4018/978-1-5225-2371-0.ch003

Tang, M., & Karunanithi, A. T. (2018). Visual Logic Maps (vLms). In *Advanced Concept Maps in STEM Education: Emerging Research and Opportunities* (pp. 108–149). Hershey, PA: IGI Global. doi:10.4018/978-1-5225-2184-6.ch005

Taşçıoğlu, M., & Yener, D. (2018). The Value and Scope of GIS in Marketing and Tourism Management. In S. Chaudhuri & N. Ray (Eds.), *GIS Applications in the Tourism and Hospitality Industry* (pp. 189–211). Hershey, PA: IGI Global. doi:10.4018/978-1-5225-5088-4.ch009

Tianming, G. (2018). Food Security and Rural Development on Emerging Markets of Northeast Asia: Cases of Chinese North and Russian Far East. In V. Erokhin (Ed.), *Establishing Food Security and Alternatives to International Trade in Emerging Economies* (pp. 155–176). Hershey, PA: IGI Global. doi:10.4018/978-1-5225-2733-6.ch008

Tiftikçigil, B. Y., Yaşgül, Y. S., & Güriş, B. (2017). Sustainability of Foreign Trade Deficit in Energy: The Case of Turkey. In N. Ray (Ed.), *Business Infrastructure for Sustainability in Developing Economies* (pp. 94–109). Hershey, PA: IGI Global. doi:10.4018/978-1-5225-2041-2.ch005

Tiwari, S., Vaish, B., & Singh, P. (2018). Population and Global Food Security: Issues Related to Climate Change. In I. Management Association (Ed.), Climate Change and Environmental Concerns: Breakthroughs in Research and Practice (pp. 41-64). Hershey, PA: IGI Global. doi:10.4018/978-1-5225-5487-5.ch003

Toujani, A., & Achour, H. (2018). A Data Mining Framework for Forest Fire Mapping. In I. Management Association (Ed.), Information Retrieval and Management: Concepts, Methodologies, Tools, and Applications (pp. 771-794). Hershey, PA: IGI Global. doi:10.4018/978-1-5225-5191-1.ch033

Tripathy, B., & K., S. B. (2017). Rough Fuzzy Set Theory and Neighbourhood Approximation Based Modelling for Spatial Epidemiology. In I. Management Association (Ed.), *Public Health and Welfare: Concepts, Methodologies, Tools, and Applications* (pp. 1257-1268). Hershey, PA: IGI Global. doi:10.4018/978-1-5225-1674-3.ch058

Trukhachev, A. (2017). New Approaches to Regional Branding through Green Production and Utilization of Existing Natural Advantages. In I. Management Association (Ed.), Advertising and Branding: Concepts, Methodologies, Tools, and Applications (pp. 1758-1778). Hershey, PA: IGI Global. doi:10.4018/978-1-5225-1793-1.ch081

Tsobanoglou, G. O., & Vlachopoulou, E. I. (2017). Social-Ecological Systems in Local Fisheries Communities. In G. Korres, E. Kourliouros, & M. Michailidis (Eds.), *Handbook of Research on Policies and Practices for Sustainable Economic Growth and Regional Development* (pp. 306–316). Hershey, PA: IGI Global. doi:10.4018/978-1-5225-2458-8.ch026

Tsou, M. (2010). Geographic Information Retrieval and Text Mining on Chinese Tourism Web Pages. *International Journal of Information Technology and Web Engineering*, *5*(1), 56–75. doi:10.4018/jitwe.2010010104

Tuydes-Yaman, H., & Karatas, P. (2018). Evaluation of Walkability and Pedestrian Level of Service. In I. Management Association (Ed.), Intelligent Transportation and Planning: Breakthroughs in Research and Practice (pp. 264-291). Hershey, PA: IGI Global. doi:10.4018/978-1-5225-5210-9.ch012

Uddin, S., Chakravorty, S., Ray, A., & Sherpa, K. S. (2018). Optimal Location of Sub-Station Using Q-GIS and Multi-Criteria Decision Making Approach. *International Journal of Decision Support System Technology*, *10*(2), 65–79. doi:10.4018/IJDSST.2018040104

Uddin, S., Chakravorty, S., Sherpa, K. S., & Ray, A. (2018). Power Distribution System Planning Using Q-GIS. *International Journal of Energy Optimization and Engineering*, *7*(2), 61–75. doi:10.4018/IJEOE.2018040103

Uzun, F. V. (2018). Natural Resources Management. In A. Eneanya (Ed.), *Handbook of Research on Environmental Policies for Emergency Management and Public Safety* (pp. 1–21). Hershey, PA: IGI Global. doi:10.4018/978-1-5225-3194-4.ch001

V, M., Agrawal, R., Sharma, V., & T.N., K. (2018). Supply Chain Social Sustainability and Manufacturing. In I. Management Association (Ed.), *Technology Adoption and Social Issues: Concepts, Methodologies, Tools, and Applications* (pp. 226-252). Hershey, PA: IGI Global. doi:10.4018/978-1-5225-5201-7.ch011 van der Vliet-Bakker, J. M. (2017). Environmentally Forced Migration and Human Rights. In C. Akrivopoulou (Ed.), *Defending Human Rights and Democracy in the Era of Globalization* (pp. 146–180). Hershey, PA: IGI Global. doi:10.4018/978-1-5225-0723-9.ch007

Vaskov, A. G., Lin, Z. Y., Tyagunov, M. G., Shestopalova, T. A., & Deryugina, G. V. (2018). Design of Renewable Sources GIS for ASEAN Countries. In V. Kharchenko & P. Vasant (Eds.), *Handbook of Research on Renewable Energy and Electric Resources for Sustainable Rural Development* (pp. 1–25). Hershey, PA: IGI Global. doi:10.4018/978-1-5225-3867-7.ch001

Vázquez, D. G., & Gil, M. T. (2017). Sustainability in Smart Cities: The Case of Vitoria-Gasteiz (Spain) – A Commitment to a New Urban Paradigm. In L. Carvalho (Ed.), *Handbook of Research on Entrepreneurial Development and Innovation Within Smart Cities* (pp. 248–268). Hershey, PA: IGI Global. doi:10.4018/978-1-5225-1978-2.ch012

Wahab, I. N., & Soonthodu, S. (2018). Geographical Information System in Eco-Tourism. In S. Chaudhuri & N. Ray (Eds.), *GIS Applications in the Tourism and Hospitality Industry* (pp. 61–75). Hershey, PA: IGI Global. doi:10.4018/978-1-5225-5088-4.ch003

Weiss-Randall, D. (2018). Cultivating Environmental Justice. In *Utilizing Innovative Technologies to Address the Public Health Impact of Climate Change: Emerging Research and Opportunities* (pp. 110–143). Hershey, PA: IGI Global. doi:10.4018/978-1-5225-3414-3.ch004

Weiss-Randall, D. (2018). Cultivating Resilience. In *Utilizing Innovative Technologies to Address the Public Health Impact of Climate Change: Emerging Research and Opportunities* (pp. 204–235). Hershey, PA: IGI Global. doi:10.4018/978-1-5225-3414-3.ch007

Weiss-Randall, D. (2018). Climate Change Solutions: Where Do We Go From Here? In *Utilizing Innovative Technologies to Address the Public Health Impact of Climate Change: Emerging Research and Opportunities* (pp. 236–268). Hershey, PA: IGI Global. doi:10.4018/978-1-5225-3414-3.ch008

Whyte, K. P., List, M., Stone, J. V., Grooms, D., Gasteyer, S., Thompson, P. B., . . . Bouri, H. (2018). Uberveillance, Standards, and Anticipation: A Case Study on Nanobiosensors in U.S. Cattle. In I. Management Association (Ed.), Biomedical Engineering: Concepts, Methodologies, Tools, and Applications (pp. 577-596). Hershey, PA: IGI Global. doi:10.4018/978-1-5225-3158-6.ch025

Wulff, E. (2017). Data and Operational Oceanography: A Review in Support of Responsible Fisheries and Aquaculture. In P. Diviacco, A. Leadbetter, & H. Glaves (Eds.), *Oceanographic and Marine Cross-Domain Data Management for Sustainable Development* (pp. 303–324). Hershey, PA: IGI Global. doi:10.4018/978-1-5225-0700-0.ch013

Yener, D. (2017). Geographic Information Systems and Its Applications in Marketing Literature. In S. Faiz & K. Mahmoudi (Eds.), *Handbook of Research on Geographic Information Systems Applications and Advancements* (pp. 158–172). Hershey, PA: IGI Global. doi:10.4018/978-1-5225-0937-0.ch006

About the Contributors

Prasenjit Chatterjee is currently working as an Associate Professor of Mechanical Engineering Department at MCKV Institute of Engineering, India. He has over 75 research papers in various international journals and peer reviewed conferences, published by ELSEVIER, Springerlink, Emerald, Inderscience etc. He has received numerous awards including Best Track Paper Award, Outstanding Reviewer Award, Best Paper Award, Bright Researcher Award, Outstanding Researcher Award and University Gold Medal to name a few. He is the Guest Editor of special issues in different Scopus and Emerging Sources Citation Index (Clarivate Analytics) indexed journals. He has edited and authored books on decision-making theories and sustainability.

Morteza Yazdani is a PhD in Business and Management sciences and currently works at Universidad Loyola Andalucia. He worked in University of Toulouse and European University of Madrid as post-doctoral and part time lecturer, respectively. He participates as editorial board of International Journal of Decision support system technology and at the same time reviewer of some journals like expert systems with applications, journal of cleaner production, soft computing, applied soft computing among others. His main research areas are decision making modeling and fuzzy decision system in application of supply chain and energy systems. He has published in journals like Expert systems with applications, soft computing, journal of cleaner production, management decision, clean technologies and environmental policy, etc.

Shankar Chakraborty is a Professor in Production Engineering Department of Jadavpur University. He had been graduated in 1986 from University of Calcutta and had obtained his postgraduate degree from Jadavpur University in 1989. He had been awarded with Ph.D. (Engg.) degree from Jadavpur University in 1994. His research interests include applications of different multi-criteria decision-making methods in manufacturing environment, control chart pattern recognition, and development of MIS and ERP systems for diverse engineering applications. He has guided several

M.E. and Ph.D. (Engg.) theses, and published numerous papers in international journals. He is also a regular reviewer of several journals of international repute.

Dilbagh Panchal is currently working as an Assistant Professor in the Department of Industrial and Production Engineering, Dr. B. R. Ambedkar National Institute of Technology, Jalandhar, Punjab, India. His area of research is reliability and maintenance engineering, performance optimization, fuzzy decision making and process safety. He obtained his Bachelor's degree (Hons.) in Mechanical Engineering from Kurukshetra University, Kurukshetra, India in 2007, Master's (Gold Medalist) in Manufacturing Technology in 2011 from Dr. B. R. Ambedkar National Institute of Technology Jalandhar, India and Phd from Indian Institute of Technology-Roorkee, India in the year of 2016. He has published/presented 18 research articles in reputed international journals and conferences. He is reviewer of many international journals.

* * *

Soma Bandyopadhyay received her Master Degree (M.E) in Computer Science & Engineering in the year 2008 from BESUS. Mrs. Bandyopadhyay has been a faculty member in the Department of Computer Science & Engineering, at MCKV Institute of Engineering, Liluah, Howrah since July 2008. She is an Associate member of Institution of Engineers India. Her research interest includes Embedded System and Data Mining and Soft Computing. She has 15 research publications in different journal and conferences.

Siddhartha Bhattacharyya did his Bachelors in Physics, Bachelors in Optics and Optoelectronics and Masters in Optics and Optoelectronics from University of Calcutta, India in 1995, 1998 and 2000 respectively. He completed PhD in Computer Science and Engineering from Jadavpur University, India in 2008. He is the recipient of the University Gold Medal from the University of Calcutta for his Masters. He is the recipient of the coveted National Award Adarsh Vidya Saraswati Rashtriya Puraskar for excellence in education and research in 2016. He is the recipient of the Distinguished HoD Award and Distinguished Professor Award conferred by Computer Society of India, Mumbai Chapter, India in 2017. He is the recipient of the coveted Bhartiya Shiksha Ratan Award conferred by Economic Growth Foundation, New Delhi in 2017. He received the NACF-SCRA, India award for Best Faculty for Research in 2017. He received the Honorary Doctorate Award (D. Litt.) from The University of South America and the South East Asian Regional Computing Confederation (SEARCC) International Digital Award ICT Educator of the Year in 2017.He also received the Rashtriya Shiksha Gaurav Puraskar from Center for Education Growth and Research, India in 2017. He has been appointed as the ACM

Distinguished Speaker for the tenure 2018-2020. He received the Young Scientist (Science & Technology) Award from CSERD, India in 2018. He is currently serving as a Senior Research Scientist in the Faculty of Electrical Engineering and Computer Science of VSB Technical University of Ostrava, Czech Republic. He is also the Principal of RCC Institute of Information Technology, Kolkata, India. In addition, he is also serving as the Professor of Computer Application and Dean (Research and Development and Academic Affairs) of the institute. Prior to this, he was the Professor of Information Technology of RCC Institute of Information Technology, Kolkata, India. He served as the Head of the Department from March, 2014 to December, 2016. Prior to this, he was an Associate Professor of Information Technology of RCC Institute of Information Technology, Kolkata, India from 2011-2014. Before that, he served as an Assistant Professor in Computer Science and Information Technology of University Institute of Technology, The University of Burdwan, India from 2005-2011. He was a Lecturer in Information Technology of Kalyani Government Engineering College, India during 2001-2005. He is a co-author of 5 books and the co-editor of 26 books and has more than 220 research publications in international journals and conference proceedings to his credit. He has got two PCTs to his credit. He was the convener of the AICTE-IEEE National Conference on Computing and Communication Systems (CoCoSys-09) in 2009. He was the member of the Young Researchers' Committee of the WSC 2008 Online World Conference on Soft Computing in Industrial Applications. He has been the member of the organizing and technical program committees of several national and international conferences. He served as the Editor-In-Chief of International Journal of Ambient Computing and Intelligence (IJACI) published by IGI Global, Hershey, PA, USA from 17th July 2014 to 06th November 2014. He was the General Chair of several international conferences like ICCICN 2014, ICRCICN 2015, ICRCICN 2016 and ICRCICN 2018 organized by the Department of Information Technology, RCC Institute of Information Technology, Kolkata and held at Kolkata, India. He is the Associate Editor of International Journal of Pattern Recognition Research and the founding Editor in Chief of International Journal of Hybrid Intelligence; Publisher: Inderscience. He is the member of the editorial board of International Journal of Engineering, Science and Technology and ACCENTS Transactions on Information Security (ATIS). He is also the member of the editorial advisory board of HETC Journal of Computer Engineering and Applications. He is the Associate Editor of the International Journal of BioInfo Soft Computing, IEEE Access and Evolutionary Intelligence, Springer. He is the member of the editorial board of Applied Soft Computing, Elsevier, B. V. He was the Lead Guest Editor of the Special Issue on Hybrid Intelligent Techniques for Image Analysis and Understanding of Applied Soft Computing, Elsevier, B. V. He was the Lead Guest Editor of the Special Issue on Computational Intelligence and Communications in International Journal of

Computers and Applications (IJCA); Publisher: Taylor & Francis, UK in 2016. He is the Editor of International Journal of Pattern Recognition Research since January 2016. He was the General Chair of the 2016 International Conference on Wireless Communications, Network Security and Signal Processing (WCNSSP2016) held during June 26-27, 2016 at Chiang Mai, Thailand. He was also the General Chair of First International Conference on Advanced Computational and Communication Paradigms (ICACCP 2017) organized by the Department of Computer Science and Engineering, Sikkim Manipal Institute of Technology, Sikkim, India and held at Sikkim during September 08-10, 2017. He is serving as the Series Editors of IGI Global Book Series Advances in Information Quality and Management (AIQM), De Gruyter Book Series Frontiers in Computational Intelligence (FCI), CRC Press Book Series Computational Intelligence and Applications, Wiley Book Series Intelligent Signal and Data Processing and Elsevier Book Series Hybrid Computational Intelligence for Pattern Analysis and Understanding. His research interests include soft computing, pattern recognition, multimedia data processing, hybrid intelligence and quantum computing. Dr. Bhattacharyya is a life fellow of Optical Society of India (OSI), India, life fellow of International Society of Research and Development (ISRD), UK, a fellow of Institute of Electronics and Telecommunication Engineers (IETE), India a fellow of Institution of Engineers (IEI), India and a fellow of Institute For Engineering Research and Publications (IFERP), India. He is also a senior member of Institute of Electrical and Electronics Engineers (IEEE), USA, International Institute of Engineering and Technology (IETI), Hong Kong and Association for Computing Machinery (ACM), USA. He is a life member of Cryptology Research Society of India (CRSI),Computer Society of India (CSI), Indian Society for Technical Education (ISTE), Indian Unit for Pattern Recognition and Artificial Intelligence (IUPRAI), Center for Education Growth and Research (CEGR), Integrated Chambers of Commerce and Industry (ICCI), and Association of Leaders and Industries (ALI).He is a member of Institution of Engineering and Technology (IET), UK, International Rough Set Society, International Association for Engineers (IAENG), Hong Kong, Computer Science Teachers Association (CSTA), USA, International Association of Academicians, Scholars, Scientists and Engineers (IAASSE), USA, Institute of Doctors Engineers and Scientists (IDES), India, The International Society of Service Innovation Professionals (ISSIP) and The Society of Digital Information and Wireless Communications (SDIWC). He is also a certified Chartered Engineer of Institution of Engineers (IEI), India. He has been inducted in the Board of Directors of International Institute of Engineering and Technology (IETI), Hong Kong.

Bhaskar Bhowani is an Assistant Professor in the Department of Financial Management, at Xavier Institute of Social Service, XISS, Ranchi. He has done his doctoral studies from Ranchi University, Ranchi in the area of Commerce and Business Management and is also an M.Com from Ranchi University and an MBA (Finance) from BIT, Mesra. He has PG teaching experience of about 15 years at XISS, Ranchi and about 2 years of UG teaching at St. Xavier's College Ranchi. Dr. Bhaskar Bhowani is also a visiting faculty at NIT, Rourkela (School of Management Studies). Dr.Bhaskar Bhowani has presented research papers in several national and international conferences and has quite a few publications to his credit in acclaimed journals. He is also a regular member in the recruitment board of Jharkhand State Cooperative Bank, Jharkhand Tribal Development Society and Dept. of Urban Development GoJ. He has also contributed as a member in the team of consultants from XISS, Ranchi to the Government of Jharkhand (MHRD) in its recent project on restructuring of the District Rural Development Agency (DRDA).

Chiranjib Bhowmik works as a research scholar at the Department of Mechanical Engineering, National Institute of Technology, Silchar, India.

Sumit Bhowmik is currently working as an Assistant Professor in the Department of Mechanical Engineering in National Institute of Technology. His research interests are Composite material, Fracture mechanics, Structural analysis, Advanced Manufacturing, Optimization, Renewable Energy. He has published number of research paper in reputed journal.

Agnimitra Biswas is an Assistant Professor at the Department of Mechanical Engineering, National Institute of Technology Silchar, Assam, India. He obtained Ph.D. in Mechanical Engineering with specialization in Manufacturing Engineering from National Institute of Technology Silchar and Post graduation in Thermal Engineering from National Institute of Technology Silchar, India. His areas of interests are Renewable Energy, Thermal Engineering, Solar Energy, and Vertical Axis Wind Turbine. He has over 9 years of teaching and research experience. He has published more than 45 research papers and 3 book chapters. Currently, he is doing research in solar energy collectors, water turbine, and Vertical axis wind turbine.

Tapas Kumar Biswas is currently working as an Assistant Professor of Automobile Engineering Department at MCKV Institute of Engineering. He has published numerous research papers in peer reviewed journals and conferences. His research interest is focused on applications of decision-making theories in automotive environment.

Soumava Boral received his B. Tech Degree in Mechanical Engineering from MCKV Institute of Engineering, Liluah, India, in 2013 and Master of Engineering from Department of Production Engineering, Jadavpur University, Jadavpur, India, in 2016. He is currently pursuing the Ph. D. degree from Subir Chowdhury School of Quality and Reliability, Indian Institute of Technology Kharagpur, India. His current research interests include Case-Based Reasoning, Multi-Criteria Decision Making, Maintenance Engineering, Reliability Engineering.

Sanjay Chaturvedi is currently working as a Professor at Subir Chowdhury School of Quality and Reliability (formerly Reliability Engineering Centre), Indian Institute of Technology, Kharagpur (WB) India. He received his Ph. D. degree from Reliability Engineering Centre, IIT, Kharagpur (India) in year 2003. He did his Bachelor's degree in electrical engineering and Master's degree in system engineering and operations research, both from Indian Institute of Technology, Roorkee (erstwhile University of Roorkee), U. P., India. He has research interest in the area of reliability modeling and analysis, network reliability, life-data analysis, maintenance and optimization. He has published papers in several international journals such as IEEE Transactions on Reliability, Performability Engineering (IJPE), Quality and Reliability Management, Quality, Reliability and Safety Engineering, Uncertain Systems, Risk and Reliability Part 'O', Quality Technology and Quantitative Management, Quality and Reliability International etc. and executed several consultancy projects of private and Govt. organizations. He's guided 6 Ph. D., 50+ post graduate students and written two books. He is also on the review panel of IEEE transactions on Reliability, IJPE, IJQRM, IJ System Science, IJ Failure Analysis. He's on the editorial board of International Journal of Mathematical, Engineering and Management Sciences (IJMEMS). He's also a senior member to IEEE.

Jyoti Chawla received her Ph.D in Chemical Sciences from Guru Nanak Dev University(GNDU) where she conducted research as junior research fellow under collaborative IUC-DAE project of GNDU, Amritsar and Bhabha Atomic Research Centre (BARC) Mumbai. She is NET qualified. Currently, Dr. Jyoti Chawla is working as a Professor& Head, Department of Chemistry, Manav Rachna International Institute of Research & Studies, Faridabad Haryana, India. Her research interest includes surface Chemistry, and risk assessment of nanomaterials. She has contributed more than 45 papers in journals and conferences proceedings. She has also contributed 2 chapters in Books. She serves as reviewer for many National and International journals.

Biplab Das is presently working as an Assistant Professor at Department of Mechanical Engineering, National Institute of Technology Silchar (An Institute of National Importance, under the Ministry of HRD, Government of India). Dr. Das completed his Ph.D. from NERIST, Itanagar, India, in the year of 2014. Later, he pursued his Post-Doctoral research from University of Idaho, USA. He is recipient of prestigious Bhaskara Advance Solar Energy (BASE) fellowship form IUSSTF and DST, Govt. of India. He is also awarded with 'DBT Associateship' by Department of Biotechnology, Govt. of India. He has 10+ years of experience in teaching and research, and published more than 30 nos. of International and National Journal and conference papers. Dr. Das completed 02 nos. of sponsored project worth 06 million INR from Department of Science and Technology (DST), Govt. of India, to establish a biodiesel reactor system at NERIST, Itanagar, India, which 4th of its kind in India. Presently Dr. Das is involved in 06 nos. of ongoing sponsored project to develop a solar thermal system for North East India, worth 0.268 billion INR, sponsored by DST, Govt. of India, Ministry of Power, Govt. of India, and Ministry of Climate Change, Govt. of India. He has ongoing research activities in collaboration with Jadavpur University, India, IIT Guwahati, India, University of Idaho, USA, Ulster University, UK.

Sumita Debbarma is an Assistant Professor at the Department of Mechanical Engineering, National Institute of Technology Silchar, Assam, India. She obtained Ph.D. in Mechanical Engineering with speciliazation in Thermal Engineering from National Institute of Technology Silchar. Her areas of interests are Biofuel, nao fuel, alternative fuwl and applied soft computing techniques. She has over 5 years of experience in the field of teaching and research experience. She has published more than 5 research papers.

Rishi Dwivedi is currently working as an Assistant Professor in the Department of Finance of XISS, Ranchi. Before joining XISS, Ranchi, he has worked as an Assistant Professor in Management Department at Central University of Jharkhand. He has obtained his PhD degree from Jadavpur University, Kolkata. Dr. Rishi has been awarded with esteemed "UGC-BSR Research Fellowship in Science for Meritorious Students". He has also almost one year of corporate experience as an Assistant Manager at ICICI Bank Limited, Kolkata, West Bengal. He has published several research papers in international journals of high repute as well as presented numerous papers in international conferences. He is regular reviewer of many international journals of high repute like, International Journal of Production Research, Journal of Cleaner Production, OPSEARCH etc. His research interests include development of integrated activity based costing and quality function deployment models for various industries in order to attain competitive edge. Of late he is also interested in exploring the ways in which MCDM model can be applied to achieve sustainable advantage for assorted industries.

Rajat Gupta is an eminent professor in Mechanical Engineering Department of National Institute of Technology Silchar, Assam, India. He did his PhD from Indian Institute of Technology Delhi in 1996. He is presently the Director of National Institute of Technology Mizoram, Aizawl, India. Earlier he served as the Director of National Institute of Technology Srinagar, J & K, India from 2011-16. He is a member of various MHRD constituted committees in India. He has got vast teaching and research experience spanning over more than 25 years. He has published more than 120 papers in various referred International Journals and Conferences. His research experience includes Thermal Engineering, Renewable Energy, Energy and Exergy analysis of Energy Systems.

Ian Howard graduated from the University of Western Australia with a Bachelor's degree and Doctorate in Mechanical Engineering in 1984 and 1988 respectively. He then worked for the Defence Science and Technology Organization for five years conducting research into helicopter gear health monitoring using vibration. In 1994 he joined Curtin University of Technology as a Lecturer in the area of Applied Mechanics and Dynamic systems. In 1998 he was promoted to Senior Lecturer, in 2004 to Associate Professor and to full Professor in 2016. He continues to supervise research in the dynamic behavior of rotating machinery for fault detection and classification, developing power harvesting and smart sensor devices for industrial applications, and investigating fault tolerant control techniques for Wind Turbine systems.

Jagadish is presently working as Assistant Professor in the Department of Mechanical Engineering, National Institute of Technology Raipur Chhattisgarh, India. He obtained Ph.D. in Mechanical Engineering with specialization in Manufacturing Engineering from National Institute of Technology Silchar and Post graduation in Product Design and Development from National Institute of Technology Warangal, India. His areas of interests are advanced manufacturing process, polymer machining, green manufacturing, computer aided design, and applied soft computing techniques. He has over 3 years of industrial experience in the field of design and analysis and 6 years of teaching and research experience. He received an Institutional Award (Gold Medal) by institution of engineers India and best innovative award by springer for his outstanding research contribution and published more than 35 papers includes journals/conferences/book and book chapters. Currently, he is doing research in advanced machining process, green manufacturing, polymer machining, and decision-making tools.

Rajeev Kumar received hisPh.D in Chemistry from IIT Roorkee. Dr. Rajeev Kumar has more than 9 years of experience related to teaching, research and administration. He has qualified NET conducted by University Grants Commission and Council of Scientific and Industrial Research (CSIR), New Delhi, India in 2003. He did Ph.D. from IIT Roorkee on synthesis of organo-metallic compounds in 2009. He was awarded Atama Ram gold medal in M.Sc. He has more than 35 international and national papers in journals and conferences. The broad area of his current research is synthesis of materials and removal of contaminants from water.

Vijay Kumar is a Professor & Head of Department of Mathematics at Manav Rachna International Institute of Research and Studies, India. He received his Ph.D in Applied Mathematics. His major areas of research interests are fuzzy information theory and decision making. He has published many papers in peer reviewed journals and conferences.

Jyotsna Kumar Mandal has done his M.Sc. degree in Physics from Jadavpur University in 1986, and M.Tech. in Computer Science from University of Calcutta. He was awarded PhD. (Engg.) in Computer Science & Engineering by Jadavpur University in 2000. Presently he is working as Professor of Computer Science & Engineering and former Dean, Faculty of Engineering, Technology and Management, Kalyani University, Kalyani, Nadia, West Bengal for two consecutive terms. He has teaching and research experience of 30 years. His area of research is Coding theory, Data and Network Security, Remote Sensing & GIS based Applications, Data Compression, Error correction, Visual Cryptography and steganography. He has produced 23 Ph. D degrees three submitted (2015) and 8 are pursuing. He is chief editor of various National/International Journals of AIRCC, editors of various National/International Conferences and proceedings. He has published more than 400 articles out of which 160 articles in various National and International Journals and 7 books from Springer and LAP Germany.

Debesh Mishra is a research scholar in KIIT Deemed to be University. He has also published more than 8 research articles in different reputed journals.

V. N. A. Naikan is currently Professor of the Subir Chowdhury School of Quality and Reliability (formerly, Reliability Engineering Centre) at the Indian Institute of Technology Kharagpur, India. He is also Professor in the Department of Industrial and Systems Engineering at IIT. He did his M.Tech. and Ph.D. degrees in Reliability Engineering. He graduated in Mechanical Engineering with silver medal from the University of Kerala. Before joining IIT, he also worked with Indian Space Research Organization, Chinese University of Hong Kong, Indian Institute of Management, Ahmedabad and Union Carbide India Limited in various capacities. He has published more than 120 research papers in international journals and conferences, a book on Reliability Engineering and Life Testing and a chapter on SPC in the Handbook on Performability Engineering, and a book chapter on Multiple Fault Classification using Support Vector Machines in a Machinery Fault Simulator, in the book "Vibration Engineering and Technology of Machinery". He was the Co-editor-in-chief of the International Journal of Performability Engineering, advisory and editorial board member of several other journals. He had been the technical chair of several international conferences. He has guided about 100 students for their B.Tech., M.Tech., and Ph.D. theses. He is Fellow of Institution of Engineers, and member of professional societies including, IEEE, IEI, SREI, System society of India, etc. He has been doing consultancy and research projects for organizations like ISRO, BARC, BHEL, Defence Forces, and Ministry of Textiles. He has organized several short-term courses and seminars for engineers and officers in industry. He also worked as a visiting professor at CALCE, University of Maryland, USA. He was awarded "Emerald Literati Network Award for Excellence (2016)". His paper has been selected for the "best paper award in the international conference IEOM Conference 2016", Detroit, USA. He has been awarded "Certificate of Outstanding Contribution in Reviewing" by the International Journal of Reliability Engineering and System Safety (ELSEVIER, 2018).

Dragan Pamucar is an Assistant Professor at University of Defence in Belgrade, Department of Logistics, Serbia. Dr. Pamucar received a PhD in Applied Mathematics with specialization of Multi-criteria modeling and soft computing techniques, from University of Defence in Belgrade, Serbia in 2013 and an MSc degree from the Faculty of Transport and Traffic Engineering in Belgrade, 2009. His research interests are in the field of Computational Intelligence, Multi-criteria decision making problems, Neuro-fuzzy systems, fuzzy, rough and intuitionistic fuzzy set theory, neutrosophic theory. Application areas include wide range of logistics problems. Dr. Pamucar has authored/co-authored over 80 papers published in refereed International Journals including Experts Systems with Applications, Computational intelligence, Technical gazette, Sustainability, Symmetry, Water, Asia-Pacific Journal of Operational Research, Operational Research, Journal of intelligent and fuzzy systems, Land use policy, Environmental impact assessment review, International Journal of Physical Sciences, Economic computation and economic cybernetics studies and research and so on, and many more.

Kanika Prasad is an Assistant Professor at National Institute of Technology, Jamshedpur, India. She graduated from Jadavpur University in Production Engineering after completing her undergraduate study in Mechanical Engineering from SMIT, Sikkim in India, earning distinction in both degrees. She earned prestigious DST INSPIRE fellowship to pursue her doctoral research from Jadavpur University.

With many publications in journal and conferences of international repute so early in her academic career, much is expected of her research career. She specializes in application of quality function deployment technique and multi-criteria decision-making to develop expert systems for selection and design problems in manufacturing. Of late sustainability in manufacturing domain interests her and she is working in that area. She has guided several UG and PG dissertations. She is also a regular reviewer of several journals of international repute.

Rajeev Ranjan is currently an Assistant Professor of Doon Business School, India. His areas of research interests are multi-criteria decision-making, sustainability modelling, production and operations management. He has published numerous papers in reputed peer reviewed journals and conferences.

P. Kritee Rao is currently pursuing her PGDM in Finance from Xavier Institute of Social Service, Ranchi, India. She has already published few papers on the domain of multi-criteria decision making (MCDM). She stands among top five students in the class. She completed her B.Sc in Mathematics hons with first division and chose finance as her domain to pursue higher studies.

Amitava Ray is currently working as a Principal, Jalpaiguri Government Engineering College. His research area of interests is Production & Operations management, Applied soft computing, Green manufacturing, Composite machining.

Suchismita Satapathy is working as associate professor in mechanical engineering department of KIIT Deemed to be University. She has published more than 25 research articles in different referred journals.

Željko Stević is Assistant professor at Faculty of Transport and Traffic Engineering Doboj, University of East Sarajevo. He received PhD in Transport and Traffic Engineering from University of Novi Sad, Faculty of Technical Sciences 2018. Interests: logistics; supply chain management; transport; traffic engineering; multi-criteria decision making problems; rough set theory; sustainability; fuzzy set theory. Hehas published over 80 papers from the area of his interest. He has contributed outstanding research in the mentioned fields. In all his researches he has provided a very good application studies and practical contributed solving different problems in transportation, logistics, supply chain management, traffic engineering etc. In his doctoral dissertation he developed few MCDM models including one universal model for supplier selection in different fields. His published studies are very well cited in other research that can be seen in ResearchGate and Google Scholar. Dr. Stević has authored/co-authored of papers published in refereed International Journals including Applied Soft Computing, Sustainability, Symmetry, Engineering Economics, Transport, Neural Computing and Applications, Information, ECECSR, Technical gazette, SIC and other. Awards: Jan 2018 Award: Medal merit for the people in the field of education and science. Nov 2017 Award: The best young researcher of the 3rd cycle (Doctoral) studies-(Festival of Science 2017).

Subro Thakur is a Research Engineer with 10+ years' in Research & Development and 11+ years' experience in Teaching and administrative responsibilities. Skilled at Networking, developing Condition Monitoring based systems & software and meeting the delivery timelines. Eager to take on new challenges with a company ideally in computer systems or product development/project management/coordination role. Position would have a focus on creativity and Innovation balanced with speed to the market.

Index

I

L

M

N

O

P

R

S

T

U

V

W

Ensure Quality Research is Introduced to the Academic Community

Become an IGI Global Reviewer for Authored Book Projects

The overall success of an authored book project is dependent on quality and timely reviews.

In this competitive age of scholarly publishing, constructive and timely feedback significantly expedites the turnaround time of manuscripts from submission to acceptance, allowing the publication and discovery of forward-thinking research at a much more expeditious rate. Several IGI Global authored book projects are currently seeking highly qualified experts in the field to fill vacancies on their respective editorial review boards:

Applications may be sent to:

development@igi-global.com

Applicants must have a doctorate (or an equivalent degree) as well as publishing and reviewing experience. Reviewers are asked to write reviews in a timely, collegial, and constructive manner. All reviewers will begin their role on an ad-hoc basis for a period of one year, and upon successful completion of this term can be considered for full editorial review board status, with the potential for a subsequent promotion to Associate Editor.

If you have a colleague that may be interested in this opportunity, we encourage you to share this information with them.

www.ingramcontent.com/pod-product-compliance
Lightning Source LLC
LaVergne TN
LVHW081537300926
849161LV00006B/1465